Casing and Liners for Drilling and Completion

Ted G. Byrom

Casing and Liners for Drilling and Completion

HOUSTON, TX:
Gulf Publishing Company
2 Greenway Plaza, Suite 1020
Houston, TX 77046

ISBN-10: 1933762063
ISBN-13: 9781933762067

10 9 8 7 6 5 4 3 2

Library of Congress Cataloging-in-Publication Data

Byrom, Ted G.

Casing and liners for drilling and completions / Ted G. Byrom.

p. cm.

Includes bibliographical references and index.

ISBN 1-933762-06-3 (978-1-933762-06-7 : alk. paper)

1. Oil well casing--Design and construction. I. Title.

TN871.22.B97 2007

622'.3381—dc22

2006038254

Printed in the United States of America
Printed on acid-free paper. ∞
Text design and composition by TIPS Technical Publishing, Inc.

This book is dedicated to a very special group of drilling engineers in Houma, Louisiana, 1962–1965, who both taught and tolerated me as I began my career: Henry Arnaud, Warren Sexton, Gary Chenier, C. J. Scott, Don Hecker, George Drenner, Randy Newcomer, Sheldon Roberdeau, Cleve Gilley, and Joe Prosser.

Contents

This book was prepared under the auspices of the IADC Technical Publications Committee, but has not been reviewed or endorsed by the IADC Board of Directors. While the committee strives to include the most accurate and correct information, IADC cannot and does not warranty the material contained herein.

The mission of the IADC Technical Publication Committee is to publish a comprehensive, practical, and readily understandable series of peer-reviewed books on the petroleum drilling industry known as the *Gulf Drilling Series* in order to educate and guide industry personnel at all levels.

This book has been peer-reviewed in accordance with this mission by:

Ross Kastor

Tommy Warren

Bob Pilko

Leon Robinson, Committee Chairman

Preface

Hardly anyone reads a Preface. Please read this one, because this book is a bit different and what is written here is the actual introduction to the book. I never read a textbook that I really liked when I was a student. The main reason is that most authors seemed more interested in presenting the information with the goal of impressing colleagues rather than instructing the reader as a student of the subject. For a long time, I thought they were so smart that they could not relate to the ordinary student. I now know that is rarely true. You should know that I have reached a point in my career where no one is important enough that I need to impress, and certainly no money is to be made writing a textbook. My reason for accepting the task of writing this text is that I truly wanted to attempt to explain this subject in an understandable manner to the many petroleum engineers who need or want to understand it but at best received a couple of classroom lectures and a homework assignment on the subject from someone who never designed or ran a real string of casing in his life. I was in that same position some 44 years ago. This book is also intended for those coming into the oil field from other disciplines and needing to understand casing design.

This book is not written in the style of most textbooks. That is because it its main purpose is to teach you, the reader, about casing and casing design without need of an instructor to "explain" it to you. I would like you to read this as if you and I were sitting down together as I explain the material to you. While some of the material requires a little formality, I have tried to put it on a readable level that progresses through the various processes in a logical manner. I have also tried to anticipate, pose, and answer some of the questions you might ask in the process of our discussion.

The first five chapters of this book lay a foundation in basic casing design. It is, if you will, a recipe book for basic casing design. It does go into some detail at times, but overall its purpose is to actually teach an understanding of basic casing design. If you are not an engineer, and many casing strings are designed by nonengineers, do not be discouraged by the many equations you see. The information in this part should be sufficient to design adequate casing strings for the vast majority of the wells drilled in the world, and although the chapter on hydrostatics contains some calculus, none of it is beyond the capabilities of a second-year engineering student. The sixth chapter is about running and landing casing. Most of it is common sense, but there are some practical insights that are worth the time it takes to read.

Chapter 7 begins the discussion of slightly more advanced material. Some of this material is not covered in universities, except on a graduate level, but I have tried to present it so that any undergraduate engineering student should be able to understand it. The remaining chapters continue in the same vein.

I have not tried to cover everything about casing or casing design in this book. I have never had any aspirations of writing the definitive text on casing or any other subject, mostly because some aspects hold no interest at all for me. I have personally run and cemented close to a couple of hundred casing strings as a field drilling engineer, designed several hundred more, and been involved with several thousand casing strings over my career. These have ranged from very shallow strings to a few over 23,000 ft. Never have I designed a string for a geothermal well, and my corrosion and sour gas experience is limited. Consequently, little is said about those subjects in this book. There are much better sources for that than what I could write on those particular topics.

Opinions: you will note that there are a few opinions in this text. You will easily recognize them. My meager defense is that they come from experience, much of it unpleasant. You are free to disagree.

Errors: Anyone who has ever written a book or had a paper published has been humbled by the appearance of errors that have somehow mysteriously materialized in the carefully written and edited work. While I and the editors tried diligently to ferret them out of this book, I know they are present. I apologize in advance. May you derive pleasure in discovering them.

Information or Knowledge: finally, one should understand that what we now call the *information age* is fast becoming an age of ignorance. Make no mistake; information is not knowledge. Knowledge connotes understanding, and information alone is of little value without understanding. Daily we are barraged with analysis of information with no perception of understanding. Now that we rely heavily on computers, we often do something comparable with casing design. We plug in some numbers, and out comes a multipage page printout of the results. If you do not understand it, and many do not, then it is merely information. Are you going to trust it? It is of little solace to know that the people who wrote the software understand it. It is my best hope that this book will lend a bit to a better overall understanding of this one small topic. If some of my explanations are not perfectly lucid; you have my permission to read them more than once; no one else will ever know.

CHAPTER 1

Oil-Field Casing

1.1 Introduction

We begin with a chapter on the usual obligatory information for a book on oil-field casing. The real book starts in Chapter 2, but in the rare case that the subject is totally new to you, you will find a rudimentary coverage of what casing is in this chapter. The steel tubes that become a permanent part of an oil or gas well are called casing, and the tubes that are removable, at least in theory, are referred to collectively as tubing which are not covered in this book. Oil-field casing is manufactured in various diameters, wall thicknesses, lengths, strengths, and with various connections. The purpose of this text is to examine the process of selecting the type and amount we need for specific wells. But first, a question: What purpose does casing serve in a well? There are three:

- Maintain the structural integrity of the bore hole.
- Keep formation fluids out of the bore hole.
- Keep bore hole fluids out of the formations.

It is as simple as that, though we could list many subcategories under each of those. Most are self-evident. Additionally, there are some cases where the casing also serves a structural function to support or partially support some production structure, as in water locations.

1.2 Setting the Standards

By necessity oil-field tubulars are standardized. Until recent times, the standards were set by the American Petroleum Institute (API) through various committees and work groups formed from personnel in the industry. Now, the International Organization for Standardization (ISO) is seen as taking on that role. Currently, most of the ISO standards are merely the API standards, but that role may expand in the future. In this text, we refer primarily to the API standards, but it should be understood that there are generally identical standards, and in some cases more advanced standards, under the ISO name.

It is important that some degree of uniformity and standardization is in force and that manufacturers be held to those standards through some type of approval or licensing procedure. In times of casing supply shortages, a number of manufacturers have entered the oil-field tubular market with substandard products. Some of these have led to casing failures where no failure should have occurred. The important point here is that any casing purchased for use in oil or gas wells should meet the current standards as set for oil-field tubulars by the API or ISO.

Some casing is not covered by API or ISO standards. Some of this non-API casing is for typical applications, some for high-pressure applications, high-temperature applications, low-temperature applications, and some for applications in corrosive environments. Most of this type of casing meets API standards or even higher standards, but one must be aware that the standards and quality control for these types of casing are set by the manufacturer. It probably should not be mentioned in the same paragraph with the high-quality pipe just referred to, but it should also be remembered that there are some low-quality imitations of API products on the market as well.

1.3 Manufacture of Oil-Field Casing

There are two types of oil-field casing manufactured today: seamless and welded. Each has specific advantages and disadvantages.

1.3.1 Seamless Casing

Seamless casing accounts for the greatest amount of oil-field casing in use today. Each joint is manufactured in a pipe mill from a solid cylindrical piece of steel, called a billet. The billet is sized so that its volume is equal to that of the joint of pipe that will be made from it. The manufacturing process involves

- Heating the billet to a high temperature.
- Penetrating the solid billet through its length with a mandrel such that it forms a hollow cylinder.
- Sizing the hollow billet with rollers and internal mandrels.
- Heat treating the resulting tube.
- Final sizing and straightening.

The threads may be cut on the joints by the manufacturer or the plain-end tubes may be sent or sold to other companies for threading. The most difficult aspect of the manufacture of seamless casing is that of obtaining a uniform wall thickness. For obvious reasons, it is important that the inside of the pipe is concentric with the outside. Most steel companies today are very good at this. A small few are not, and that is one reason that API and ISO standards of quality were adopted. Current standards allow a 12.5% variation in wall thickness for seamless casing. The straightening process at the mill affects the strength of the casing. In some cases, it is done with rollers when the pipe is cool and other cases when the pipe is still hot. Seamless casing has its advantages and also a few disadvantages.

- Advantages of Seamless Casing.
    - No seams to fail.
    - No circumferential variation of physical properties.
- Disadvantages of Seamless Casing.
    - Variations in wall thickness.
    - More expensive and difficult manufacturing process.

1.3.2 Welded Casing

The manufacturing process for welded casing is quite different from that of seamless casing. The process also starts with a heated steel slab that is rectangular in shape rather than cylindrical. One process uses a relatively small slab that is rolled into a flat plate and trimmed to size for a single joint of pipe. It is then rolled into the shape of a tube and the two edges are electrically flash welded together to form a single tube. Another process uses electric resistance welding (ERW) as a continuous process on a long ribbon of steel from a large coil. The first stage in this process is a milling line in the steel mill:

- A large heated slab is rolled into a long flat plate or ribbon of uniform thickness.
- Plate is rolled into a coil at the end of the milling line.

The large coils of steel "ribbon" are then sent to the second stage of the process, called a forming line.

- Steel is rolled off the coil and the thickness is sized.
- Width is sized to give the proper diameter tube.
- Sized steel ribbon is formed into a tubular shape with rollers.
- Seam is fused using electric induction current.
- Welding flash is removed.
- Weld is given an ultrasonic inspection.
- Seam is heat treated to normalize.
- Tube is cooled.
- Tube is externally sized with rollers.
- Full body of pipe is ultrasonically inspected.
- Tube is cut into desired lengths.
- Individual tubes are straightened with rollers.

This is the same process by which coiled tubing is manufactured, except coiled tubing is rolled onto coils at the end of the process instead of being cut into joints. Note that, in the welding process, no filler material is used; it is solely a matter of heat and fusion of the edges.

Welded casing has been available for many years, but there was an initial reluctance by many to use it because of the welding process. Welding has always been a matter of quality control in all applications, and a poor-quality weld can lead to serious failure. Today, it is both widely accepted and widely used for almost all applications except high-pressure and/or high-temperature applications. It is not used in the higher yield strength grades of casing.

- Advantages of Welded Casing.
    - Uniform wall thickness.
    - Less expensive than seamless.
    - Easier manufacturing process.
    - Casing is inspected during manufacturing process (ERW) and the defective sections removed.
    - Uniform wall thickness is very important in some applications, such as the newer expandable casing.

- Disadvantages of Welded Casing
    - High temperatures of welding process.
    - Possible variation of material properties due to welding.
    - Possible faulty welds.
    - Possible susceptibility to failure in weld.

Welded casing has been used for many years now. Many of the so-called disadvantages are perhaps more a matter of perception than actuality.

1.3.3 Strength Treatment of Casing

When a cast billet or slab is formed into a tube it is done at quite high temperature. The deformation that takes place in the forming process is in a plastic or viscoplastic regime of behavior for the steel. As it cools its crystalline structure begins to form. Once the crystalline structure forms or begins to form, any additional plastic deformation to which we subject the tube will change its properties. The change may be minor or significant, depending on the constituents of the steel, the amount of deformation, and the temperature. Heating a tube above certain temperatures and cooling slowly allows the crystals to form more uniformly with fewer structural imperfections, called *dislocations*, in the lattice structure. The properties of the steel can modified by the addition of certain constituents to the alloy and, to some extent, by controlling the cooling rate. One common process for enhancing the performance properties of casing is to heat the tube above a certain temperature then quickly cool it by spraying it with water or some other cool fluid to change its properties, especially near the surface. This process is called *quench and temper*, or QT for short, and is an inexpensive alternative to adding more expensive alloys.

Some steels are said to get "stronger" when they are deformed plastically at ambient temperatures. This is part of the manufacturing process in some steels and is called *cold working*. Cold working typically increases the steel's yield strength, however it does not, in general, increase the ultimate strength. Straightening casing joints in the latter stages of the manufacturing process can also have an effect on the properties of the tube depending on whether it is done at "cool" temperatures or "warm" temperatures.

1.4 Casing Dimensions

Casing comes in odd assortment of diameters ranging from 4½-in. to 20-in. that may seem quite puzzling at first encounter, such as 5½-in., 7-in., 7⅝-in., 9⅝-in., 10¾-in. Why such odd sizes? All we can really say about that is that they stem from historical sizes from so far back that no one knows the reasons for the particular sizes any longer. Some sizes became standard and some vanished. Within the different sizes, there are also different wall thicknesses. These different diameters and wall thicknesses were eventually standardized by the API (and now ISO). The standard sizes as well as dimensional tolerances are set out in API Specification 5CT (2001) and ISO 11960 (2004).

1.4.1 Outside Diameter

The size of casing is expressed as a nominal diameter, meaning that is the designated or theoretical diameter of the pipe. API and ISO allow for some tolerance in that measurement, and the specific tolerance differs for different size pipe. The tolerance for the pipe being under sized is typically less than for the pipe being oversized. This is to ensure that the threads cut in the pipe are full threads. The tolerances for non-upset casing 4½ in. and larger are given as fractions of the outside diameter, $+0.01d_o$, $-0.005d_o$. For upset casing, Table 1–1 shows the current API and ISO tolerances measured 5 in. or 127 mm behind the upset.

Table 1–1 API/ISO Tolerances for Upset Casing Outside Diameter (API 5CT, ISO 11960)

Nominal Outside Diameter, d_o (in.)	Tolerances (in.)		Tolerances (mm)	
	+	–	+	–
>3½ to 5	7/64	$0.0075d_o$	2.78	$0.0075d_o$
>5 to 8⅝	⅛	$0.0075d_o$	3.18	$0.0075d_o$
>8⅝	5/32	$0.0075d_o$	3.97	$0.0075d_o$

Table 1–2 Minimum Drift Mandrel Dimensions (API Spec 5CT, ISO 11960)

Casing Outer Diameter (in.)	Mandrel Length		Mandrel Diameter	
< 9⅝	6 in.	152 mm	d_i – ⅛ in.	d_i – 3.18 mm
9⅝ to 13⅜	12 in.	305 mm	d_i – 5⁄32 in.	d_i – 3.97 mm
>13⅜	12 in.	305 mm	d_i – 3⁄16 in.	d_i – 4.76 mm

1.4.2 Inside Diameter and Wall Thickness

The inside diameter of the casing determines the wall thickness or vice versa. Rather than a specific tolerance for the amount at which the internal diameter might exceed a nominal value, the tolerance specified by API and ISO is given in terms of minimum wall thickness. The minimum wall thickness is 87.5% of the nominal wall thickness. The maximum wall thickness is given in terms of the nominal internal diameter, however. It specifies the smallest diameter and length of a cylindrical drift mandrel that must pass through the casing (see Table 1–2).

The internal diameter of casing is a critical dimension. It determines what tools and so forth may be run through the casing. It is not uncommon to have to select a casing for a particular application such that the drift diameter is less than the diameter of the bit normally used with that size casing, even though the bit diameter is less than the nominal internal diameter of the pipe. In cases like this, it is a practice to drift the casing for the actual bit size rather than the standard drift mandrel. This may be done with existing pipe in inventory, and those joints that will not pass the bit are culled from the proposed string. Or it may be done at special request at the steel mill, in which case there will be an extra cost. This procedure applies only to casing where the desired bit diameter falls between the nominal internal diameter and the drift diameter of the casing.

1.4.3 Joint Length

The lengths of casing joints vary. In the manufacture of seamless casing, it all depends on the size of the billet used in the process. Usually, there is some difference in weight of the billets, and this results in some variation in the length of the final joints. One could cut all the joints to the same length, but that would be a needless expense and, in fact, would not be desirable. (Wire line depth correlation for perforating and other operations in wells usually depends on an electric device to correlate the couplings with a radioactive formation log; so if all the joints are the same length, it can cause errors in perforating or packer setting depths.) For ERW casing, it is much easier to make all of the joints the same length, but there may still be some waste if that is done. Even if the joints vary in length, they need to be sorted into some reasonable ranges of lengths for ease of handling and running in the well. Three ranges of length are specified by API Recommended Practices 5B1 (1999), Ranges 1, 2, and 3 (see Table 1–3).

Most casing used today is in either Range 2 or 3, with most of that being Range 3. Range 1 is still seen in some areas where wells are very shallow, and the small rigs that drill those wells cannot handle longer pipe.

Table 1–3 Length Ranges of Casing (API RP 5B1, 1999)

Range 1		Range 2		Range 3	
(ft)	(m)	(ft)	(m)	(ft)	(m)
16 to 25	4.88 to 7.62	25 to 34	7.62 to 10.36	34 to 48	10.36 to14.63

1.4.4 Weights of Casing

The term *casing weight* usually refers to the specific mass of casing expressed as mass per unit length, such as kg/m or lb/ft. The use of the term weight is so common that we are going to use that term for now, but it should be understood that we are not talking about weight but mass. In Chapter 2, we discuss this more thoroughly. So the casing weight is determined by the density of the steel and the dimensions of the casing body. For instance, we may have a joint of 7-in. 26 lb/ft casing. We might reasonably assume from that that our joint actually weighs 26 lb/ft. Our assumption would be wrong. This is the nominal casing weight not the actual weight. For some reason, the nominal weight of casing is based on a joint that is 20 feet in length. It includes the weight of the plain-end pipe plus the weight of a coupling, minus the weight of the casing cut away to make the threads on each end and divided by 20 feet to give the nominal weight in terms of pounds per foot (or kilograms per meter). In other words, casing almost never weighs the same as its nominal weight. Fortunately the difference is small enough that in most cases it does not really matter. API Spec 5CT has formulas for calculating the actual weight of a joint, but it requires specification of the thread dimensions and so forth, and we are not going to concern ourselves with that here. One particular formula in API Spec 5CT or ISO 11960 sometimes is useful though, and that is the formula for calculating the nominal casing weight of the plain pipe without threads or couplings:

$$w = C(d_o - t)t$$

or

$$w = C\left(\frac{d_o^2 - d_i^2}{4}\right) \tag{1.1}$$

where

- w = plain-end "casing weight," mass per unit length, lb/ft or kg/m.
- C = conversion factor, 10.69 for oil-field units; 0.0246615 for metric units
- d_o = outside diameter, in. or mm.
- d_i = inside diameter, in. or mm.
- t = wall thickness, in. or mm.

These formulas appear in several API/ISO publications accompanied by a statement that martensitic chromium steels (L-80, Types 9Cr and 13Cr) have densities different from carbon steels and a correction factor of 0.989 should be applied. Interestingly though, the density of carbon steel is nowhere to be found in those publications except by backing it out of these formulas. From these formulas, the density of API carbon steel is 490 lb/ft^3 or 7850 kg/m^3. Again, we discuss units of measure, mass and weight, and so forth in Chapter 2.

1.5 Casing Grades

Casing is manufactured in several different grades. Grade is a term for classifying casing by strength and metallurgical properties. Some of the grades are standardized and manufacture is licensed by the API; others are specific to the particular manufacturer.

1.5.1 API Grades

The API grades of casing are manufactured under a license granted by the API. These grades must meet the specifications listed in API Spec 5CT (2004) or ISO11960. These grades have yield strengths ranging from 40,000 lbf/in.^{2} to 125,000 lbf/in.^{2}. These grades are listed in Table 1–4.

The letter designations are essentially arbitrary, although there may be some historical connotation. The numbers following the letters are the minimum yield strengths of the metal in 10^3 lbf/in.^{2}. The minimum yield strength is the point at which the metal goes from elastic behavior to plastic behavior. And it is specified as a "minimum," meaning that all joints designated as that particular grade should meet that minimum strength requirement, although it is allowed to be higher. Typically, we use the minimum yield strength as the design limit in most casing design. The yield strength also may be higher than the minimum, and a maximum value also is listed in the table. The reason for the maximum value is to assure that the casing sold in one particular grade category does not have tensile and hardness properties that may be undesirable in a particular application. Some years back, it was common practice to downgrade pipe that did not meet the minimum specifications for which it was manufactured. In other words, if a batch of casing did not meet the minimum specifications for the grade it was intended, it could be downgraded and sold as the next lower grade. There also were cases where one grade was sold as the next lower grade to move it out of inventory. Some of the consequences of this practice were disastrous. One typical example was the use of N-80 casing for tieback strings and production strings in high-pressure gas wells in the Gulf Coast area of the United States. Many of these wells drilled in the 1960s used lignosulfonate drilling fluids and packer fluids, which over time degraded to form hydrogen sulfide (H_2S). As it turned out, some wells that were thought to have N-80 grade casing, actually had P-110, and there were a number a serious casing failures due to hydrogen embrittlement. Many of these "N-80" casing strings had P-110 grade couplings on them, and in some wells almost every coupling in the entire string cracked and leaked.

You will also notice in the chart that some different grades have the same minimum yield strength. Again, this is a case where the metallurgy is different. For instance both N-80 and L-80 have a minimum yield strength of 80,000 lbf/in.^{2}, but their other properties are different. L-80 has a maximum Rockwell hardness value of 22 but N-80 does not. N-80 actually might be a down-rated P-110 but L-80 cannot be. The grades with the letter designation L and C have maximum hardness limitations

Table 1–4 API Casing Grades, Tensile Strength and Hardness Specifications (API Spec 5CT, 2001)

Grade	Yield Strength (ksi)		Minimum Tensile Strength (ksi)	Hardness	
	Minimum	Maximum		HRC	HBW/HBS
H-40	40	80	60		
J-55	55	80	75		
K-55	55	80	95		
N-80	80	110	100		
M-65	65	85	85	22	235
L-80	80	95	95	23	241
C-90	90	105	100	25.4	255
C-95	95	110	105		
T-95	95	110	105	25.4	255
P-110	110	140	125		
Q-125	125	150	135		

and are for specific applications where H_2S is present. Those hardness limits are also shown in the table.

The ultimate strength value listed is the minimum strength of the casing at failure. In other words, it should not fail prior to that point. This is based on test samples and does not account for things like variations in wall thickness, pitting, and so forth. It is not really possible to predict actual failure strength, because there are too many variables, but this value essentially means that the metal should fail at some value higher than the minimum. Also shown in the table are values for minimum elongation. This is specified as the minimum percent a flat sample will stretch before ultimate failure. When you consider that K55, for instance, yields at an elongation of 0.18%, then you can imagine that nearly 20% elongation is considerable. But one should not be misled into thinking that, if we design casing with the yield strength as a design limit, there is necessarily a considerable additional "strength" remaining before the casing actually fails.

Once the material is loaded beyond the elastic limit (yield) the incremental stress required to stretch it to failure can often be quite small. We discuss more on plastic behavior in Chapter 7.

1.5.2 Non-API Grades

Non-API grades of casing are not licensed by the API and consequently do not necessarily adhere to API or ISO specifications. This is not to imply that they are inferior, in fact, the opposite is true in many cases. Most non-API grades are for specialized applications to meet requirements not covered in the API or ISO specifications. Examples are high-temperature and or high-corrosive environments and high-collapse and tensile-strength requirements. In these cases, one must rely on the specifications, quality control, and reputation of the manufacturer. For extremely critical wells, many operators elect to do a number of qualification tests and inspections on the specific casing that will be used in a particular application. For instance, one operating company has invested a very large amount of money and research into qualifying connections for use in high-pressure wells (Valigura and Cernocky 2005).

It should also be mentioned that a number of manufacturers make casing that supposedly meets API/ISO specifications but are not licensed to do so. Typically, this casing is sold below the market price for API/ISO casing. While some of this pipe has been found to be acceptable, much of it is not. This was a particular problem in the late 1970s, when the demand for casing far exceeded the supply, and similar situations reoccur from time to time. It is a case of "buyer beware."

1.6 Connections

Many types of connections are used for casing. These are threaded connections, and there are three basic types: coupling, integral, and weld-on.

The most common type is a threaded pipe with couplings. A plain joint of pipe is threaded externally on both ends and an internally threaded coupling, or collar as it is sometimes called, joins the joints together. A coupling usually is installed on one of the threaded ends of each joint immediately after the threads are cut. The end of the coupling that is installed at the threading facility (usually at the steel mill) is called the *mill end*. The other end of the coupling typically is called the *field end*, since it is connected in the field as the casing is run into the well. An integral connection is one in which one end of the pipe is threaded externally (called the *pin end*) and the other end is threaded internally (called the *box end*). The joints are connected by screwing the pin end of one joint into the box end of another. In most cases, an integral connection requires that the pipe body be thicker at the ends to accommodate both internal and external threads and still have a tensile strength reasonably close to that of the pipe body. The increased wall thickness in this case is called an *upset*, and it may be an increase of the external diameter, *external upset* (EU), a reduction in the internal diameter, *internal upset* (IU), or a combination of both (IEU). Most integral joint casing is externally upset, so as to have a uniform internal diameter to accommodate drilling and completion tools. Finally, the weld-on connection is one in which the threaded ends are welded onto the pipe instead of being cut into the pipe body itself. This type of connection typically is used for large-diameter casing (20 in. and more), where the difficulty of cutting threads on the pipe body becomes more pronounced due to the large size, uniformity of diameter, and roundness.

Of the three types of connections mentioned, there are also different ways in which threaded connections bring about a seal. These primary sealing methods are interference and metal-to-metal seals.

Interference sealing relies on the compression of the individual threads against one another to cause a seal. Typically, this is the sealing mechanism of "V" or wedge-shaped threads that are forced tight against one another as the connection is made. The threaded area is tapered so that the more it is made up the greater the contact force between the threads due to the circumferential stress in the pin and box. Despite all the force though, interference alone does not cause a total seal, because there has to be some tolerance in the thread dimensions for the connections to be made. There always is some small gap in the cross-sectional profile of a connection. In the case of wedge-shaped threads, there is a small gap between the crest of one thread and the valley of the other. These connections require a thread lubricant to fill this small gap and effect a true seal. For that reason, it is necessary to use a good-quality thread lubricant. Another aspect of this type of seal is that the contact force must be great enough to resist any pressure force tending to press fluids into the contact area.

Metal-to-metal seals rely on the contact of metal surfaces other than the threads to cause a seal. This may be a tapered surface on the pin and box that contact each other in compression, a shoulder contact, or a combination of the two. These types of seals are strictly metal-to-metal contact and do not rely on thread lubricant to bring about the seal. For this reason, it is extremely important that the connections are protected during handling and running to avoid damaging the seal surfaces. And, since these seals are also dependent on the compression of the metal surfaces, the type of thread lubricant is important to achieve the desired makeup torque.

There is a secondary type of sealing mechanism, called *resilient seals* or *rings*. Resilient seals typically are polymer rings inserted into a special recess in the threaded area to provide additional seals to keep gas or corrosive fluids out of the thread gaps. They usually are not considered a primary seal but only an additional seal to improve the quality of an interference seal or a corrosion prevention mechanism for some metal-to-metal seals.

1.6.1 API 8-rd Connections

The most common type of casing connection in use is the API 8-rd connection, where 8-rd means 8-round or eight threads per inch and a slightly rounded profile. The profile is a V or wedge-shape but slightly rounded at the crest and valleys of the threads. There is also an API 11½-V thread, which has 11½ threads per inch and a sharp V profile. This typically is called a *line pipe thread* and is seldom used in down-hole applications today. The API 8-rd connection is made in either ST&C (short thread and coupling) or LT&C (long thread and coupling). These two threads are the workhorses of the industry and sufficient for most normal applications. Like most connections, these are not as strong in tension as the pipe body itself because of the reduced net cross-sectional area of the tube, resulting from the threads being cut into the pipe body wall in the absence of an upset. These are interference-seal-type connections. The threads are wedge-shaped, cut on a tapered profile, and made up until a prescribed torque is attained. At full makeup torque, the threads do not achieve a pressure seal, because the threads do not meet in the base of the groove, leaving two small channels at the base of the thread in both pipe and the coupling. How then do they seal and prevent pressure leaks? They form a pressure seal with the use of thread lubricant that fills the voids between the thread roots. The gap is very small and its length is quite long due to the number of turns at a pitch of eight per inch, so the lubricant forms a good seal in most cases. However, one must always use an approved thread lubricant, one that ages and shrinks in time and temperature eventually will leak. Although these connections often are used in gas well applications, they generally are not recommended, because they rely on the thread lubricant for a seal. Another precaution is that because the threads are wedge shaped, they tend to override each other when subjected to high tension or compression. This override mechanism is often referred to as *jump out*. Because of this jump out tendency, ST&C and

LT&C connections generally are not recommended for wells that have high bending stress due to well-bore curvature or applications where temperature fluctuations cause high axial tensile and compressive loads.

1.6.2 Other Threaded and Coupled Connections

A number of types of threaded and coupled connections have different profiles from the API 8-rd. Instead of wedge-shaped threads, many have a square profile or something similar to give them greater tensile and bending strengths. Examples of this type of thread is the Buttress (now an API thread), 8-Acme, and the like. These threads typically are used where higher tensile strengths are needed in the joints. In general, they also rely on thread lubricant to form a seal and are prone to leak in high-pressure gas applications. Most of these connections require less makeup torque than API 8-rd connections. This is an advantage but also can be a disadvantage, because the maximum makeup torque usually is less than that required to rotate the casing in the hole. Where rotation is planned for cementing or orienting precut windows for multilateral wells, these types of connections are to be avoided. Also, because the makeup torque is relatively low, most of these joints have a "makeup mark" on the pipe. When the pipe is made up properly, the coupling is aligned with the makeup mark. If the maximum torque is attained before the coupling reaches the makeup mark, it is an indication that the thread lubricant is the wrong type, the connections have not been properly cleaned, the pipe is not round, or the connection has been damaged. If the makeup mark is reached before the optimum torque is achieved, that is an indication the connections are either worn or the threads were not properly cut. Although not as common, some threaded and coupled pipe also has metal-to-metal seals.

1.6.3 Integral Connections

Another type of connection used for casing is one in which a metal-to-metal seal is achieved that is independent of the threaded area. These usually are integral-type connections cut into both ends of the pipe with no separate coupling. Some have a smooth tapered seal that seats very tightly when the proper makeup torque is achieved, others have a shoulder type seal, and as mentioned previously, still others have a combination of both. These types of connections give both high tensile strengths (some greater than the pipe body itself), greater bending strengths for curved well bores, and greater pressure sealing for high-pressure gas wells. Some of these threads may be cut in non-upset pipe for use as liners, typically called *flush-joint connections* because both the inner and outer diameters are the same in both the tube and connection. Most integral and metal-to-metal sealing connections often are referred to as *premium connections*, but this often is a misnomer. With the exception of API X-line, these should be referred to as *proprietary threads*. They are patented, and their dimensions and properties are strictly those specified by the manufacturer, even though they usually are on API specific tubes.

Many of the proprietary connections are designed for special applications, where the loading exceeds typical casing design loads. High-tensile loads and high pressures come to mind, but there are other types of loading we often do not consider. One of these is high torsion. If a casing string is to be rotated (for cementing or drilling), the frictional torque often is much higher than the recommended maximum makeup torque of most connections. Additionally, in some wells, where temperatures cycle significantly between flowing and shut-in times, severe compressive loading can take place. That a particular connection may be strong in tension does not necessarily mean that it is as strong in compression. For these applications, special connections have been designed. One proprietary thread is of an interlocking design, so that it may used in high-torque situations, curved well bores where bending due to bore-hole curvature is a possible cause of connection failure, and situations where axial compressive loading is significant. The interlocking-type thread is somewhat unique in that it is wider at the crest than at the base, and its width also is tapered along its length.

One should always consult the individual manufacturer for properties such as strengths and makeup torque. Another important point is that one should follow the manufacturer's recommendation as to thread lubricant, as some lubricants used with API 8-rd connections can result in loss of pressure seal in some of the proprietary connections. And, on the subject of thread lubricants, it should be mentioned that some connections are coated with special clear lubricants at the mill to avoid the need for field lubrication. This is not a labor-saving process but one of avoiding possible environmental and formation damage from conventional lubricants.

1.7 Strengths of Casing

The strengths of the many sizes, weights, and grades of casing are given in various sources. API casing strengths and dimensions are given in API Bulletin 5C2, and formulas used for calculating those values are given in API Bulletin 5C3 or ISO/DIS 10400. These values are also published in many other sources. We discuss later the formulas used to calculate casing design strengths.

1.8 Closure

In this chapter, we covered a few of the basics of oil-field casing, and it was assumed that the reader has some general knowledge of casing and its use in oil and gas wells. This chapter was not intended to be a comprehensive description of the manufacturing, metallurgy, and specifications of casing. The interested reader should refer to other publications for those types of information, such as the API and ISO publications mentioned in the References of this chapter as well as the published information of several casing manufacturers as well as the manufacturers of proprietary connections.

In the next chapter, we look at some of the basics of calculations and hydrostatics for casing design.

1.9 References

API Recommended Practices 5B1. (August 1999). *Threading, Gauging, and Thread Inspection of Casing, Tubing, and Line Pipe Threads.* Washington, DC: American Petroleum Institute.

API Bulletin 5C2. (October 1999). *Bulletin on Performance Properties of Casing, Tubing, and Drill Pipe.* Washington, DC: American Petroleum Institute.

API Specification 5CT. (2004). *Specifications for Casing and Tubing.* Washington, DC: American Petroleum Institute.

ISO/DIS 10400. (2004 draft). *Petroleum and Natural Gas Industries—Formulae and Calculations for Casing, Tubing, Drill Pipe, and Line Pipe Properties.* Geneva: International Organization for Standardization.

ISO 11960. (2004). *Petroleum and Natural Gas Industries—Steel Pipes for Use as Casing or Tubing for Wells.* Geneva: International Organization for Standardization.

Valigura, G. A., and P. Cernocky. (2005). *Shell Exploration and Production Company List of Connections Tested for Well Service Pressures >8000 Based on Standardized API/ISO Qualification Testing Procedures.* SPE 97605. Richardson, TX: Society of Petroleum Engineers.

CHAPTER 2

Basic Calculations and Hydrostatics

2.1 Introduction

Hydrostatics is a subject so simple we should not even have to devote space to it. If you actually believe that, then you have not worked in the oil field for very long. The truth is that hydrostatics is a relatively simple subject, but the problem is that its simplicity is often deceptive. Most texts and courses on fluid mechanics devote very little space or time to hydrostatics because the interesting part of fluid mechanics is fluid dynamics not fluid statics, consequently many authors and instructors treat fluid statics as trivial. And I suppose it might justifiably be considered trivial relative to hydrodynamics and other more difficult topics, but the net result is that too many engineers get through it with little more than a superficial understanding. Too often, some of the simplifying assumptions become imbedded as axioms of truth. For instance many engineers (and even a few distinguished professors) go through their entire careers believing that water is incompressible. And in our particular discipline, the concept of buoyancy, for example, has led to all sorts of incredible nonsense, some of which has even been published in textbooks and peer-reviewed literature. Take a look at the example in Figure 2–1 to illustrate a bit about buoyancy.

The figure shows a smooth tube suspended vertically in an idealized, vertical well. The tube has a sliding seal assembly on the bottom that allows frictionless vertical motion in a concentric packer. It is supported entirely at the surface, where a weight indicator measures its suspended

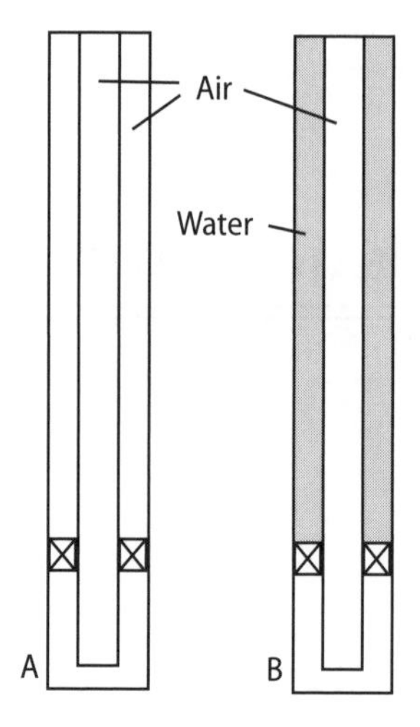

Figure 2–1 *Smooth tube suspended in a vertical well*

weight. Both cases are identical, except in case A the annular space is filled with air and in case B the annular space is filled with water. Now,

1. Will the weight indicator show that the weight of A is greater than B, less than B, or the same?
2. Suppose the tube in both cases is 4½-in. API ST&C casing, will the weight indicator show that the weight of A is greater than B, less than B, or the same?

The answer to the first question is "the same." The answer to the second question is "greater than B." Did you get both answers correct? If you did not, do not feel bad, because many engineers miss one or both of these (in addition, no one else will know you missed the correct answers). But let us suppose you answered both correctly, can you explain the reasons for the answers?

Now, let us pose two more questions. Suppose we have a joint of API casing (no defects) on a test rack in a lab and we also have a high-priced vacuum pump capable of pulling a near perfect vacuum[1] on the inside of the casing joint. Under a near perfect internal vacuum,

- Will the joint of casing collapse?
- Do we have enough information to make that determination?

1. A perfect vacuum is a theoretical concept that does not actually exist in the known universe.

The answer to the first question is "no," it will not collapse. The answer to the second is "yes," we do have enough information to determine that it will not collapse. If you missed any of these questions or if you cannot explain the answers, you need to read this chapter. If you correctly answered them all and you can actually explain the reasoning, you should still read this chapter.

2.2 Units of Measure

Before we get into any specific discussion of hydrostatics or any other calculations in this text, we should comment a bit on units of measure. This is a subject that one would like to avoid, but in an industry like ours, which actually spans the globe, some comments are necessary. There is no doubt that the world's acceptance of the Système International d'Unités (SI) as the world standard system in the latter half of the last century was a step in the right direction, and frankly no other system of units should have ever seen the light of the 21st century. But, for various reasons, local adoption of the SI system was not universal; consequently, our industry still is encumbered with a relic system, which we diplomatically refer to as oil-field units. A decision as to whether to continue perpetuating its use in a text such as this was a difficult one. My own preference is to use the SI system exclusively as a step in the right direction. However, such a choice would severely limit the utility of this text, since the vast majority of engineers and practitioners in our industry still use oil-field units, or possibly even worse, some hybrid combination of oil-field and SI units. Therefore, I chose to yield to the overwhelming popularity of the conventional oil-field units system for use in this text. But, rather than totally bow to convention, I am going to attempt to keep formulas and equations as independent of units as possible to enhance this text's usefulness for those using SI units. However, I make no concession at all to those who still use non-SI metric units such as bars, kilograms-force, and the like.

It should not be necessary to discuss units of measure any further, but if one were to read the discussions and arguments that took place in petroleum publications back in the days when a conversion to the SI system appeared imminent, one would realize that there was considerable confusion back then. To think that confusion has since been resolved is a mistake. I know this to be true, because I see it in the classroom all the time. The general confusion is not with the SI system, but with the one that has been used historically in the oil field since its beginning. The primary confusion is between units of mass and force, especially *pound-mass* and *pound-force* and their seeming interchangeability. (Those who use SI units should not be too smug here, because it was not that long ago that engineers working in the metric system used both kilograms-mass and kilograms-force.) In the English system of units, there are actually three distinct systems: the absolute system, the British gravitational system, and the engineering system. They differ in how they account for the units of mass and force. Table 2–1 summarizes the three English systems.

In the table, g is the local acceleration of gravity and g_c is a conversion factor taken to be 32.17405 (lb·ft)/(lbf·s^2), or for practical purposes 32.2 (lb·ft)/(lbf·s^2). Of the four possible systems in that table, the one used for "oil-field units" is the engineering system, which is the most confusing. One can see that, in the engineering system, it is often thought that pound (a unit of mass) and pound-force are interchangeable when it comes to mass and weight, and in fact, when the local gravity is equal to a standard gravity of 32.17405 ft/s^2, they are numerically equal (but their units never are). Note though, that they are not numerically equal in Newton's second law. Now, the gravitational system notwithstanding, the pound (avoirdupois) is a unit of mass and was defined by the National Bureau of Standards in 1959 as

$$1 \text{ pound} \equiv 0.45359237 \text{ kilogram} \tag{2.1}$$

That definition was also adopted by other countries that utilize that unit of measure. Since that is the standard definition of the *pound*, we use that definition as standard in this text and *pound* or lb always means a unit of *mass* not force. We do not use the term *pound-mass* or lbm, since that is redundant to the standard definition, and there is no system where such terminology is necessary. When we refer to force in the English engineering system we use *pound-force* or lbf.

2.2.1 Weight and Mass in Oil-Field Context

In addition to the preceding discussion of units, we should also clarify the confused use of the terms weight and mass in common oil-field language. Properly, *weight* means only one thing, force. However, that definition notwithstanding, it is quite common in the oil field and the world of commerce in general to refer to mass as *weight*. We often speak of casing, for instance, in terms of "size, weight, and grade," such as "7 inch, 26 pound, N-80." The "weight" referred to is not weight in its proper sense, but rather mass per unit length. In that example, it is in lb/ft; and in SI units it would be expressed as kg/m. Weight is a derived property, the force

Table 2–1 Common systems of units

System/ Quantity	SI	Absolute	Gravitational	Engineering
Time	second	second	second	second
Length	meter	foot	foot	foot
Mass	kilogram	pound	slug	pound
Force	Newton	poundal	pound-force	pound-force
Weight (a force)	$w = mg$	$w = mg$	$w = mg$	$w = m(g/g_c)$
Newton's second law	$f = ma$	$f = ma$	$f = ma$	$f = m(a/g_c)$

exerted on a body due to local gravity, and it is not the same on the moon as it is on earth, for example. But mass is a fundamental property and constant whether on the moon or on earth.[2] If a man jumps off a roof on the moon, his landing will be relatively soft compared to earth, because the gravitational acceleration on the moon is so low, his velocity when he hits the ground will be much less than if he jumped from the same height on earth. However, if a jogger is running along the moon's surface at 5 m/s and hits a brick wall it is going to hurt just as badly as on earth, because his mass is the same and his velocity is the same.

We should also point out that it has become common practice in SPE, API, and other publications to employ the term *pound-mass* and use the abbreviation, lbm, in addition to pound-force and lbf. This is quite unnecessary because the pound (lb) is the single standard unit of mass in both the English engineering and absolute systems. In the British gravitational system, the unit of mass is the slug, but the unit of force is the pound-force (lbf), so there should never be any confusion as to what a pound (lb) is. A pound (lb) always refers to mass in any standard system of units. Every engineer is taught that as part of his or her education (or certainly should have been taught). Therefore, it is only through ignorance and sloppiness that such a term as *pound-mass* (lbm) comes into engineering usage. It is not used here because a pound (lb) is defined as a unit of mass.

2. We are considering only Newtonian mechanics here. In special and general relativity, such a statement is false.

In casing design, unless we are calculating surge loads, we almost always use weight instead of mass. So we adopt the following convention here and for the rest of this text:

$$w \equiv \text{specific weight of casing (force per unit length)}$$

$$w = \frac{g}{g_c}\left(\frac{\text{Mass}}{\text{Length}}\right)\left(\frac{\text{lb}}{\text{ft}}\right) = \left(\frac{\text{Force}}{\text{Length}}\right)\left(\frac{\text{lbf}}{\text{ft}}\right) \tag{2.2}$$

$$w = g\left(\frac{\text{Mass}}{\text{Length}}\right)\left(\frac{\text{kg}}{\text{m}}\right) = \left(\frac{\text{Force}}{\text{Length}}\right)\left(\frac{\text{N}}{\text{m}}\right)$$

Specific weight refers to a weight/volume. That is essentially what was just defined, because the specific weight for a particular casing is determined by its outside diameter and inside diameter as well as its length. We also use a standard gravity in that definition, and that brings us to the next topic.

2.2.2 Standard Gravity

In the preceding formula, the specific weight is determined from the specific mass using the local acceleration of gravity, g. The acceleration of gravity is not a constant and varies even on earth. For instance, a string of casing with a mass of 500,000 lb might weigh 498,725 lbf at the equator and 501,275 lbf at one of the poles. We are not going to consider local variations of g in this text, but instead use an average value, called the *standard gravity.* Such an average value was adopted by the Conférence Générale des Poids et Mesures (CGPM) in 1901 and defined as the *standard gravity* to be used in legal weights and measures. The GCPM defined value is

$$g_s \equiv 9.80665\,\text{m/s}^2 \approx 9.81\,\text{m/s}^2 \tag{2.3}$$

In English units, it is

$$g_s \equiv \frac{196133}{6096}\,\text{ft/s}^2 \approx 32.17405\,\text{ft/s}^2 \approx 32.2\,\text{ft/s}^2 \tag{2.4}$$

It is not likely that the local acceleration of gravity will deviate enough from the standard gravitational acceleration to significantly affect our calculations, but it is important to understand these definitions and assumptions.

2.2.3 Fluid Density

Throughout this text, we continually use the density of fluids in well bores to calculate hydrostatic pressures. The common oil-field measure of liquid density is pound per gallon (lb/gal), which almost always requires conversion factors every time it is used. Almost anyone who uses that measure has long since committed to memory the necessary conversion factors, but it is still an awkward measure, especially for those accustomed to SI units and in regions where the common engineering density measure, pound per cubic foot, is used instead. Therefore, in the interest of making this text more universal, I use specific gravity in referring to the densities of well bore liquids. This specific gravity is easily multiplied by the density of water in whatever units one may want to use to give the density of the liquid in those units.

That, of course, brings up the question as to what is the density of water to which our measure is specific. Pure water at 4° C has a density of 1000 kg/m^3, and at 20° C, it has a density of 998 kg/m^3. Respectively then, those give us densities of 8.345 lb/gal and 8.33 lb/gal or 62.43 lb/ft^3 and 62.32 lb/ft^3. Some immediately will begin to question, then, to what temperature do we refer when we have to convert the specific gravity into a value of density for a calculation. It is interesting that this question arises in this context, because it almost never arises when one reads a drilling fluid density from a drilling report and uses it in hydrostatic calculations. At what temperature was the density measurement made on the rig, and how does that relate to the temperature and density down hole? Do we ever even think about that? In this text, I use the density of water at 20° C for sake of consistency and possibly some notion that it is perhaps closer to the average temperature at which most rig density measurements are made. But I also add that it does not make any difference, because nobody knows the density of the drilling fluid at 4° C or 20° C in the first place. Furthermore, if I work in SI units, I use water at 4° C just to make my life easier, since the density in kg/m^3 is simply 1000 times the specific gravity rather than 998 times.

2.2.4 Units in Formulas and Equations

In this text, I try to adhere to a long-standing conviction that numerical conversion factors have no place in equations and formulas. The use of numerical conversion factors in formulas only confuses understanding and often renders a formula useless for many applications. To give you an example, here is a formula for maximum stress due to planar bending of a tube (which we study in a later chapter):

$$\sigma = \pm E \frac{r_o}{R}$$

In this formula, the maximum axial bending stress, σ, is equal to Young's modulus, E, times the outside radius of the tube, r_o, divided by the radius of curvature, R, of the bore hole. The plus-or-minus sign indicates it is positive on the side of the tube that is stretched and negative on the side that is compressed in the curved configuration. This formula is valid for any system of units. Furthermore, in this form, it is easy to visualize that it is merely an expression of Hooke's law in one dimension, that is,

$$\sigma = E\varepsilon = \pm E \frac{r_o}{R}$$

We can see from this formula that, for simple bending, the strain (which is dimensionless) is the ratio of the radius of the pipe to the radius of the curvature of the bore hole. Note also that strain can be positive or negative but radii are always positive, hence, the necessity of the plus-or-minus sign in the last part. In this form, we can actually derive some insight into the nature of the bending stress just by looking at the formula. Any engineer should know that, to use this (or any) formula, one must use consistent units. For example, when this formula is written in terms of basic quantities (force and length in this case) it looks like this:

$$\left(\frac{\mathrm{F}}{\mathrm{L}^2}\right) = \left(\frac{\mathrm{F}}{\mathrm{L}^2}\right)\left(\frac{\mathrm{L}}{\mathrm{L}}\right)$$

It should be obvious that all length units should be the same as well as all force units. Now let us look at the same formula with some numerical conversion factors included:

$$\sigma = \pm \frac{E r_o}{12R} \quad \text{or} \quad \sigma = \pm \frac{E r_o}{1000R}$$

The first one uses oil-field units, where the stress is in lbf/in.^{2} and Young's modulus is also in lbf/in.^{2}. It should be obvious that the radius of the pipe and the radius of curvature of the well bore must be in the same units, either inches or feet. Typically, in the oil field, we express the pipe radius in terms inches and radius of curvature of the well bore in feet. Clearly, one of these has to be converted and the number in the first formula is a conversion factor, 12 in./ft. The second formula is in SI units where the stress and Young's modulus are in Pascals (Pa or kPa or MPa), and we measure casing radius in millimeters (mm) and radius of curvature for the well bore in meters (m), then we convert one or the other radius measures into the same units. In this case, the conversion factor is 1000 mm/m. Both these formulas are unit specific and cannot be used in another system without explanation as to what the units are for each variable in the formula. These two versions are fairly easy to decipher, but look at a version of this same formula that often appears in the petroleum literature:

$$\sigma = 218 D\theta$$

What on earth does this one mean? Is it even the same formula? In this version, D is the pipe diameter in inches and θ is the bore-hole curvature expressed as degrees per 100 ft. Let us see how this one is the same as our original version:

$$\sigma = \pm E\left(\frac{r_o}{R}\right) = \pm\left(30 \times 10^6 \frac{\text{lbf}}{\text{in}^2}\right) \frac{\left(\frac{D}{2}\text{in}\right)}{\left(\left(\frac{12\,\text{in}}{\text{ft}}\right)\left(\frac{180\,\text{deg}}{\pi\,\text{rad}}\right)\left(\frac{1\,\text{rad}}{\theta \frac{\text{deg}}{100\text{ft}}}\right)\right)}$$

$$\approx \pm 218 D\theta \text{ lbf/in}^2$$

Is there anything intuitive or insightful about the final result of that process? Granted, it may be useful to someone employing those particular units, but for someone else using SI units, how long would it take to come up with a usable version, where diameter is in millimeters and the curvature is in degrees per 10 meters? Young's modulus is buried in the formula as a numerical value, and unless one knows it is there and what the units are, one is at a loss to convert it to something useful in SI units like

$$\sigma = \pm E\frac{r_o}{R} = \pm\left(207\times10^9\,\frac{\mathrm{N}}{\mathrm{m}^2}\right)\frac{\left(\dfrac{D}{2}\,\mathrm{mm}\right)}{\left(\left(\dfrac{1000\,\mathrm{mm}}{\mathrm{m}}\right)\left(\dfrac{180\,\mathrm{deg}}{\pi\,\mathrm{rad}}\right)\left(\dfrac{1\,\mathrm{rad}}{\theta\dfrac{\mathrm{deg}}{10\mathrm{m}}}\right)\right)}$$

$$\approx \pm 181 D\theta \text{ kPa}$$

Also consider the possibility that our pipe is made of aluminum or some other alloy with a different value for Young's modulus. How useful would either of these last two formulas be then? That should be evidence enough that conversion factors have no place in equations and formulas.[3]

Along the way in this text we will obviously have to make some concessions to units, but what I will try to do is keep the explanation clear and in a few cases actually provide some conversion factors.

2.2.5 Significant Figures, Rounding, and Computers

I taught an undergraduate course in numerical methods for several semesters, and one thing never failed to get my attention. For a first assignment, I often used an easy problem to get the students adapted to the campus computer system and refresh their programming skills, since many had gone two semesters since they took a beginning computer programming course. A typical assignment was a simple beam-bending problem that consisted of entering the data from a formatted data file, calculating the transverse deflection of the beam due to a single load on the end, and printing the for-

---

3. There was a period in engineering history when it was considered de rigueur to lump all known numerical values and conversion factors in a formula into a single numerical value. That greatly aided slide rule calculations, but it does not justify this confusing practice today.

matted results to an output file. The deflection equation was given so that the assignment was only a computer exercise. The beam was 10.00 ft in length, but after a few semesters, I began to state its length as 120.00 in. to not have to spend so much time helping students find their "programming error." For most students it was a relatively simple assignment, but a semester never went by that this assignment did not generate several results stating that the deflection was something like this, "$u = 2042.89568936$." No units, no comments, just "the answer." I assume it must be in inches, since all other length measurements were given in inches. Did this student even notice that the deflection was about 17 times the original length of the beam? And what astounding precision had been accomplished! I could not even make up a scenario this ridiculous. The students were taught better than that, but for some strange reason computers seem to make a lot of people dumber rather than smarter.

Significant Figures

First of all, significant figures are not inconsequential in calculations. There are some rather strict rules that, if followed to the letter, can lead to absurdities such as $6 \times 2 = 10$. But that aside, one should at least be aware of the importance of significant figures and adhere as closely to the rules as is practical:

Rule 1. All nonzero digits are considered significant.

Rule 2. All zeros between two nonzero digits are significant. For example, 300.05 has five significant figures.

Rule 3. For any decimal number whose absolute value is less that 1, all zeros immediately to the right of the decimal are not significant figures. For example. 0.0043 and –0.00051 each have two significant figures.

Rule 4. For any decimal number, zeroes after a nonzero digit to the right of the decimal are significant. For example, 0.000400 has three significant figures.

Rule 5. For numbers whose absolute value is greater than 0, all zeros the right of the last nonzero digit are not significant unless there is a decimal. Without knowledge of their source, we must assume the number 64,000 has only two significant figures whereas 64,000. has five by virtue of the decimal point.

Rule 6. A zero by itself has one significant figure.

The application of significant figures to arithmetic operations has only two rules:

> *Multiplication and division.* The result should be rounded to the number of significant figures of the factor containing the least number of significant figures. For example, $9.81 \times 7521 = 73{,}800$.
>
> *Addition and subtraction.* The result should be rounded to the position of the least significant digit in the least uncertain number in the addition or subtraction. For example, $9.806 + 0.00055 = 9.807$.

We all relax these rules a bit from time to time, often out of laziness but occasionally with some purpose. This book is an example. The rules are violated several times intentionally to show the results of intermediate steps in calculations that would not always appear if the calculations were being done in sequence on a calculator or computer. But there are probably other places where I cannot claim this as an excuse. I have tried to weed out the most flagrant violations.

Rounding

There are different ways to round numbers, and they do have an effect on calculations. Some computer operations tend to truncate numbers, whereas we manually tend to round numbers up or down depending on whether the digit just past the rounding point is 5 or greater (round up) or less than 5 (round down). This latter method works well for small amounts of data, but it has a bias toward rounding up rather than down, since the digit 5 is a median and it is always rounded up in this method. There is a round-to-even method that compensates for this. It works by the following four rules:

- Determine the last digit to retain.
- If the following digit is 6 or greater or 5 followed by nonzero digits, then round up by 1.
- If the following digit is 4 or less, round down by 1.
- If the following digit is 5 with no following digits or all zero following digits, then round to the nearest even digit (if it is odd the nearest even digit is above, if it is even it stays the same). For example, if we are rounding to the second decimal place, 6.5251 rounds to 6.53, 6.5248 rounds to 6.52, 6.5250 rounds to 6.52, and 6.5350 rounds to 6.54.

Other methods work well on computers, but this is relatively easy for manual calculations.

Measurements

When it comes to measurements, precision and accuracy do not mean the same thing. Precision refers to repeatability and significant figures, which has nothing to do with accuracy, which refers to truth. You might possess the most precise digital scale in the world, step on it, read your body mass as 65.982812845 kg, and repeat this several times with the same reading. That scale is precise. But when you step off the scale, and its reading shows 2.032456210 kg with nothing on it, then it is not accurate. A well-head bourdon-tube pressure gauge may be more acute than a deadweight tester that has not been calibrated. Recently, three groups of physicists measured the gravitational constant (big-G, not to be confused little-g, the gravitational acceleration) with the greatest precision that had ever been achieved. All three groups got different results. What good is precision, if it is not accurate?

Another measurement transgression that seems to pop up often in the petroleum literature is the careless use of the ± symbol. "Surface casing was set at ±3000 ft" does not mean the same thing as "Surface casing was set at 3000 ± ft." In the first case, it means a 6000 ft tolerance in setting depth, so we assume the depth is measured at a datum like sea level and the location must be at least 3000 feet above sea level, right? The second case means the casing was set at about 3000 ft plus or minus some unspecified tolerance. The ± symbol is not a synonym for the word *approximate*(*ly*), especially if it appears before a measurement. In fact, when it appears after a measurement it could still mean anything if the tolerance is not specified.

Comment

The point of this brief discussion is essentially a reminder, because most of you have already been taught about significant figures, rounding, and measurements in some basic engineering course. So, if we round the standard gravity to three significant figures, then our calculations using that value are limited to the same number of significant figures. Furthermore, if local gravity varies from standard gravity in the fourth figure, which it often does, then many of our calculations are valid only to three significant figures, unless we take the variation of local gravity into consideration. Three or four significant figures are adequate for most of our casing design applications, but one should be aware of this limitation. Also, it should be mentioned, API and ISO pressure ratings are rounded to the nearest 10 lbf/in.[2] and axial load ratings to the nearest 100,000 lbf. Finally, I would be next to the last person on this planet to lament the passing of the slide rule to the electronic calculator or computer as a calculation tool, but the slide rule had two qualities computers do not possess: It forced one to live within the confines of significant figures, and since it did not place the decimal point, it forced one to become familiar enough with computations to recognize gross errors.

2.3 Fluid Statics

Our interest in casing design is primarily concerned with loads imposed by fluids; and in most cases for casing design, these fluids are static, or at least we consider the pressures and forces of the fluids only as if they are static. First of all then, we clarify a few terms and concepts. A fluid, by most definitions, is a material that cannot sustain a shearing load without continuous deformation. A fluid may be either a liquid or a gas. A static fluid is defined to be a fluid with no relative motion within the fluid itself. Hence, a fluid element does not deform, and this still allows for rigid-body motion of the fluid. (A glass of water sitting on a table is static in that there is no internal motion, but the fluid undergoes rigid-body motion in relation to the earth's axis, the sun, the galaxy, and so forth.)

Two types of forces may be exerted on a fluid, body forces and surface forces. The body force is the gravitational force. The fundamental equation for fluid statics may be written various ways. In differential form, it is

$$\mathrm{d}\vec{F}_B = \vec{g}\,\rho\,\mathrm{d}V \tag{2.5}$$

where $\mathrm{d}\vec{F}_B$ is an element body force of the fluid, $\vec{g}$ is the local acceleration of gravity, ρ is the density of the fluid, and dV is the element volume. Both the body force and gravitational acceleration are vectors and their direction is downward. The density is a scalar quantity (no direction), sometimes referred to as a field quantity, in that it is a function of position within the fluid. For simple applications where the density is nearly constant[4], we might also write the preceding equation as

4. All fluids are compressible, some more than others. The assumption that some fluids are incompressible is a convenience to aid in simplifying some calculations, which we often employ here. Unfortunately, many engineers go through their entire career believing that water actually is incompressible—it is not.

$$F = g\rho V = gm \tag{2.6}$$

where V is the volume and m is the mass of the fluid, and we omitted the vector symbols knowing that the direction of the body force (gravitational force) is always downward.

The *surface forces* on a fluid are *shear forces* and *normal forces*. A *surface* in this context is not necessarily the interface between the fluid and another substance but may be any real or imaginary surface within the fluid itself. As mentioned earlier, there are no shear forces in a static fluid, so the only surface force is a *normal* force, which is pressure, *normal* meaning perpendicular to the surface.

2.3.1 Hydrostatic Pressure

Hydrostatic pressure is more or less intuitive. It might be described as a force per unit area exerted on a static fluid or by a static fluid on some material body. The hydrostatic pressure is a result of the body force plus any additional pressure that may be applied to the fluid. We also can write the fundamental equation of fluid statics in terms of pressure instead of body force. But, before we do that, it is necessary to introduce a coordinate system to make things a little easier. In almost all texts on mechanics, the coordinate systems used for illustration show the vertical coordinate to be positive in the upward direction, however since almost everything we measure in regard to a well in the oil field is measured from the surface downward, we start out with such a coordinate system and stay with it throughout this text. Figure 2–2 shows the coordinate system we most often use.

Note that this is still a conventional right-hand coordinate system. It may look a bit awkward at first, because the positive z-axis points downward and the positive y-axis is to the right of the positive x-axis when viewed from above, but if viewed from below it is exactly what we are more accustomed to. The main advantage to us is that the z-axis is positive downward, so that our depth measurements in a well correspond in sign and direction to the z-axis. Another advantage to this coordinate system is that, if we assume the x-axis is positive in the North direction, then the y-axis is positive in the East direction, and directional azimuth and trigonometric functions are compatible since all angles are measured in the same direction. Directional azimuth in a well is measured from North in a clockwise direction, but trigonometric functions are such that the angle is measured counterclockwise from the x-axis towards the y-axis. Both of these are perfectly compatible in this coordinate system, since both appear to measure angles clockwise when viewed from above even though they are both counterclockwise in the actual coordinate system. The advantage of directional azimuth and trigonometric function compatibility is extremely important in directional drilling calculations.

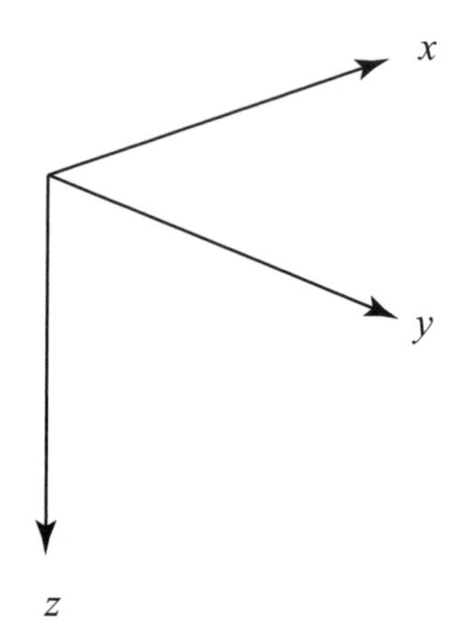

Figure 2–2 *An earth-oriented Cartesian coordinate system.*

The fundamental equation of fluid statics can be written in terms of pressure, where z is our vertical coordinate. In differential form it is

$$\mathrm{d}p = p_o + g\rho \mathrm{d}z \tag{2.7}$$

where equation dp is the element pressure and p_o is the external pressure on the element from above. Another form of the same equation for the pressure at some true vertical depth, h, is

$$p = p_o + \int_0^h g\rho \mathrm{d}z \tag{2.8}$$

If the fluid is incompressible such that the density is constant and the local gravity is constant, then

$$p = p_o + g\rho h \tag{2.9}$$

where p_o is some externally applied pressure (usually at the surface in our context). In many casing applications, we assume the pressure at the surface to be zero, and that equation becomes

$$p = g\rho h \tag{2.10}$$

The hydrostatic pressure exerted by a fluid on a material body always is normal (perpendicular) to the surface of the material body. This is very important to remember. Furthermore, the hydrostatic pressure in a fluid at any single point is equal in magnitude in all directions. And if we were to draw an imaginary surface through some part of a static fluid, the only force on that surface is a pressure normal to it. If you look back at Figure 2–1, you see that this accounts for the fact that both tubes weigh the same even though one is suspended in air and the other in water. Since the bottom of the tubes are not exposed to either annulus fluid, then those fluids have no effect on the weight of the tube as measured at the surface. The water pressure acts perpendicular to the walls of the tube and hence there is no upward or buoyant force. For a smooth vertical tube, there can be no vertical force component, due to hydrostatic pressure acting along the walls of the tube.

Before leaving this section, there is one more point we should briefly cover. We know that the units of pressure are force divided by area, F/L^2, and in the units common to the oil field, these are lbf/in.^{2} or N/m^2. Consider now a quantity of units force·length divided by volume, $F{\cdot}L/L^3$, lbf·in./in.^{3} or N·m/m^3. Note that the length dimension in the numerator cancels one of the length dimensions in the denominator leaving the same dimensions as pressure. Is it possible then that pressure might mean something in addition to our common perception? Note that this new quantity has, in the numerator, the dimension of energy, and the denominator has the dimension of volume. This might be energy divided by volume, or more correctly, *energy density*; in fact, that is another way of thinking about fluid pressure. It can be a very useful potential function from which we can derive or calculate various other quantities using energy methods rather than equilibrium formulations. While this has little application in hydrostatics, it can be a valuable tool in hydrodynamics, such as drilling hydraulics calculations. I mention this here because such a concept rarely appears in basic fluid mechanics texts but is invaluable in many fluid mechanics applications.

2.3.2 Buoyancy and Archimedes' Principle

When we mention buoyancy, Archimedes' principle automatically comes to mind. It is an essential part of hydrostatics, which states that the buoyant force on a submerged, or partially submerged, body is equal to the weight of the volume of fluid displaced by the body. This is quite handy for calculating the buoyed weight of submerged objects, such as casing, without the necessity of determining the hydrostatic forces on the body, which requires details of the body geometry and depths within the fluid. For instance, a cube with dimensions $b \times b \times b$ submerged in a liquid, as shown in Figure 2–3, has a buoyant hydrostatic force acting on the bottom cross-sectional area, $b \times b$. We can calculate the force on the bottom easily:

$$F_{\text{bottom}} = -g\rho_{\text{liquid}} h b^2 \tag{2.11}$$

The force on the top is

$$F_{\text{top}} = g\rho_{\text{liquid}} (h - b) b^2 \tag{2.12}$$

The net buoying force is the sum of those two. And the buoyed weight, $\hat{W}$, of the cube can be calculated as

$$\hat{W} = g\rho_{\text{solid}} b^3 + g\rho_{\text{liquid}} (h - b) b^2 - g\rho_{\text{liquid}} h b^2 \tag{2.13}$$

$$\hat{W} = g\left[\rho_{\text{solid}} b^3 + \rho_{\text{liquid}} \left(hb^2 - b^3 - hb^2\right)\right] \tag{2.14}$$

$$\hat{W} = g\left(\rho_{\text{solid}} - \rho_{\text{liquid}}\right) b^3 \tag{2.15}$$

We see that the depth is irrelevant (assuming the liquid is incompressible) and the buoyed weight depends only on the difference in densities of the solid and the liquid and the volume of the solid (or the displaced liquid), and equation (2.15) is a statement of Archimedes' principle, which can be generalized as

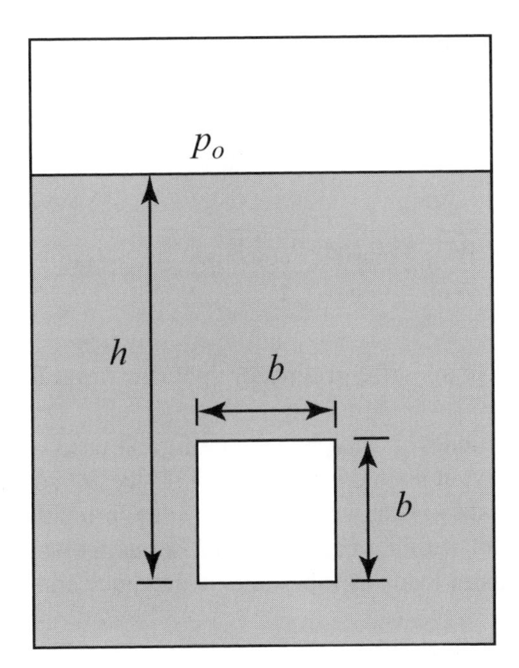

Figure 2–3 *A solid cube submerged in a liquid.*

$$\hat{W} = g\left(\rho_{\text{solid}} - \rho_{\text{liquid}}\right)V_{\text{solid}} \tag{2.16}$$

where

$\hat{W}$ = buoyed weight of solid
V_{solid} = volume of solid
ρ_{solid} = density of solid
ρ_{liquid} = density of liquid

Now suppose that, instead of the orientation in Figure 2–3, the cube had been oriented with the "bottom" face 60° to the vertical and a "side" face 30° to vertical. The calculation of the buoyed weight is much more complicated but yields the same result. So, it should be obvious that Archimedes' principle is of considerable use in simplifying calculations of buoyed weight of submerged or semi-submerged objects.

Another handy relationship we can derive for employing Archimedes' principle is the formula for a buoyancy factor for a body, which is the ratio of the buoyed weight in a liquid to the weight in air. We can start with a definition of the buoyancy factor, f_b and equation (2.16):

$$f_b \equiv \frac{\hat{W}}{W} = \frac{g\left(\rho_{\text{solid}} - \rho_{\text{liquid}}\right)V_{\text{solid}}}{g\,\rho_{\text{solid}}\,V_{\text{solid}}} = 1 - \frac{\rho_{\text{liquid}}}{\rho_{\text{solid}}} \tag{2.17}$$

We could also use specific gravity or specific weights in the formula instead of density.

While Archimedes' principle can be of great use, as we just demonstrated, it can also get us into serious trouble if we are not careful. Here is an important question: Can we use Archimedes' principle to determine a load at some point within a buoyed body? Let us look at some examples of axial and moment loads on tubes due to buoyancy and gravity.

Axial Load in a Vertical Tube

Suppose we have a smooth tube suspended vertically in a liquid to some depth, h, as in Figure 2–4. We want to know the axial load, F_a, in that tube at point a, some distance, ℓ, from the bottom. The density of the liquid (fluid) is ρ_f, the density of the tube (solid) is ρ_s, and the tube is open at both ends.

We use Newton's third law by saying the force at point a is equal to the buoyed weight of the portion of the tube below that point, and we can calculate that buoyed weight by calculating the displaced volume times the difference in the density of the solid and the fluid, in other words Archimedes' principle:

$$\sum F_z = 0$$

$$gV\Delta\rho - F_a = 0$$

$$F_a = g\pi\left(r_o^2 - r_i^2\right)\ell\left(\rho_s - \rho_f\right) \tag{2.18}$$

Now, is that correct? Is that really the axial force in the tube at point a? Let us check it using Newton's third law and the actual forces, that is, the body

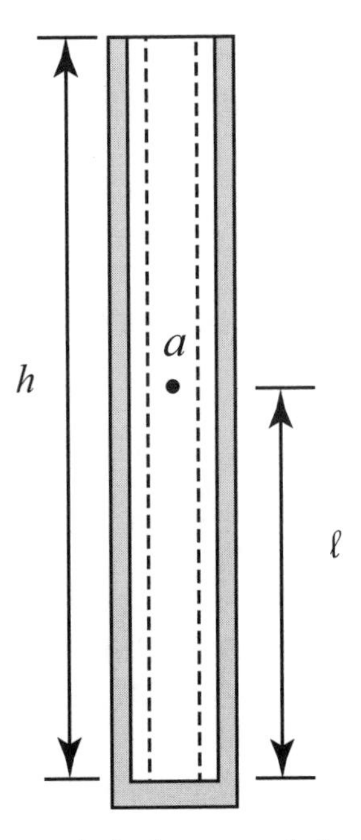

Figure 2–4 *A smooth open-ended tube suspended vertically in a liquid.*

force of the section of tube below point a and the force of the fluid on the bottom of the tube. Assuming there is no pressure at the surface, then

$$\sum F_z = 0$$

$$g\rho_s \pi \left(r_o^2 - r_i^2\right)\ell - g\rho_f \pi \left(r_o^2 - r_i^2\right)h - F_a = 0$$

$$F_a = g\pi\left(r_o^2 - r_i^2\right)\left(\ell\rho_s - h\rho_f\right) \tag{2.19}$$

The two methods give different results because ℓ and h are not the same! We know that Newton's law with the forces is correct, so what is wrong with using Archimedes' principle in Newton's law? Before we answer that, let us note one thing about the results. If $\ell = h$, then they both give the same results. In other words, the only point where Archimedes' principle can give the correct axial load in the suspended tube is at the surface. The casing in this example is being acted on by forces (at the surface) not accounted for in Archimedes' principle. The problem with Archimedes' principle is that, in general, it cannot be used on part of a body to give the loads within the body itself. This is important to remember, and it is a common mistake made by many who should know better. Newton's third law with actual forces gives us the true loads; Archimedes' principle gives us only something often termed *effective loads* in the vertical direction. Actually, there are legitimate applications for the effective loads, and we discuss those later. For now we consider how Archimedes' principle can give us misleading results.

Axial Load in a Horizontal Tube

Now let us attempt another application using Archimedes' principle. Suppose we have an open-ended tube fixed to a vertical wall and extending horizontally into a liquid, and the central longitudinal axis of the tube is at some depth, h, as in Figure 2–5. What is the longitudinal axial load in this horizontal tube (neglecting bending loads due to gravity for now)? We can see that a pressure load on the end of the tube is acting on the cross-sectional area, so we know there is a compressive axial load in the tube. Can we use Archimedes' principle to determine that load? No, we cannot, because Archimedes' principle says nothing about horizontal hydrostatic loads. Also we cannot simply multiply the pressure by the cross-sectional area, as we did previously, because it is apparent from the figure that the pressure on the cross section is not the same at all points since the pressure varies with depth. So, let us see how we calculate the load on the end due to the liquid pressure, which varies with depth.

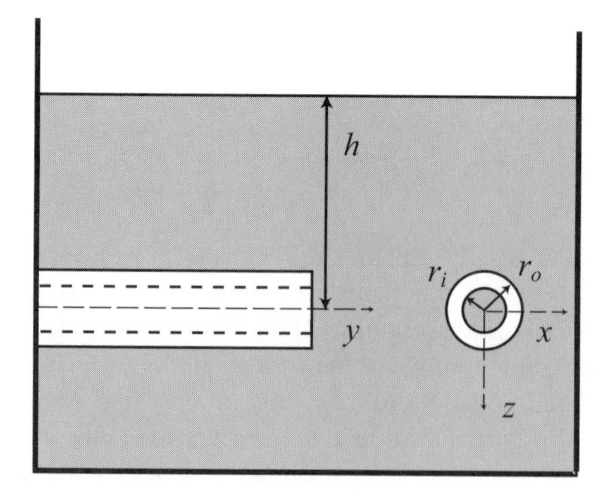

Figure 2–5 *Horizontal tube, centroid at depth,* h.

We could show this more easily if it were a solid bar with a rectangular cross section, but since our interest is in tubes, we might as well see the details of how it is done. Since pressure varies with depth, we can express the pressure at some point on the tube end as follows:

$$p = p_o + g\rho_f \left(h + r\cos\theta\right) \tag{2.20}$$

Note carefully the orientation of our coordinate system, because we have adopted the convenient system mentioned earlier for our use in well-bore calculations, and it appears to be upside down to what we are accustomed to seeing; that is, the z-axis is positive downward. The angle, θ, is measured counterclockwise from the positive z-axis. Since the pressure varies over the area of the tube, the force due to the pressure on the end of the tube is the pressure integrated over the area of the tube:

$$\sum F_y = 0$$

$$-F - \iint p\,\mathrm{d}A = 0$$

$$F = -\int_{r_i}^{r_o}\int_0^{2\pi}\left[p_o + g\rho_f\left(h + r\cos\theta\right)\right]r\,\mathrm{d}\theta\,\mathrm{d}r$$

$$F = -\int_0^{2\pi}\frac{p_o}{2}\left(r_o^2 - r_i^2\right)\mathrm{d}\theta - g\rho_f\int_0^{2\pi}\left[\frac{h}{2}\left(r_o^2 - r_i^2\right) + \frac{r_o^3 - r_i^3}{3}\cos\theta\right]\mathrm{d}\theta$$

$$F = -\left(p_o + g\rho_f h\right)\pi\left(r_o^2 - r_i^2\right) \tag{2.21}$$

From this result, we see that the force on the end of the tube is equal to the pressure at the center of the tube times the cross-sectional area of the tube. Is this a general result for any tube, or is it specific for a horizontal tube face? This can be generalized to any inclination and is an important result in fluid statics, in that the force of a fluid of constant density on a submerged flat surface is equal to the pressure at the centroid of the surface times the area of the surface.

Axial Load in an Inclined Tube

We generalize the preceding statement with the following example of an inclined tube. The tube in Figure 2–6 is inclined at some angle, α, from vertical. Note that, in oil-field applications, well bore inclination angles are always measured from vertical not horizontal. So again the pressure at the center of the tube end at its end is given by

$$p = p_o + g\rho_f h \tag{2.22}$$

where p_o is a surface pressure and h is the depth of the center of the tube end. If we use Newton's third law, again with the subscript s denoting a coordinate direction along the longitudinal axis of the tube, the axial force on the end of the tube is

$$\sum F_s = 0$$

$$-F - \iint p\,\mathrm{d}A = 0$$

$$F = -\int_{r_i}^{r_o}\int_0^{2\pi}\left[p_o + g\rho_f\left(h + r\cos\theta\sin\alpha\right)\right]r\,\mathrm{d}\theta\,\mathrm{d}r$$

$$F = -\int_0^{2\pi}\frac{p_o}{2}\left(r_o^2 - r_i^2\right)\mathrm{d}\theta - g\rho_f\int_0^{2\pi}\left[\frac{h}{2}\left(r_o^2 - r_i^2\right) + \frac{r_o^3 - r_i^3}{3}\cos\theta\sin\alpha\right]\mathrm{d}\theta$$

$$F = -\left(p_o + g\rho_f h\right)\pi\left(r_o^2 - r_i^2\right) \tag{2.23}$$

We see then that the result is the same. No matter what the inclination angle the hydrostatic force on the end of the tube is equal to the cross sectional area of the tube times the pressure at the centroid of the tube end.

Now let us look at the axial force at some point, a, some distance, ℓ, from the end of an inclined tube:

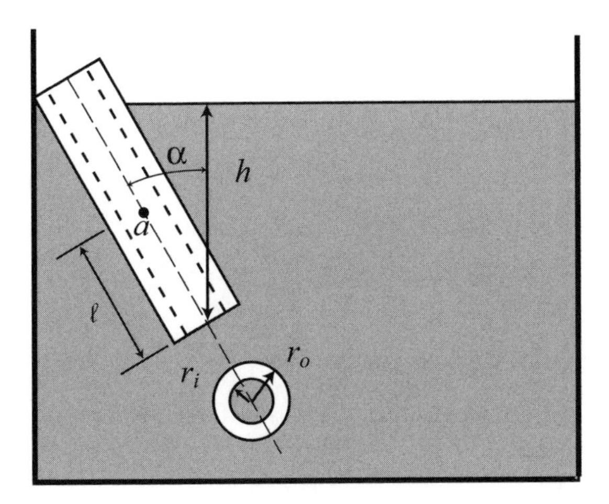

Figure 2–6 *Tube at an inclined angle in a liquid.*

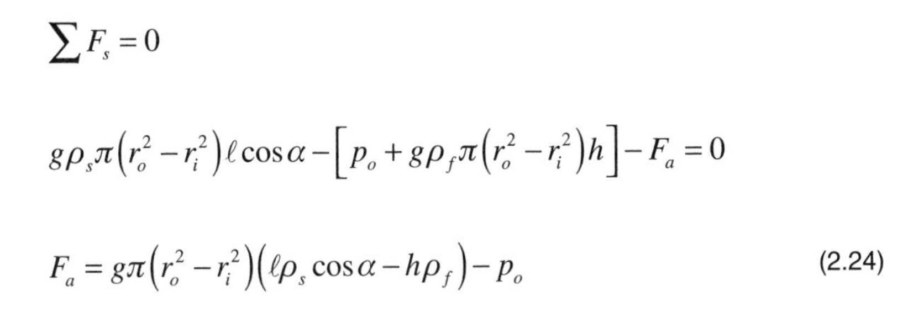

$$\sum F_s = 0$$

$$g\rho_s \pi \left(r_o^2 - r_i^2\right)\ell \cos\alpha - \left[p_o + g\rho_f \pi \left(r_o^2 - r_i^2\right)h\right] - F_a = 0$$

$$F_a = g\pi\left(r_o^2 - r_i^2\right)\left(\ell\rho_s \cos\alpha - h\rho_f\right) - p_o \tag{2.24}$$

Notice that, as the inclination angle goes to 0° (vertical), equation (2.24) is identical to equation (2.19) with the addition of a surface pressure, and when the inclination goes to 90° (horizontal), it is identical to equation (2.21).

Moment in a Horizontal Tube

Let us now look at one more example of a horizontal tube and determine the moment in the tube at some point *a*. We determine the moment about the *x*-axis, in Figure 2–7.

Again we might be tempted to use Archimedes' principle as others have before. We try that using the buoyed weight of the end segment to find the moment at point *a* about the x-axis:

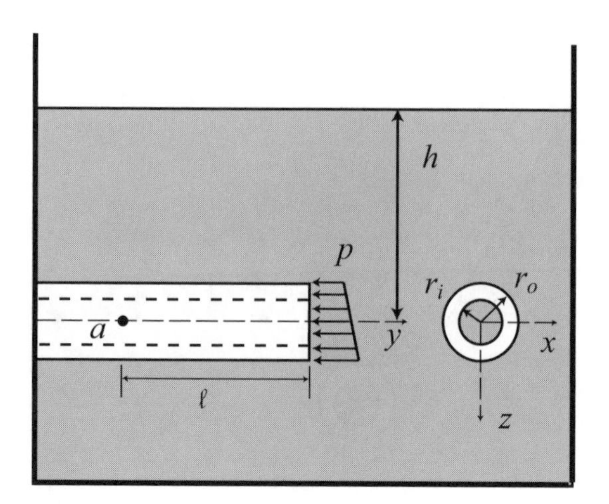

Figure 2–7 *A fixed horizontal tube.*

$$\sum M_x = 0$$

$$\int_{y_a}^{\ell} \pi\left(r_o^2 - r_i^2\right) g\left(\rho_s - \rho_f\right)\left(y - y_a\right) \mathrm{d}y - M_a = 0$$

$$M_a = \frac{\left(\ell - y_a\right)^2}{2} \pi\left(r_o^2 - r_i^2\right) g\left(\rho_s - \rho_f\right) \tag{2.25}$$

That is pretty straight forward and relatively easy. Now, let us do it using the actual forces, and to save a little bit of calculation, we use the center of gravity of the segment as the length of the moment arm:

$$\sum M_x = 0$$

$$\frac{\left(\ell - y_a\right)^2}{2} g \begin{bmatrix} \rho_s \pi\left(r_o^2 - r_i^2\right) + \rho_f \int_0^{2\pi} \left(h + r_o \cos\theta\right) r_o \cos\theta \, \mathrm{d}\theta \\ -\rho_f \int_0^{2\pi} \left(h + r_i \cos\theta\right) r_i \sin\theta \, \mathrm{d}\theta \end{bmatrix}$$

$$+ g\rho_f \int_0^{2\pi} \int_{r_i}^{r_o} \left(h + r\cos\theta\right) r^2 \cos\theta \, \mathrm{d}r \, \mathrm{d}\theta - M_a = 0$$

$$M_a = \frac{(\ell - y_a)^2}{2} g\left[\rho_s \pi (r_o^2 - r_i^2) - \rho_f \pi r_o^2 + \rho_f \pi r_i^2\right] + g\rho_f \frac{\pi}{4}(r_o^4 - r_i^4)$$

$$M_a = \frac{(\ell - y_a)^2}{2} g\pi (r_o^2 - r_i^2)(\rho_s - \rho_f) + g\rho_f \frac{\pi}{4}(r_o^4 - r_i^4) \tag{2.26}$$

We can see immediately that the results are different. The first term of the result is the moment due to gravity, and we see that it is exactly the same as the moment we derived using Archimedes' principle. But the second term is a moment at the end of the tube due to the difference in the pressure from the top of the tube to the bottom of the tube. We might call this term a pressure-end moment. Clearly, this term is finite and does contribute to the moment in the tube, so it is not something fictitious that we can arbitrarily disregard. But, how significant is this term in oil-field applications? Let us look at an example to get some idea of the magnitude of that term in an oil-field context. Say, the tube is 9⅝ in. casing with an inside diameter of 8.681 in., and the fluid is drilling fluid with a specific gravity of 2.0 (124.64 lb/ft^3 or 1996 kg/m^3):

$$g\rho_f \frac{\pi}{4}(r_o^4 - r_i^4)$$

$$= \frac{32.2\left(\frac{\text{ft}}{\text{s}^2}\right)}{32.2\left(\frac{\text{lb}\cdot\text{ft}}{\text{lbf}\cdot\text{s}^2}\right)} 124.64\left(\frac{\text{lb}}{\text{ft}^3}\right)\frac{\pi}{4}\left\{\left[\frac{\frac{9.625}{2}(\text{in})}{12\left(\frac{\text{in}}{\text{ft}}\right)}\right]^4 - \left[\frac{\frac{8.681}{2}(\text{in})}{12\left(\frac{\text{in}}{\text{ft}}\right)}\right]^4\right\}$$

$$= 0.86 \text{ lbf}\cdot\text{ft} \tag{2.27}$$

This is a relatively small value in oil-field calculations (0.86 lbf·ft or 1.16 J), so in oil-field applications, we almost always choose to ignore it. But if we choose to assume its value to be negligible, then we must certainly acknowledge at least to ourselves that we are doing so.

Moment in an Inclined Tube

That was for a horizontal tube, but what is generalized result for an inclined tube? In this example, the inclination angle is α, h is the depth at the center of the tube end, and s is a coordinate along the axis of the inclined tube. Now the pressure varies along the length of the tube.

$$\sum M_x = 0$$

$$\begin{aligned}
&\int_{s_a}^{\ell} g\rho_s\pi\left(r_o^2 - r_i^2\right)\left(s - s_a\right)\sin\alpha\,\mathrm{d}s \\
&\quad + \int_{s_a}^{\ell}\int_0^{2\pi} g\rho_f\left\{\left[h - (\ell - s)\cos\alpha\right] + r_o\sin\alpha\cos\theta\right\}r_o\left(s - s_a\right)\cos\theta\,\mathrm{d}\theta\,\mathrm{d}s \\
&\quad - \int_{s_a}^{\ell}\int_0^{2\pi} g\rho_f\left\{\left[h - (\ell - s)\cos\alpha\right] + r_i\sin\alpha\cos\theta\right\}r_i\left(s - s_a\right)\cos\theta\,\mathrm{d}\theta\,\mathrm{d}s \\
&\quad + g\rho_f\int_0^{2\pi}\int_{r_i}^{r_o} g\rho_f\left(h + r\cos\theta\sin\alpha\right)r^2\cos\theta\,\mathrm{d}r\,\mathrm{d}\theta - M_a = 0
\end{aligned}$$

$$M_a = \frac{\left(\ell - s_a\right)^2}{2} g\left[\rho_s\pi\left(r_o^2 - r_i^2\right) - \rho_f\pi r_o^2 + \rho_f\pi r_i^2\right]\sin\alpha + g\rho_f\frac{\pi}{4}\left(r_o^4 - r_i^4\right)\sin\alpha$$

$$M_a = \frac{\left(\ell - s_a\right)^2}{2} g\pi\left(r_o^2 - r_i^2\right)\left(\rho_s - \rho_f\right)\sin\alpha + g\rho_f\frac{\pi}{4}\left(r_o^4 - r_i^4\right)\sin\alpha \tag{2.28}$$

This was also a bit tedious to do, but most of the terms evaluate to zero and the results are identical to the horizontal case, except for the sine of the inclination angle. Note that, as the inclination angle goes to zero (vertical well bore), both terms, the gravitational moment and the pressure-end moment, vanish, as one would expect. Likewise, both terms are a maximum when the tube is horizontal.

2.4 Oil-Field Calculations

It is important that we become familiar with routine calculations involving hydrostatics in bore-hole applications. This is particularly important, because more often than not, we do not have measured pressures down hole, and we must rely on surface pressures and known fluid densities to calculate down-hole pressures and loads. In drilling and casing applications, we typically work with gauge pressures as opposed to absolute pressures. Atmospheric pressure is negligible in the context of these types of applications, so it typically is ignored, but at least we acknowledge here that we recognize the difference between absolute pressures, which include atmospheric pressure, and gauge pressures, which do not.

2.4.1 Hydrostatic Pressures in Well Bores

Liquid Columns

Calculating hydrostatic pressure due to a liquid column in a well bore is probably the most frequent type of calculation made in drilling, completion, intervention, and stimulation work. It is easy to do and one of the first things a field engineer learns to do. The best way to understand it is with an example. Figure 2–8 shows a simple but common well-bore situation. A tube is hanging freely in a vertical well bore with an open end. The well-bore fluid has a specific gravity of 1.5 and the depth of the tube is h. What is the hydrostatic pressure at the end of the tube?

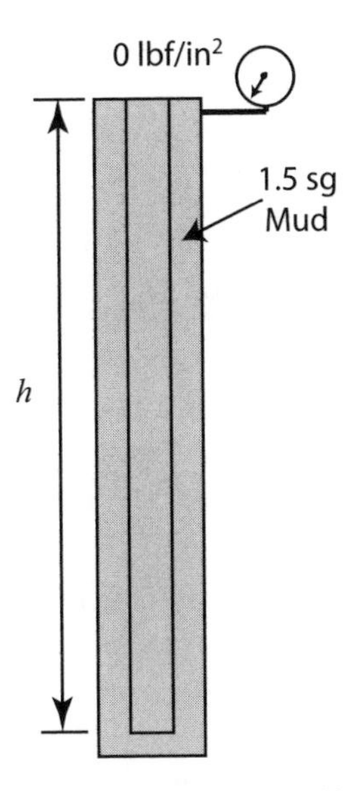

Figure 2–8 *An open-end tube suspended in a well.*

The pressure can be calculated as

$$p = p_o + g\rho_f h = p_o + g(1.5\rho_w)h$$

where ρ_w is the density of water. This calculation is quite simple in SI units, because the density and depth are always in compatible units. For instance, if the depth is 3000 m then the calculation is

$$p = 0 + \left(9.81\frac{\text{m}}{\text{s}^2}\right)(1.5)\left(998\frac{\text{kg}}{\text{m}^3}\right)(3000\,\text{m}) = 43.97\times10^6\,\frac{\text{N}}{\text{m}^2} = 43.97\,\text{MPa}$$

SI units are consistent and all these types of calculations are straight forward, however, oil-field units are not consistent and conversion factors are required. The fact that pressure is measured in lbf/in.^{2}, depth in ft, and density in lb/gal basically means that length, area, and volume are measured in three distinct and inconsistent units. So let us derive the necessary conversion factor such that we may have it available for use in all our oil-field calculations of this type. One way to do this is as follows:

$$p = p_o + \frac{g}{g_c} C_{\text{L}} \rho_f h \tag{2.29}$$

where C_{L} is a conversion factor for the length, area, and volume units and g_c is a conversion factor for the mass units as previously discussed. We can calculate the value of this factor as

$$C_{\text{L}} = \left(\frac{\text{gal}}{231\,\text{in}^3}\right)\left(\frac{12\,\text{in}}{\text{ft}}\right)$$

$$C_{\text{L}} = 0.051948\left(\frac{\text{gal}}{\text{in}^2\cdot\text{ft}}\right)$$

$$C_{\text{L}} \approx 0.052\left(\frac{\text{gal}}{\text{in}^2\cdot\text{ft}}\right) \tag{2.30}$$

This is the factor commonly used in the oil field, and since any deviation of local gravity from standard gravity is usually negligible, the term g/g_c is assumed to be unity and generally is ignored in this system of units. However, this has also led to a certain amount of misunderstanding on the part of those using these units, since the gravitational acceleration is an essential part of any equation dealing with gravitational body forces (e.g., hydrostatic pressure) and cannot be omitted. We could use the proper formula for pressure in a fluid of constant density as

$$p = p_o + g\rho_f h \tag{2.31}$$

with the understanding that, when using oil-field units for calculation, it means

$$p = p_o + \frac{g}{g_c} C_L \rho_f h \tag{2.32}$$

but we take one more step to simplify that a bit. If we use specific weight instead of density, it makes the use of conversion factors somewhat easier. Specific weight is defined as

$$\gamma \equiv g\rho \tag{2.33}$$

In the English engineering system, this means

$$\gamma = \frac{g}{g_c}\rho \tag{2.34}$$

The SI units of specific weight are N/m^3, and in oil-field units, they are lbf/gal. This makes the pressure formula slightly less cumbersome.

$$p = p_o + \gamma h \tag{2.35}$$

Remember though that, in oil-field units, we still need the conversion factor for length, area, and volume as

$$p = p_o + C_L \gamma h \tag{2.36}$$

Now, if we take the previous example, with a depth of, say, 10,000 ft, and calculate the pressure at the end of the tube it goes like this:

$$p = p_o + C_L \gamma_f h$$

$$p = 0 + \left(0.052 \frac{\text{gal}}{\text{in}^2 \cdot \text{ft}}\right)(1.5)\left(8.33 \frac{\text{lbf}}{\text{gal}}\right)(10000\,\text{ft})$$

$$p = 6487 \frac{\text{lbf}}{\text{in}^2}$$

And in SI units

$$p = p_o + \gamma_f h$$

$$p = 0 + (1.5)\left(9810 \frac{\text{N}}{\text{m}^3}\right)(3048\,\text{m})$$

$$p = 44.85\,\text{MPa}$$

In the SI unit calculation, we used the specific weight of water at 4° C to make life easier, so the results are slightly different. If you prefer, you could use the density at 20° C and the specific weight of water would be $\gamma = g\rho = 9.81(998) = 9790$ N/m^3. That type of calculation should become routine for the engineer or anyone doing hydrostatic calculations in well bores. From now on, we do not show the conversion factor in the formula, so that the formula is not unit specific, but you must remember to include it when using oil-field units.

Figure 2–9a shows another example of the same well with pressure on the surface. The formula is the same but the pressure at the surface is not zero in this case.

$$p = p_o + \gamma_f h$$

$$p = p_o + (1.5\gamma_w)h$$

If the surface pressure is 1200 lbf/in.^{2}, calculate the pressure at the bottom of the tube:

$$p = 1200 + 0.052(1.5)(8.33)(10000) = 7697\frac{\text{lbf}}{\text{in}^2}$$

Figure 2–9b shows different fluids in the same well bore. In this case, the fluid in the annulus has a specific gravity of 1.5, as before, but the fluid in the tubing has a specific gravity of 1.1.

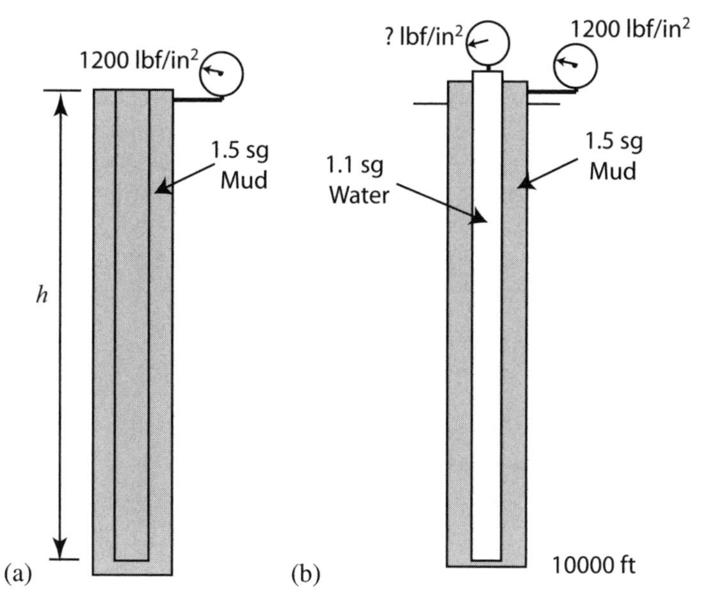

Figure 2–9 *(a) Same well with surface pressure. (b) Well with different fluids and surface pressure.*

There is a pressure of 1200 lbf/in.^{2} on the annulus, and the tubing is closed at the surface. Here, our task is to calculate the pressure at the surface in the tubing. There are various ways to do this, but the easiest way is to set up an equality knowing the pressure in both the tubing and annulus are equal at the bottom of the tubing:

$$p_{ot} + \gamma_{ft} h = p_{oa} + \gamma_{fa} h$$

$$p_{ot} = p_{oa} + \left(\gamma_{fa} - \gamma_{ft}\right) h$$

$$p_{ot} = 1200 + (0.052)(1.5 - 1.1)(8.33)(10000)$$

$$p_{ot} = 2933 \frac{\text{lbf}}{\text{in}^2}$$

For a slightly different perspective, suppose we do not know the density of the fluid in the tubing, but we measure the pressure at the surface of the tubing to be 2000 lbf/in.^{2}. All other variables are the same. What is the specific gravity of the fluid in the tubing? What is the density of the fluid in the tubing? We start out exactly the same; that is, we know the pressure at the bottom of the tubing:

$$p_{ot} + \gamma_{ft} h = p_{oa} + \gamma_{fa} h$$

$$\rho_{ft} = \frac{p_{oa} + \gamma_{fa} h - p_{ot}}{gh}$$

$$\rho_{ft} = \frac{p_{oa} - p_{ot}}{gh} + \rho_{fa}$$

$$\frac{\rho_{ft}}{\rho_w} = \frac{p_{oa} - p_{ot}}{\gamma_w h} + \frac{\gamma_{fa}}{\gamma_w}$$

$$S_{gt} = \frac{p_{oa} - p_{ot}}{\gamma_w h} + S_{ga}$$

$$S_{gt} = \frac{1200 - 2000}{0.052(8.33)10000} + 1.5$$

$$S_{gt} = 1.32$$

And the answer to the second question is

$$\rho_t = S_g \rho_w = 1.32(8.33) \approx 11.0 \ \frac{\text{lb}}{\text{gal}}$$

The only difficulty with this particular problem is where to use the conversion factors, and that is just a matter of practice and familiarity with the units.

Gas Columns

Frequently in casing design, it becomes necessary to calculate pressures at various points in the well bore in the presence of a full or partial column of gas. This is common in the design of intermediate and production strings of casing. Unlike most liquids, we cannot take the density of the gas to be constant over some interval in a well bore, because its density is a function of both pressure and temperature. There are numerous ways to calculate gas density and pressures, and all depend on the type of gas. Many charts and simple computer programs are available for the purpose, and that is what most people use. However, to keep this text as simple as possible, we use an approximation based on methane gas. The advantage to this is that we need not know anything about the composition of the gas we encounter in a particular well, because methane is the lightest of all the possible gases encountered in oil and gas wells (with the exception of helium); hence, it almost always represents the worst case load on the casing. In fact, a number of companies always use methane as the gas in casing designs that require a gas load on casing. This makes calculations easy for us because we can derive a simple formula for our use.

We start with the modified ideal gas law:

$$pV = ZnRT \tag{2.37}$$

where

$p =$ absolute gas pressure

$V =$ gas volume

$Z =$ gas compressibility factor (a correction from ideal gas behavior)

$n =$ number of moles of gas

$R =$ ideal gas constant (a conversion factor of gas units)

$T =$ absolute temperature

and the relationship

$$n = \frac{m}{M} \tag{2.38}$$

where

$n =$ number of moles of gas
$m =$ mass of the gas
$M =$ molecular mass of the gas (commonly called molecular weight)

We combine those two equations to give us an equation of state for the density, ρ, of a gas:

$$\rho = \frac{m}{V} = \frac{pM}{ZRT} \tag{2.39}$$

We substitute this into equation (2.7) without the initial pressure, which is not needed here, to give us a differential equation for the gas pressure as a function of the vertical coordinate:

$$\mathrm{d}p = g\rho\,\mathrm{d}z = g\frac{pM}{ZRT}\mathrm{d}z \tag{2.40}$$

Separating variables, we get

$$\frac{\mathrm{d}p}{p} = g\frac{M}{ZRT}\mathrm{d}z \tag{2.41}$$

The temperature is a function of the depth, but to keep things simple we assume it is a linear function and that we may use an average temperature as a constant. We also assume the compressibility factor is a constant. This gives us

$$\int_{p_1}^{p_2} \frac{\mathrm{d}p}{p} = \frac{gM}{ZRT_{\text{avg}}} \int_{h_1}^{h_2} \mathrm{d}z \tag{2.42}$$

The result of the integration is

$$\ln p_2 - \ln p_1 = \frac{gM}{ZRT_{\text{avg}}} \left(h_2 - h_1\right) \tag{2.43}$$

which, when exponentiated, gives us

$$\frac{p_2}{p_1} = e^{\frac{gM(h_2 - h_1)}{ZRT_{\text{avg}}}} \tag{2.44}$$

and finally

$$p_2 = p_1 \, e^{\frac{gM(h_2 - h_1)}{ZRT_{\text{avg}}}} \tag{2.45}$$

where

p_1 = pressure at point 1
p_2 = pressure at point 2
h_1 = vertical depth of point 1
h_2 = vertical depth of point 2
g = local accelaration of gravity (assumed constant)
M = molecular mass of gas (called molecular weight)
R = ideal gas constant (a conversion factor)
Z = compressibility factor of gas (assumed constant)
T_{avg} = average absolute temperature between points 1 and 2

One may use this formula as is, but it is a bit confusing when going from oil-field units to SI units and vice versa. The ideal gas constant is nothing more than a conversion factor for units of measure of the gas, and there are many different numerical values. The normal units are energy/mole/degree absolute temperature. That means that, in English units, the molecular mass is in lb/mole, which must be converted to lbf/mole; and in SI units, it is in kg/mole, which must be converted to N/mole. Add to that the value we normally assume to be the molecular mass; for methane, for instance, it is 16, which in English units is 16 lb/mole, but in SI units it is 0.016 kg/mole because the conventional metric description of the mole is 16 g/mole. So, to alleviate some of the confusion, we modify equation (2.45) for our use in this text as follows:

$$p_2 = p_1 e^{\frac{M(h_2 - h_1)}{ZRT_{avg}}} \tag{2.46}$$

where

p_1 = pressure at point 1

p_2 = pressure at point 2

h_1 = vertical depth of point 1

h_2 = vertical depth of point 2

M = molecular mass of gas

R = ideal gas constant at standard gravity

Z = compressibility factor of gas (assumed constant)

T_{avg} = average absolute temperature between points 1 and 2

We assume the following values for oil-field units:

$Z = 1$ (for methane)

$M = 16$ lb/mole (for methane)

$R = 1544 \dfrac{\text{lb} \cdot \text{ft}}{\text{mole} \cdot {}^{\circ}\text{R}}$ at standard gravity and for SI units

$$Z = 1 \text{ (for methane)}$$
$$M = 16 \text{ g/mole (for methane)}$$
$$R = 847.8 \frac{\text{g} \cdot \text{m}}{\text{mole} \cdot \text{k}} \text{ at standard gravity}$$

Those may not be the values of the ideal gas constant you are accustomed to seeing, but they are consistent with the units we are using. Remember, the ideal gas constant is only a conversion factor.

Note that it is inconsequential as to which points we label as point 1 and point 2, as in p_1, h_1 and p_2, h_2. As long as we are consistent in assigning the correct vertical depths to the appropriate pressures, the signs will be correct. For example, if point 1 is the upper point then $(h_2 - h_1)$ will be positive and p_2, at the deeper point, will be greater than p_1. Likewise, if point 1 is the deeper point, then the difference in depths will be negative and the pressure at point 2 will be less than at point 1. The correct pressure in this formula is absolute pressure rather than gauge pressure, but for all practical purposes, it makes an insignificant difference considering the magnitudes of casing load pressures and the approximate nature of the formula itself and our application. For those reasons, we will use gauge pressure in our casing design calculations. While this formula is easy to use and is acceptable to most casing designers, many prefer to use a more sophisticated computation done with a computer.

2.4.2 Buoyed Weight of Casing

To calculate the axial loads on casing, we have to find the weight of the casing in the fluid that surrounds it. We already stated that we work with the specific weight of casing as opposed to the specific mass, which is listed in the published tables.

Weight of Casing in Air

In some casing designs, we use what we call the weight of the casing string in air. However, strictly speaking, we actually mean the weight of casing in a vacuum, since air is a gas and there is a difference in the pressure on the top and on the bottom of a casing string suspended in a bore hole containing only air. However, this buoyant force is relatively small compared to the weight of the casing and usually ignored in practice. So *when we speak of the weight of casing in air, we actually ignore all buoyant forces on the casing string.* To calculate the weight of a casing string suspended vertically in air, we merely multiply the weight per unit length times the length to get the weight of the string and the axial load at the surface.

Example 2–1 Axial Load of Vertical Casing in Air

Table 2–2 shows a 7 in. casing string with four sections having different specific weights and wall thickness. We wish to calculate the weight of this casing string in air.

Table 2–2 Example 7 In. Casing String

Section Number	Length, ft	Specific Weight, lbf/ft	Internal Diameter, in.
4	1000	32	6.093
3	4000	23	6.366
2	3000	26	6.276
1	1500	29	6.184

We can write a formula for calculating the axial force at the top any section in the string like this:

$$F_k = F_o + \sum_{i=1}^{k} w_i L_i \quad \text{for } k = 0,1,2,\ldots n \tag{2.47}$$

where n is the number of sections in the string. This formula provides for the possible presence of some load on the bottom, F_o, in case there is a tensile or compressive load at the base of the string, although this value is usually zero. (Note that there is no summation when $k < 1$.) For this example then, the calculation is as follows:

$$F_k = F_o + \sum_{i=1}^{4} w_i L_i$$

$$F_o = 0 \text{ lbf}$$

$$F_1 = 0 + 29(1500) = 43500 \text{ lbf}$$

$$F_2 = 43500 + 26(3000) = 121500 \text{ lbf}$$

$$F_3 = 121500 + 23(4000) = 213500 \text{ lbf}$$

$$F_4 = 213500 + 32(1000) = 245500 \text{ lbf}$$

The last value is the total weight of the unbuoyed casing string.

Weight of Casing in a Liquid

When the casing is hanging in a liquid, we must include the buoyant forces on the tube to determine the weight of the string and the axial loads. We already distinguished between the true axial load and something called the effective load. We saw that Archimedes' principle gave us the correct load at the surface but did not give us the true axial load at any other point within the string, rather what we called the effective axial load. Another factor to consider is that, in most casing string designs, the wall thickness is not the same for the entire string, and that contributes further to the inaccuracy of using Archimedes' principle. To get the true axial load, we must use the actual forces attributable to the weight of the tube and the hydrostatic pressure at the points where it acts along a vertical axis. We can illustrate this with a simple figure, Figure 2–10.

In the figure, we show a string of casing with different wall thicknesses. At each point where there is a vertical force due to the fluid pressure, we labeled it with a node number, starting with 0 at the bottom of the string. The true axial load changes at each node where the internal

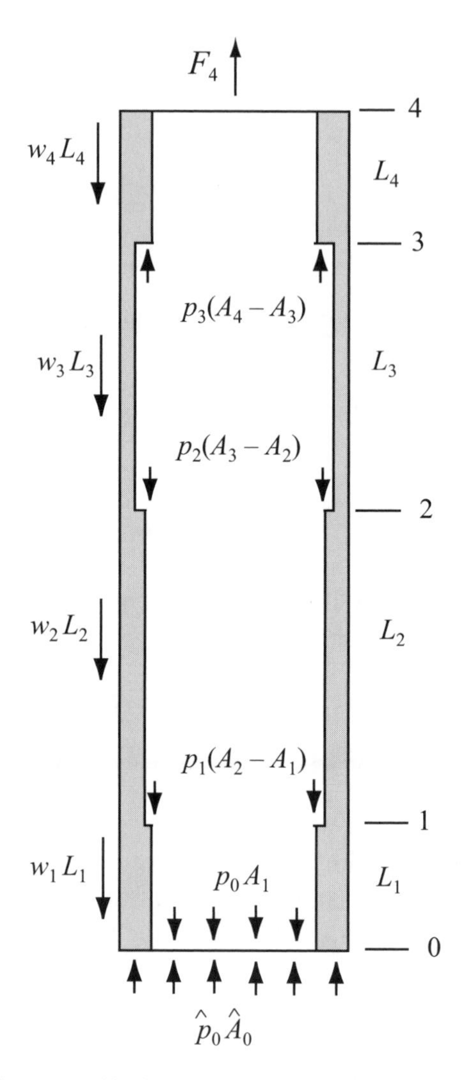

Figure 2–10 *Schematic of hydrostatic forces on a casing string.*

diameter changes. The true axial load at the bottom of each section (just above each node) is given by

$$F_k = -\hat{p}_o \hat{A}_o + p_0 A_1 + \sum_{i=1}^{k} p_i \left(A_{i+1} - A_i\right) + \sum_{i=1}^{k} w_i L_i \qquad k = 0,1,2,\ldots,n-1 \tag{2.48}$$

The true axial load at the top of each section (just below each node where the internal diameter of the tube changes) is given by

$$\hat{F}_k = -\hat{p}_o \hat{A}_o + p_0 A_1 + \sum_{i=1}^{k-1} p_i \left(A_{i+1} - A_i\right) + \sum_{i=1}^{k} w_i L_i \qquad k = 1,2,\ldots,n \quad (2.49)$$

where

i,k = node or section number
n = number of sections in string
F_i = axial load (force) in section just above node i
$\hat{F}_i$ = axial load (force) in section just below node i
$\hat{p}_o$ = external pressure at bottom of casing
p_i = internal pressure at node i
$\hat{A}_o$ = total exterior cross sectional area of casing
A_i = internal cross sectional area of casing section i
L_i = length of section i
w_i = specific weight of casing in section i

Note that, in these formulas where summations take place over a range from $i = 1$ to k or $i = 1$ to $k-1$, the summation terms are zero for $k < 1$ or $(k-1) < 1$, respectively. These two formulas can be posed in a number of ways, but this form works well for a computer algorithm or a spreadsheet. The easiest way to understand these two formulas is with an example.

Example 2–2 True Axial Load of Vertical Casing in a Liquid

Using Figure 2–10 as our example, let us assume all the casing in that figure is 7 in. diameter. We assume that the fluid inside and outside the casing has a specific gravity of 1.5 and there are no other sources of pressure. Use the same casing string from Example 2–1.

We begin by calculating some of the values we will need in the formulas. First, we calculate the values of the cross-sectional areas:

$$\hat{A}_o = \frac{\pi}{4} d_o^2 = \frac{\pi}{4} (7.0)^2 = 38.485 \text{ in}^2$$

$$A_1 = \frac{\pi}{4} d_i^2 = \frac{\pi}{4} (6.184)^2 = 30.035 \text{ in}^2$$

$$A_2 = \frac{\pi}{4} (6.276)^2 = 30.935 \text{ in}^2$$

$$A_3 = \frac{\pi}{4} (6.366)^2 = 31.829 \text{ in}^2$$

$$A_4 = \frac{\pi}{4} (6.094)^2 = 29.167 \text{ in}^2$$

Next, we calculate the pressures at nodes 0, 1, 2, and 3:

$$\hat{p}_o = h_o \gamma_o = (1000 + 4000 + 3000 + 1500)(0.052)(1.5)(8.3) = 6150 \frac{\text{lbf}}{\text{in}^2}$$

$$p_0 = h_0 \gamma = (1000 + 4000 + 3000 + 1500)(0.052)(1.5)(8.3) = 6150 \frac{\text{lbf}}{\text{in}^2}$$

$$p_1 = h_1 \gamma = (1000 + 4000 + 3000)(0.052)(1.5)(8.3) = 5179 \frac{\text{lbf}}{\text{in}^2}$$

$$p_2 = h_2 \gamma = (1000 + 4000)(0.052)(1.5)(8.3) = 3237 \frac{\text{lbf}}{\text{in}^2}$$

$$p_3 = h_3 \gamma = (1000)(0.052)(1.5)(8.3) = 647 \frac{\text{lbf}}{\text{in}^2}$$

Of course, we see that the pressure at the bottom of the casing is the same on the inside as on the outside, but we showed the calculation just for illustration. Next, we calculate the forces at the bottom of each section at a point just above each node, where the area changes.

$$F_{1\downarrow} = -\hat{p}_o\hat{A}_o + p_0A_1 = -(6150)(38.485) + (6150)(30.035) = -51968 \text{ lbf}$$

$$\begin{aligned} F_{2\downarrow} &= -\hat{p}_o\hat{A}_o + p_0A_1 + p_1(A_2 - A_1) + w_1L_1 \\ &= -51968 + 5179(30.935 - 30.035) + 29(1500) = -3806 \text{ lbf} \end{aligned}$$

$$\begin{aligned} F_{3\downarrow} &= -3745 + p_2(A_3 - A_2) + w_2L_2 = -3806 + 3237(31.829 - 30.935) \\ &\quad +26(3000) = 77087 \text{ lbf} \end{aligned}$$

$$\begin{aligned} F_{4\downarrow} &= 77136 + p_3(A_4 - A_3) + w_3L_3 = 77087 + 647(29.167 - 31.829) \\ &\quad +23(4000) = 167365 \text{ lbf} \end{aligned}$$

Then, we calculate the force at the top of each section (just below each node):

$$F_{1\uparrow} = -\hat{p}_o\hat{A}_o + p_0A_1 + w_1L_1 = -51968 + 29(1500) = -8468 \text{ lbf}$$

$$\begin{aligned} F_{2\uparrow} &= -8468 + p_1(A_2 - A_1) + w_2L_2 = -8468 + 5179(30.935 - 30.035) \\ &\quad +26(3000) = 74193 \text{ lbf} \end{aligned}$$

$$\begin{aligned} F_{3\uparrow} &= 72426 + p_2(A_3 - A_2) + w_3L_3 = 74193 + 3237(31.829 - 30.935) \\ &\quad +23(4000) = 169087 \text{ lbf} \end{aligned}$$

$$\begin{aligned} F_{4\uparrow} &= 169087 + p_3(A_4 - A_3) + w_4L_4 = 169087 + 647(29.167 - 31.829) \\ &\quad +32(1000) = 199365 \text{ lbf} \end{aligned}$$

To satisfy our curiosity let us now calculate the axial load using Archimedes' principle in the form of a buoyancy factor:

$$f_b = 1 - \frac{1.5}{7.85} = 0.809$$

We can write a formula for the buoyancy calculation as follows:

$$F_k = f_b \sum_{i=1}^{k} w_i L_i \qquad \text{for } k = 1, 2, \ldots, n \tag{2.50}$$

Using this formula we can now calculate the weight of the buoyed sting using a buoyancy factor based on Archimedes' principle:

$$F_0 = 0$$

$$F_1 = f_b w_1 L_1 = 0.809(29)(1500) = 35192 \text{ lbf}$$

$$F_2 = F_1 + f_b w_2 L_2 = 35192 + 0.809(26)(3000) = 98294 \text{ lbf}$$

$$F_3 = F_2 + f_b w_3 L_3 = 98294 + 0.809(23)(4000) = 172722 \text{ lbf}$$

$$F_4 = F_3 + f_b w_4 L_4 = 172722 + 0.809(32)(1000) = 198610 \text{ lbf}$$

Note first that, at node 0, the buoyant force is zero. This is a result of using Archimedes' principle and is something we discussed earlier. This is the effective load and not the true load in the pipe. But, look at something else. We said that both methods would give exactly the same results at the surface, but in this case, they did not. Why not? They differ by 755 lbf or about 0.4%, which is basically negligible for practical purposes. One might think the difference in the two methods is due to rounding off in the calculations, but it is not. The primary source for the difference is the way we used the specific weight of the casing. The method we used to calculate the true axial load assumes all the weight is in the tube body and pressure acts on the pipe where there is a change in diameter. What this method fails to account for are the presence of the external couplings. The weight of the coupling is averaged into the linear weight of the pipe, so it is not missing from the formulation, but there is a net buoyant force in an upward direction on each coupling, because the pressure on the bottom of the coupling is slightly greater than the pressure on the top of the coupling. If this is taken into account, the two methods give the same results at the surface. But, remember that the weight per unit length of API casing is calculated based on 20 ft joints with standard couplings, so unless the joints exactly conform to that length and coupling type, then either method will differ from the actual weight by some small amount. Figure 2–11 shows a plot of the true axial load and the effective axial load for this example.

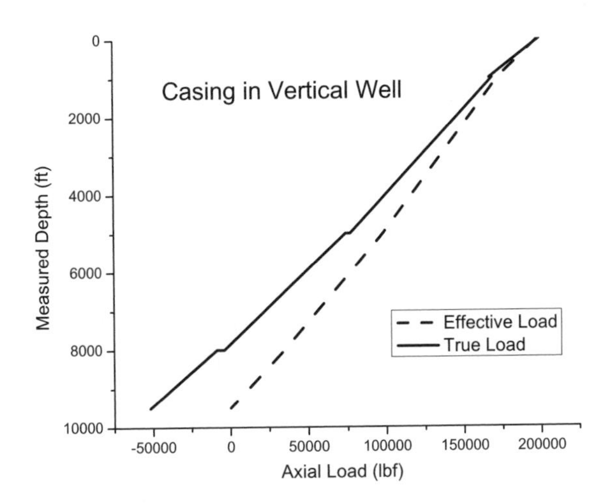

Figure 2–11 *True and effective axial loads for the example casing string.*

2.4.3 Buoyed Weight of Casing in Inclined and Curved Well Bores

Casing rarely hangs vertically in an actual well bore. Most wells are inclined to some extent, and the well bores are also curved. The casing is partially or sometimes totally supported by the walls of the borehole. Therefore, the axial load is that component of the gravitational body force not supported by the bore-hole wall plus the friction along the bore hole.

In this discussion, we ignore friction, which will be covered in detail in Chapter 9. So, for our purposes here, the only force acting on the casing are the body forces of the casing due to gravity and the hydrostatic forces of the fluid in the well bore. Our particular interest for the moment is in determining the axial load in the casing, but later, when we discuss friction, we also determine the contact force of the casing with the bore-hole wall. If we look at a segment of casing in an inclined well bore, as in Figure 2–12, we can see that it is easy to resolve the gravitational body force into an axial force and a force normal to the bore-hole wall.

The axial force at the top of the segment is

$$F_s = F_o + Lw\cos\alpha \tag{2.51}$$

where

F_s = axial load at top of segment

F_o = axial load at the bottom segment

L = length of segment

w = specific weight of segment

α = inclination angle (always measured from vertical)

In previous calculations, we always labeled the axial force with a subscript z to denote that it was along the vertical axis of our coordinate system, but now we are going to use a subscript, s, to denote a curvilinear coordinate or measure along the axis of the pipe, which may or may not coincide with the vertical coordinate. This is fairly straightforward so far, there are no buoyant forces to contend with. But, if we add buoyant forces, which we must always do in a real well, how do they affect that equation? Actually, in that case, it is fairly simple, in that we can put the buoyant force into the value, F_o, since no buoyant force along the walls of the segment acts on the axial load. In fact, all we have to do is use equations (2.48) and (2.49) and substitute $w_i\cos\alpha$ for w_i in those equations. The problem with this, though, is that the inclination angle, α, is not likely to be a constant. All we have accomplished is a formula for a straight section of well with a constant inclination angle. We can generalize this for a curved well bore, but it starts to get a bit messy. This is where a good application for the effective load comes in. If, instead of the specific weight of the casing, we use a buoyed specific weight, we calculate an

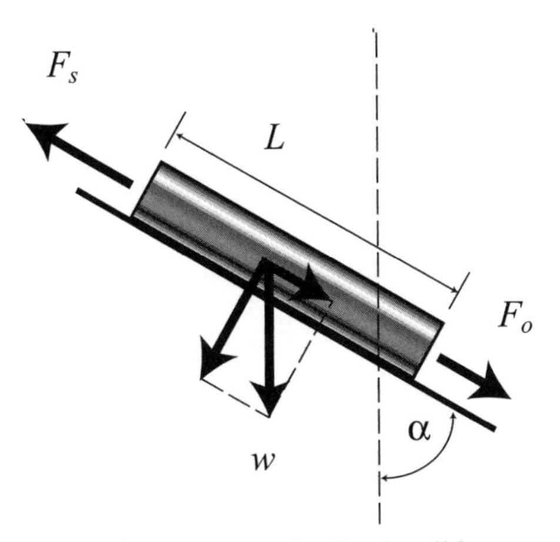

Figure 2–12 *A segment of casing in an inclined well bore.*

effective axial load, which is easily converted to the true axial load. We define the buoyed specific weight as

$$\hat{w} \equiv f_b w \tag{2.52}$$

Then, the differential axial load at any point is

$$\frac{d\hat{F}}{ds} = \hat{w}\cos\alpha \tag{2.53}$$

and the effective axial load is

$$\hat{F}_s(L) = \hat{F}_o + \int_0^L \hat{w}\cos\alpha\, ds \tag{2.54}$$

where

$\hat{F}_s(L) =$ effective axial load at a measured distance, L, from the bottom

$\hat{F}_o =$ effective axial load on bottom, if present (usually weight set on bottom)

$s =$ curvilinear coordinate measured along the axis of the pipe

$\hat{w} =$ buoyed specific weight of pipe

$\alpha =$ inclination angle

To convert the effective axial load to the true axial load, we use the following relationships. When the pressure is the same on the inside of the casing as on the outside,

$$F_s = \hat{F}_s - p\pi\left(r_o^2 - r_i^2\right) = \hat{F}_s - pA_t \tag{2.55}$$

and, where the inside pressure is different from the outside pressure,

$$F_s = \hat{F}_s - \pi\left(p_o r_o^2 - p_i r_i^2\right) \tag{2.56}$$

where for both formulas

$F_s =$ true axial load at some point
$\hat{F}_s =$ effective axial load at that point
$A_t =$ cross sectional area of the casing at that point
$p =$ hydrostaic pressure at that point (when inside and outside are equal)
$p_i =$ inside hydrostaic pressure at that point
$p_o =$ outside hydrostaic pressure at that point
$r_r =$ inside radius of casing
$r_o =$ outside radius of tube

It may seem that it will be a bit difficult to integrate equation (2.54) for a real bore hole, and while that is true, we address numerical methods for doing that in Chapter 9. For an idealized well in a vertical plane, we may make some assumptions, which we illustrate in an example.

Example 2–3 Buoyed Weight of Casing in a Curved Well Bore

Suppose we have a curved well bore, as shown in Figure 2–13, and are going to use the casing string used in the previous examples and a mud with 1.5 sg.

Before we start calculating axial casing loads, we need to get some measured depths. We are given that the kickoff point (where the build or curvature starts) is at 4000 ft and we call that length the length of the ver-

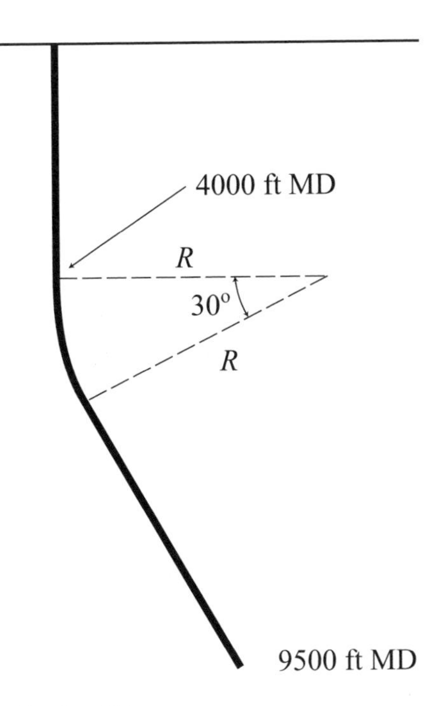

Figure 2–13 *An inclined well bore.*

tical section, L_{vert} The well builds angle at a constant rate until it is inclined at 30° then remains constant until total depth. The measured depth of the well is 9500 ft, as measured along the well bore path. What is the length of the curved section? The inclination changed by 30° so the angle between a radius at the beginning of the curve and at the end of the curve also is 30°. We are given that the radius is 1000 ft, so we can calculate the length of the curved section of the well bore:

$$L_{crv} = R\alpha = 1000\left(\frac{\pi}{180}\right)(30) \approx 524 \text{ ft}$$

The length in the inclined section is then the measured depth less the length in the vertical section and the length in the curved section. Note that we had to change the angle in degrees to radians for this formula. Also note that, whenever an angle appears outside a trigonometric function in a formula, it is almost always in radians not degrees. That will always be the case in this text.

$$L_{\text{incl}} = 9500 - 4000 - 524 = 4976 \text{ ft}$$

To calculate the load, it is necessary to know the specific weights and lengths of casing in each of the three sections of the well. We might best show that in a figure, as in Figure 2–14.

We determine the effective load in the casing string first, then convert it to the true axial load. (Remember that we are not considering friction in this example.) We start at the bottom and assume the casing is not sitting with any weight on bottom, so that $\hat{F}_o = 0$. Our buoyancy factor for 1.5 sg mud is

$$f_{\text{b}} = 1 - \frac{1.5}{7.85} = 0.809$$

Our bottom section of casing is 1500 ft of 29 lb/ft, and it is all in the inclined section of the bore hole. If we integrate equation (2.54) over a straight but inclined section, it gives us exactly equation (2.51). Using the buoyed specific weight then, we get

$$\hat{F}_1 = L_1 \hat{w}_1 \cos\alpha = 1500(0.809)(29)\cos 30 = 30477 \text{ lbf}$$

Section 2 is 3000 ft of 26 lb/ft casing, and all of it is in the inclined section of the well:

$$\hat{F}_2 = \hat{F}_1 + L_2 \hat{w}_2 \cos\alpha = 30477 + 3000(0.809)(26)\cos 30 = 85125 \text{ lbf}$$

Now Section 3 has a portion in the inclined section, 476 ft, a portion in the curved section, 524 ft, and a portion in the vertical section, 3000 ft. Before we set up this calculation, we must do something with equation (2.54) to integrate it over this interval. Since the curvature of the curved section is constant (it has a constant radius), we can substitute the relationship $\mathrm{d}s = R\,\mathrm{d}\alpha$ and change the limits of integration, so that it becomes

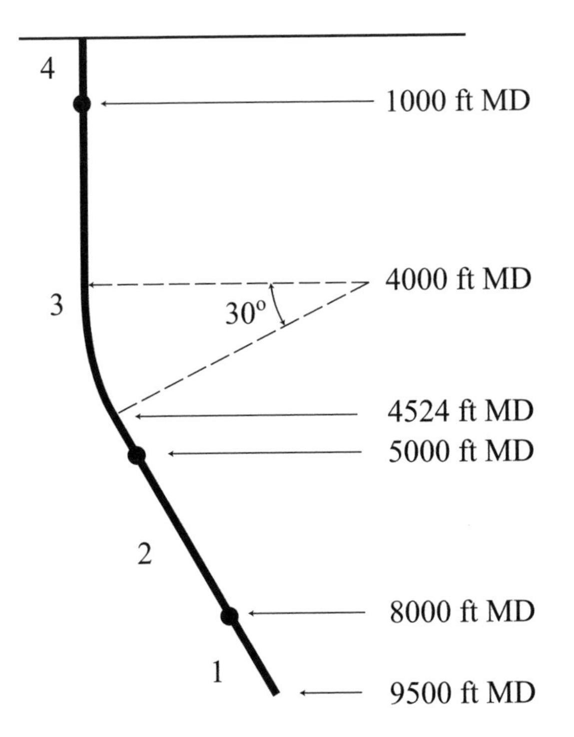

Figure 2–14 *Well sections and casing string sections.*

$$\hat{F}_s = \hat{F}_o + \int_{\alpha_1}^{\alpha_2} \hat{w} R \cos\alpha \, d\alpha = \hat{F}_o + \hat{w} R \int_{\alpha_1}^{\alpha_2} \cos\alpha \, d\alpha = \hat{F}_o + \hat{w} R \left(\sin\alpha_2 - \sin\alpha_1\right) \tag{2.57}$$

The substitution we made is legitimate, but in this equation, it can lead to different results, depending on how we label the inclination angles. Curvature can be positive or negative and is defined as

$$\kappa \equiv \frac{d\alpha}{ds} = \pm \frac{1}{R}$$

But the radius of curvature can only be positive, hence the ± sign. So, for equation (2.57) to work in our application, we must take the absolute value of the difference in the sine functions. For a constant curvature and constant buoyed specific weight, equation (2.57) is rewritten as

$$\hat{F}_s = \hat{F}_o + \hat{w}\,R\left|\sin\alpha_2 - \sin\alpha_1\right| \tag{2.58}$$

Now the effective load at the top of Section 3 is

$$\hat{F}_3 = \hat{F}_2 + L_{3\ \mathrm{incl}}\hat{w}\cos\alpha + \hat{w}\,R\left|\sin\alpha_2 - \sin\alpha_1\right| + L_{3\ \mathrm{vert}}\hat{w}$$

$$\begin{aligned}\hat{F}_3 &= 85125 + \left[476(0.809)(23)\cos 30\right] + \left[0.809(23)(1000)\left|\sin 0 - \sin 30\right|\right] \\ &\quad + \left[3000(0.809)(23)\right] \\ &= 157920 \text{ lbf}\end{aligned}$$

Finally, at the top of Section 4, the effective load is

$$\hat{F}_4 = \hat{F}_3 + L_4\,\hat{w} = 157920 + 1000(0.809)(32) = 183808 \text{ lbf}$$

We can then calculate the true axial loads at the tops of each section. To do this, however, we need to know the true vertical depth at the top of each section, and that involves a bit of trigonometry. The top of Section 4 is at the surface, $h_4 = 0$; the top of Section 3 is in the vertical section, $h_3 = 1000$ ft. The top of Section 2 must be calculated:

$$\begin{aligned}h_2 &= 4000 + R\sin(\Delta\alpha) + L_{3\mathrm{incl}}\cos\alpha = 4000 + 1000\sin 30 + 476\cos 30 \\ &= 4912 \text{ ft}\end{aligned}$$

The top of Section 1 is at

$$h_1 = h_2 + L_2\sin\alpha = 4912 + 3000\cos 30 = 7510 \text{ ft}$$

And, finally, the bottom of Section 1 is

$$h_o = h_1 + L_1 \cos\alpha = 7510 + 1500\cos 30 = 8809 \text{ ft}$$

We can now calculate the pressures at each depth:

$$p_4 = 0 \text{ lbf/in}^2$$

$$p_3 = 1000(0.052)(1.5)(8.33) = 650 \text{ lbf/in}^2$$

$$p_2 = 4912(0.052)(1.5)(8.33) = 3192 \text{ lbf/in}^2$$

$$p_1 = 7510(0.052)(1.5)(8.33) = 4880 \text{ lbf/in}^2$$

$$p_0 = 8809(0.052)(1.5)(8.33) = 5724 \text{ lbf/in}^2$$

Since the pressure on the inside and outside are the same, we need the cross-sectional area of each section of casing:

$$A_4 = \frac{\pi}{4}\left(7.00^2 - 6.094^2\right) = 9.317$$

$$A_3 = \frac{\pi}{4}\left(7.00^2 - 6.366^2\right) = 6.655$$

$$A_2 = \frac{\pi}{4}\left(7.00^2 - 6.276^2\right) = 7.549$$

$$A_1 = \frac{\pi}{4}\left(7.00^2 - 6.184^2\right) = 8.449$$

Next we calculate the true axial load at the top of each section:

$$F_{4\uparrow} = 183808 - 0(9.317) = 183808 \text{ lbf}$$

$$F_{3\uparrow} = 157920 - 650(6.655) = 153594 \text{ lbf}$$

$$F_{2\uparrow} = 85125 - 3192(7.549) = 61029 \text{ lbf}$$

$$F_{1\uparrow} = 30477 - 4880(8.449) = -10754 \text{ lbf}$$

Finally, we calculate the true axial load at the bottom of each section:

$$F_{4\downarrow} = 157920 - 650(9.317) = 151864 \text{ lbf}$$

$$F_{3\downarrow} = 85125 - 3192(6.655) = 63882 \text{ lbf}$$

$$F_{2\downarrow} = 30477 - 4880(7.549) = -6362 \text{ lbf}$$

$$F_{1\downarrow} = 0 - 5724(8.449) = -48362 \text{ lbf}$$

We can plot the true axial load at the top and bottom of each section and the effective axial loads. Keep in mind that the load between the top and bottom of Section 3 is not a straight line, as we show in Figure 2–15, because of the well bore curvature, but we calculated the values only at the top and the bottom of that section. Although Figure 2–15 and Figure 2–11 appear similar, the values are different.

We did not account for friction in our calculations, meaning the sliding friction running the pipe into the hole or picking it up from bottom. If we could rotate the pipe several revolutions to remove the sliding friction, the axial loads would look like the figure.

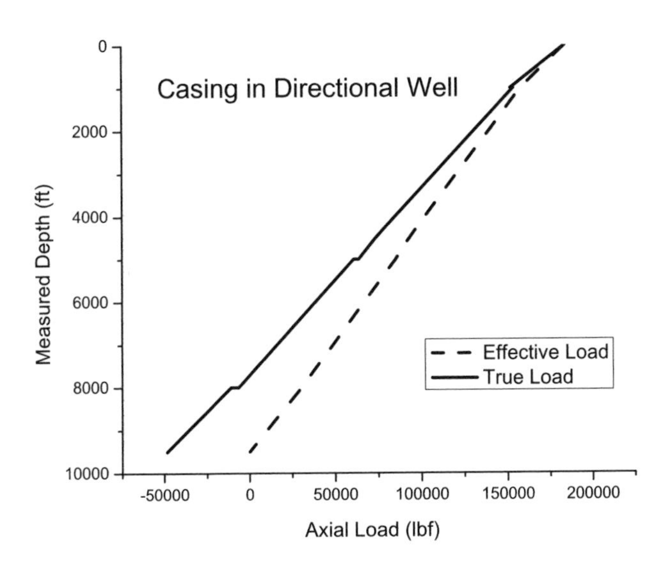

Figure 2–15 *True axial load and effective axial load in curved well bore example (without friction).*

2.4.4 The Ubiquitous Vacuum

This section should not even be here, except that it seems to pop up in oil-field thinking time and again. A perfect vacuum is zero absolute pressure and does not exist. In oil-field terms, that is roughly 15 lbf/in.^{2} less than atmospheric pressure. A near perfect vacuum does not cause casing to collapse. It does not suspend a significant column of fluid in an annulus. I will not mention any specific instances (and I know of several) because, to this day, those involved are embarrassed. If you are close enough to an oil-field disaster that you have to consider a vacuum, then you are too close to worry about it.

2.5 Closure

In this chapter, we covered several definitions and the conventions we use for units and formulas in the remainder of this text. You may wish to refer back to this chapter from time to time, because I do not intend to remind you of those conventions in later chapters.

We may have spent an inordinate amount of time on basic hydrostatics, especially as the subject applies to oil-field casing in well bores. One reason for this is that it is an important topic: a secondary reason is that there seems to exist a certain degree of confusion in the oil field about hydrostatics, especially in regard to buoyancy.

2.6 References

NIST Special Publication 330. (2001). *The International System of Units, the National Institute of Standards and Technology.* Washington, DC: U.S. Department of Commerce (free online).

NIST Special Publication 811. (1995). *Guide for the Use of the International System of Units.* Gaithersburg, MD: Physics Laboratory, National Institute of Standards and Technology (free online).

CHAPTER 3

Casing Depth and Size Determination

3.1 Introduction

Arguably the most critical step in casing design is determining the setting depths for the various casing strings. Setting the wrong size casing at the wrong depth can preclude the well ever reaching its objective. Figure 3–1 is a schematic of a typical well showing four strings of casing: conductor casing, surface casing, intermediate casing, and production casing. Why does this well require four strings of casing? How is that determination made? How are the setting depths determined? How are the casing sizes determined? This chapter addresses those questions.

3.2 Casing Depth Determination

Casing depth is determined by a number of parameters, most of which we cannot control. What are those parameters?

3.2.1 Depth Parameters

When we make a determination of the setting depths for the various casing strings in our well several parameters must be considered, including

- Pore pressures (formation fluid pressures).
- Fracture pressures.
- Experience in an area.
- Bore-hole stability problems.
- Corrosive zones.
- Environmental considerations.
- Regulations.
- Company policy.

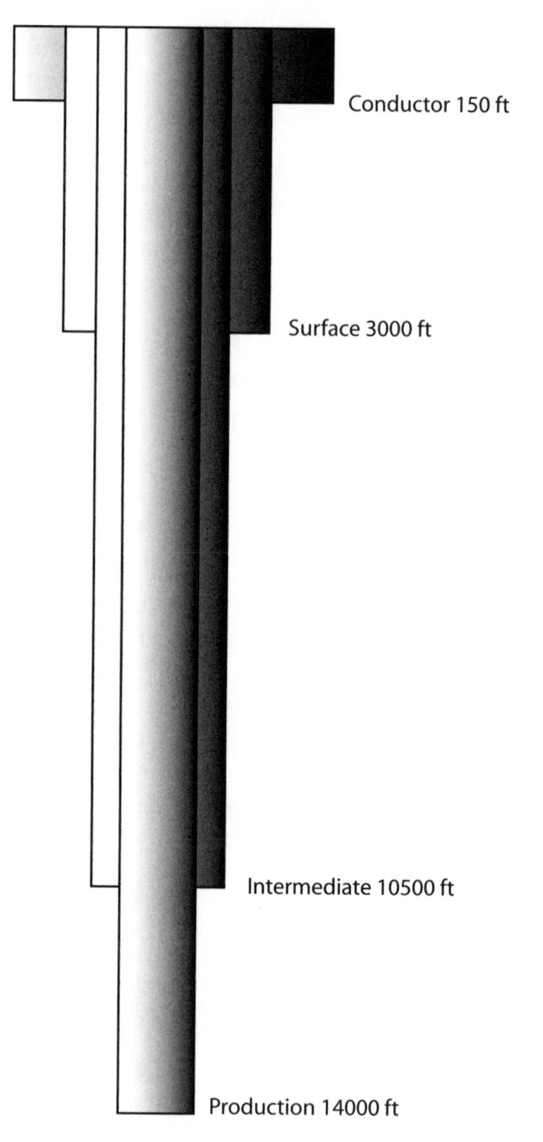

Figure 3–1 *A typical casing installation.*

Some of these criteria may overlap in practice. For instance, many regulations to protect fresh water sources near the surface might also be considered environmental parameters. While this is a text primarily about casing, two of these criteria, pore pressures and fracture pressures, are so important that we discuss them in detail to understand their importance and what they represent.

Pore Pressure

Pore pressure (also called *formation pressure*) is the pressure of the fluid that fills the pore spaces (or voids) in the rock. This pressure normally determines the lower limit of the drilling fluid density (exceptions are underbalanced drilling operations or bore-hole stability problems). All sedimentary rock contains some type of fluid in the pore spaces and this fluid may be in the form of liquid or gas. In general, the pressure of the fluid depends on the depth of the rock and the density of the fluid, in particular, it depends on the connectivity, if any, of the pore spaces to the surface. Figure 3–2 illustrates the depositional process. It is a simple schematic similar to one used by Terzaghi and Peck (1948) to illustrate the nature of soil compaction.

The cylinder contains porous rock and liquid. A downward force is applied to the piston, representing the weight of subsequent layers of rock (called *overburden*) deposited above the sample. As the force or weight is increased with increasing deposition, the rock in the cylinder is compressed, forcing some of the fluid into the drain tube, whose length represents the depth of the rock below the earth's surface. If the fluid can flow freely into the tube and to the surface, it can be seen that, no matter how deep the rock or the weight of the overburden, the pressure of the fluid in the pore spaces are equivalent to the hydrostatic pressure of the fluid column in the tube. This closely models the situation in many formations that have porosity or channels that freely connect to the surface (though the connection may be far from obvious). In this instance, the pore pressure of the formation is equal to the hydrostatic pressure of the fluid column between the formation and the surface. Typically, such a fluid contains dissolved minerals, such as salt, that make the density heavier than fresh water.

In general, the path through pore spaces of various formations from some depth to the surface is not quite so direct, and there may be considerable variation if the deposition, for instance, is occurring at a faster rate than the fluid can escape. This is illustrated in Figure 3–3, where we have placed a valve or choke in the drain tube.

If we restrict the rate at which the fluid is allowed to escape as the formation is compacted by the increasing overburden, then the pressure of the fluid in the cylinder will be higher than in the first illustration because of the restriction of the rate at which the fluid is allowed to escape. Such a restriction could be caused by very low permeability formations between the sample and the surface, for instance. If the depositional rate slows or stops, that pore pressure might eventually equalize to the hydrostatic head

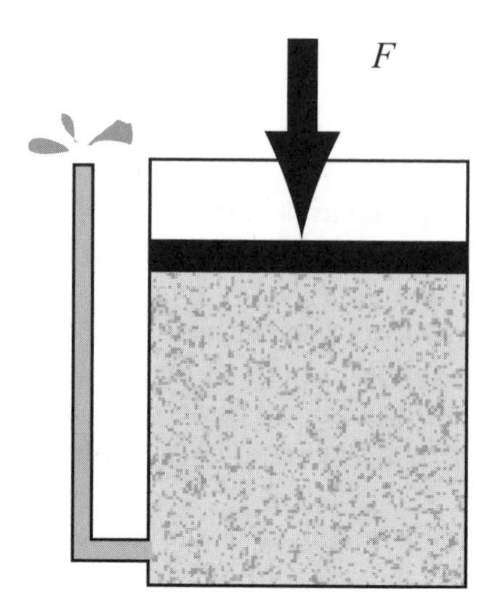

Figure 3–2 *Soil compaction model.*

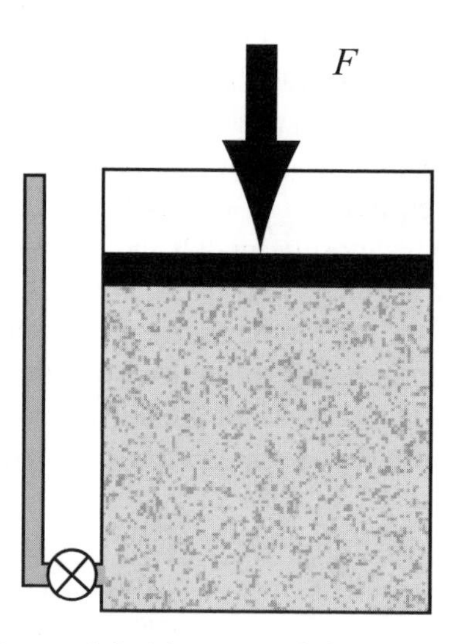

Figure 3–3 *Compaction model with restricted drainage.*

of the fluid column, but such a process might require several hundred thousand years or so. In that case, we might consider, for all practical purposes, that the escape tube is totally sealed, depending on where in the age of the process we drill into the formation. Then, where the fluid has been trapped or its escape severely restricted, we consider the pore pressure to be abnormally high or overpressured. It can be considerably higher than in the first model. We call such a formation an *undercompacted formation*, meaning that it has not compacted to a normal density for the given amount of overburden pressure, because some of the fluid has been trapped and is also contributing to the support of the weight of the overburden. The fluid pressure could vary considerably from slightly above normal to an upper limit, which would be a pressure equivalent to the weight of the overburden. We could write this formation pressure as

$$p \leq \int_0^h g\rho \, dz \tag{3.1}$$

where p is the formation pressure, g is the local acceleration of gravity, ρ is the mass density of the rock, z is a vertical coordinate axis, and the limits of integration are from the surface to some depth, h. Four obvious consequences of these two illustrations are that

1. The density of an undercompacted formation is less than that of a normally compacted formation, since it contains a greater percentage of fluid.
2. If the fluid is saltwater, the resistivity measurement from an electric log is lower than the normal formation, again because it contains a greater percentage of fluid.
3. The sonic travel time in the undercompacted formation is reduced.
4. If we are drilling through the two formations, the drilling rate is faster through the undercompacted formation, because it is less dense, and the differential pressure between the drilling fluid and pore pressure is less, hence the drilled cuttings are more easily removed from the formation face.

These four mechanisms are the primary means we have to detect the presence of abnormally high pressured formations. While this fluid entrapment mechanism explains most of the abnormally high pressured formations in the world, there are other causes, such as a long gas column in a steeply dipped reservoir, artesian flow, a reservoir that has been pressured by flow from a higher-pressure reservoir during uncontrolled subsurface flow in a blowout, or flow in an uncemented casing annulus. We have no means to detect these types of situations other than direct measurements or previous knowledge of the situation.

On the other hand, there are also situations where the formation pore pressures are abnormally low, that is, less that the gradient of saltwater. These cases generally are caused by depletion of the normal pore pressure by production of the fluids from the formation. This type of situation usually is known in advance of drilling—but not always.

Another quite important point is this. Much of what we know about formation pore pressures in a given area seems to fail miserably at shallow depths. A number of rigs have been lost because of drilling into shallow pockets of trapped gas. The uncertainty and unpredictability of the presence or absence of these small shallow gas accumulations often require that small diameter pilot holes be drilled in some areas to ascertain their presence prior to starting a well. At the other extreme of low pressure, there was at least one occurrence of a rig drilling into a shallow cavern that swallowed the entire rig and all the water from a small lake (the "cavern" in that case happened to be a human-made mine shaft). The data that one has available for casing point selection seldom includes data near the surface, and one should never assume that the absence of such data means that pore pressures near the surface are of no consequence. There are various methods for determining or estimating the magnitude of pore pressures in well bores, and we are not going into those methods. For this text, we assume that we already have access to reasonable pore pressure estimates for our bore hole, and after reading this, we have some fundamental understanding of what it means.

Fracture Pressures

The subject of fracture pressures for drilling mud programs and casing design is quite complicated—a lot more so than many realize. This is primarily because there is considerable confusion as to what is meant by *fracture pressure*. One true definition is that the fracture pressure is the pressure at which a formation matrix opens to admit whole liquids through an actual crack in the matrix of the rock as opposed to invasion through the natural porosity of the rock. This sounds straight forward, but some of the things we often hear called *fracture pressures* are not true fracture pressures by that definition.

The pressure at which the matrix of a rock physically fractures and admits the entrance of whole liquids depends on a number of things. Here is what we normally expect when we think about that definition of fracture pressure:

- The well bore liquid cannot enter the formation pore spaces prior to fracture (e.g., a filter cake building mud or an impermeable formation such as shale).
- The formation is in a state of compression from an in situ stress field (due to overburden, lateral constraints, or tectonic activity).
- The formation matrix has some amount of cementation and, hence, some amount of tensile strength (may be relatively small though).

For this formation to be hydraulically fractured, the pressure of the liquid in the well bore has to exceed both the near-well-bore stress field that is compressing the rock at the point of fracture and also the tensile strength of the rock matrix at that point. The near-well-bore stress field usually is not the same as the in situ stress field because the rock deforms slightly when some of it is removed to form the bore hole; consequently, the stress field near the bore hole changes as well. Once the fracture is initiated, the fluid enters the fracture and the fracture propagates at a lower pressure than the initial fracture pressure. The reason that the propagation pressure is lower than the original fracture pressure is that, once the fracture is open, the fluid pressure acts like a wedge. In other words, mechanical advantage is gained as the fracture length grows (up to a limit).

That definition is widely accepted. It is normally the way we interpret "fracture pressure," and that is what we usually assume when we design casing. However, the "fracture pressure" we utilize at the time of casing depth selection and casing design often comes from several sources, such as leak-off tests, integrity tests, fracture gradient curves and correlations, pilot hole mini-fracture tests, and production-stimulation fracture jobs. Often the values obtained from these sources are not what we assumed from the preceding definition. Additionally, there are certain consequences of that definition of fracture pressure that we do not always anticipate, such as

- Once fractured, a formation as just described can never be pressured again to the original fracture pressure in that well bore, because it has no tensile strength at that point. On repressuring, the formation will open at the fracture closure pressure, because the tensile strength, if any, is now permanently lost. The fracture closure pressure is a function of the in situ stress field and not the rock itself. The original tensile strength at the point of fracture is never restored. The rock is broken, and it stays broken. (Note: Some mud systems may form plugs that can sometimes effectively divert the point of fracture to a different point in the rock matrix and cement can sometimes restore some tensile strength at a point.)

- The fracture pressure also depends on the orientation and inclination of the bore hole in most cases. The reason for that is that the principal in situ stress components generally are not equal in magnitude and their orientation in relation to the well bore can vary. That is a complicating factor, in that the fracture pressure is commonly reduced as the bore-hole inclination increases. The fracture pressure in an inclined well bore also varies with the direction of the well bore as well as its inclination.

Here are some other situations to consider about fracture pressure:

- Some formations have no tensile strength; for example, unconsolidated sandstones, formations with micro fractures, and faults.
- A well-bore fluid actually can enter the pore spaces under pressure prior to fracture (e.g., fracturing a porous formation with a clear fluid such as brine water).

In the case of no tensile strength, the fracture pressure depends entirely on the in situ stress field. It opens at the same pressure each time, so that fracturing it does not reduce its strength, as in the previous case. This is why, in many areas, it is perfectly "safe" to fracture a zone while drilling, because the "strength" of the formation has not been reduced. In actuality, the formation has no tensile strength; the compressive in situ stress field is what holds it together.

In the case of the pressured fluid entering the pore spaces before fracture occurs, the fracture pressure usually is lower than in the earlier definition, because there is mechanical advantage to the pressurized fluid in the pore spaces.

So we have seen that fracture pressure depends on well-bore orientation to the stress field, the type of fracturing fluid we use, and whether or not the formation has any tensile strength. If we are aware of those situations, then we are better able to understand exactly what the fracture pressure means. However, fracture pressure values come from various sources and we have to understand the nature of the sources to understand what the values mean.

One source of fracture pressures comes from fracture gradient curves and similar correlations. These sources usually are accurate enough for casing design and mud programs in vertical wells, but many of them are based on erroneous assumptions about the mechanics of rocks. They typically assume that the horizontal in situ stress is uniform in all directions and the fracture gradient can be calculated from knowledge of the overburden density (from density logs or estimated density gradients) and Poisson's ratio of the rock. This assumes that the rock is in a state of plane strain, in that it is perfectly constrained laterally from expansion due to the weight of the overburden. First of all, it is rare that the horizontal principal stress components of the in situ stress field are equal in magnitude, the mere presence of faults and fractures precludes this. And in those cases, the values for Poisson's ratio used in those calculations are correlation values not actually Poisson's ratio. To further complicate matters, these methods calculate the fracture pressure based on the in situ stresses. When a bore hole is drilled in rock, the rock around the well bore deforms, and the in situ stress field in the proximity of the well bore changes. It then depends on whether the fluid can enter the pore spaces prior to fracture as to how the fracture pressure is calculated. Typically these methods use the in situ stress field prior to the bore hole being drilled. The consequence of this is a fracture pressure based on a non-filter-cake-building fracture fluid, which is not what is used for most drilling applications. But, despite the technical problems with these methods, they are fairly successful in vertical and near vertical wells in unconsolidated formations and give good approximations, but beware that the engineering assumptions on which they are based in general are not true.

We come to the often-used "fracture pressure" called the *leak-off pressure*. Because there are differing opinions as to what constitutes a leak-off pressure, Figure 3–4 is an illustration of a pressure recording of a mini-fracture test (three or four barrels) done in a vertical pilot hole.

In the figure, a small fracture test was done in open hole by pumping a few barrels of mud into a formation at constant rate with an open-hole DST tool and a down-hole pressure recorder. The test was done in a vertical well bore, so that the fracture closure is the same magnitude as the minimum horizontal in situ stress component. This particular formation is an example of our first definition: It has a definite tensile strength (the difference between the fracture pressure and fracture closure pressure) and is impermeable to the fracture fluid (the bleed-down after pump shutoff was

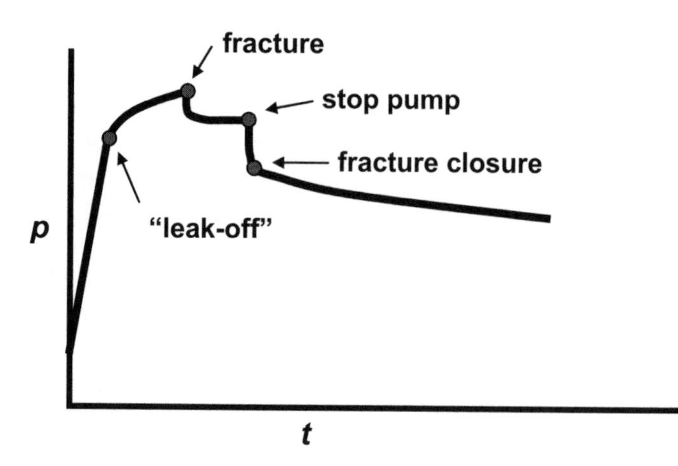

Figure 3–4 *A mini-fracture test for bore-hole stability analysis.*

accomplished by opening a valve at the surface). This test was performed to determine the fracture pressure of the formation (*B*), the fracture closure pressure (*D*), and the fracture directional orientation (from an imaging log run later). That is all the useful information derived from a test like this, but note a point (*A*) on the curve that is often called the *leak-off pressure*. Most leak-off tests are conducted in a similar manner, except the pressure is often recorded at the surface and pumping usually is stopped before reaching the actual fracture pressure of the formation. The pertinent question then is, What is the physical significance of the "leak-off pressure"? The truth of the matter is that, in cases like this, there is no particular significance to this point, as it is unrelated to both the fracture pressure and the fracture closure pressure. The only true thing one can say of this definition of a leak-off point in a test like this is that it is the point at which the pressure-volume relationship of the bore hole becomes significantly nonlinear. If the borehole is linearly elastic, the pressure-volume relationship cannot possibly be linear anyway, since the radial strain is a linear function of the pressure, but the volume cannot be. But that nonlinearity is quite small in the elastic range, and we usually can ignore it. You will hear several different explanations for this nonlinear point or "leak-off point." One is that it is the point at which "whole mud begins to enter the formation." That is pretty hard to believe, as whole mud (which builds a filter cake) cannot enter a formation except by fracture—the particles are too large. And once a fracture initiates, it gets big very quickly. There are cases where permeabilities on the order of several Darcies allow the entrance of "whole mud," but in that case, lost circulation is present before the test even begins. Another popular explanation says that it is the time during which the fracture is propagating "down the hole" before it begins to propagate radially from the bore hole. That is a bit difficult to imagine also, in that a surface fracture under constant pressure propagates with the speed of a Rayleigh wave or at about 25% of the velocity of sound in the rock; so it might propagate 60 ft of open hole in about 0.02 sec. Can you read that time interval on any chart? The problem is that, somewhere along the way, an idea has become entrenched that rock behaves like a linear elastic material all the way up to the point of failure. Possibly, this comes from seeing yield points on uniaxial stress-strain curves for some metals. But those metals usually do not fail at the onset of nonlinearity and, in general, neither does rock. This does not address the possible nonlinear compressibility of the mud used in the test. As a consequence, a leak-off test is unfortunately meaningless as far as determining anything about the rock properties or the stress field; however, it works for drilling, because the leak-off pressure is a safe value for maximum mud densities. In other words, the leak-off pressure is less than the actual fracture pressure, therefore it is a safe upper limit on the drilling fluid density, so there should be no reason not to use it in well planning or casing depth selection, as long as we understand what it is and is not.

As mentioned earlier, the subject of fracture in rocks is exceedingly complicated and far beyond what we need for casing design. The point of all this discussion is to instill awareness such that, when we speak of fracture pressure in casing design, we have a clear idea of what we mean and realize that the values we often use are not necessarily what they are supposed to represent.

It is not within the scope of this text to detail the various methods for quantifying pore pressures or fracture pressures. This brief description at least may give a little insight into the subject for those whose backgrounds do not include any study of geology or reservoir mechanics. A number of reliable sources are available for quantifying pore pressures and fracture pressures for well planning. Unfortunately, the majority of those are local to the Louisiana and Texas Gulf Coast. The methods work anywhere, but the correlations are quite restricted geographically, and my reluctance to include any here is motivated strictly by a desire to keep any part of this text from being restricted to a limited geographical area. A readable source of information on rock mechanics is the book by Fjær et al. (1992). Another good source for understanding and quantifying pore pressures and fracture pressures is the book by Fertl (1976). In any event, one should be aware that the pore pressure and fracture pressure data are only approximate.

Other Parameters

The other parameters listed previously are self-explanatory and need little elaboration. However, a few comments may be in order.

Experience in an area should never be casually tossed aside in favor of pore pressure and fracture pressure data. There usually are good reasons that casing setting depths have become standardized in a particular area. Before making any changes, one should investigate those reasons thoroughly.

Bore-hole stability problems exist in many areas. Casing is not the only solution in but a few cases, so all possibilities should be evaluated.

Regulations should never be violated. That should not have to be said, but many would be surprised to learn how frequently violations actually occur. In the long run, the environment is a lot more important to the future of human existence on this planet than we seem to appreciate today.

Casing setting depth is determined by the requirements to maintain the integrity of the bore hole and protect the environment. Yes, it is that simple. Or perhaps we should say it is that complicated.

Casing String Configuration

Once we determine the depths of the casing strings, we still have several alternatives. Figure 3–5 shows three possible configurations for a well similar to the one in Figure 3–1. The first shows a production casing string. The second shows a production liner, where the intermediate string also serves as part of the production string. The third shows a tieback string inside the intermediate string and connected to a liner at the bottom of the intermediate string. One can see that the second option might save the operator money by eliminating a full production string, but why would an operator elect to choose the third option as opposed to the first or second? One reason might be to reduce the weight of the final string and save money using a lower-tensile-strength casing. Of course, that has to be more saving than the additional cementing and equipment cost and additional rig time required. However, here is a typical situation for choosing the third option. We are drilling a high-pressure well and the intermediate casing is required to contain the high-density mud while drilling the lower part of the hole. Suppose it takes a few weeks to drill the hole below the intermediate casing, so there may be considerable wear from the drill string on the intermediate string. This means we have to rule out option two because the intermediate casing may not be able to contain the pressures required of a production string due to loss of wall thickness from the wear. In this case, the first option usually is cheaper than the

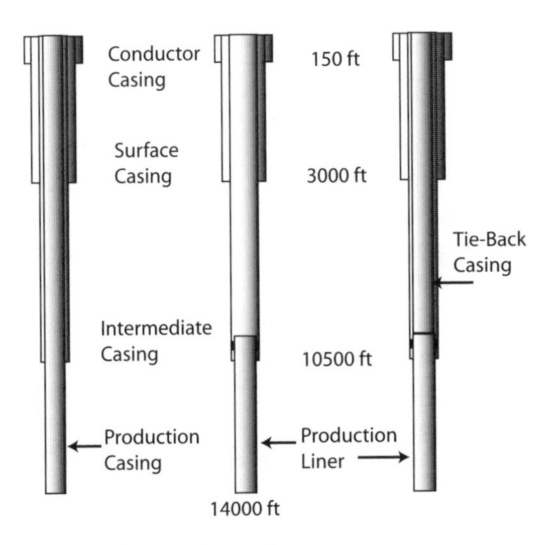

Figure 3–5 *Three possible configurations.*

third, which requires more time, more cement, and more equipment, so we still see no reason for selecting the third option. Consider two more things though. Remember that we said that it was a high-pressure well. The operator wants to be assured that the casing above the cement does not leak and the best way to assure this is to hydrostatically test the casing connections as the casing is being run in the hole. This cannot be done with a full string of pipe because the static time required to test each connection probably would allow the casing to become stuck before it gets to bottom. This would then be an extremely costly situation that would require another liner of a smaller diameter than the production casing. So, while the third option is not common, there often are very good reasons for doing it. Further, many wells require two liners instead of one, and the tieback string is always a preferred option in that case. There are many possibilities. Well conditions and costs dictate the actual choices. We discuss those choices in more detail later. For now, we discuss the requirements particular to the individual types of casing strings.

3.2.2 Conductor Casing Depth

The conductor casing is the largest diameter casing run in the well. As already mentioned, it often supports the weight of the subsequent tubes run in the well bore and maintains some minimal amount of bore-hole integrity while drilling the surface hole for the surface casing. Individual wells may require two conductors, one a structural conductor to support the wellhead and casing, another to provide bore-hole integrity while drilling the surface hole.

Conductor casing may require drilling a hole in the ground and cementing it in place, or it may be driven into the ground with a diesel pile-driving hammer. The criteria for selecting the depth of the conductor can be very simple or very complicated. On the simple side, we want the conductor deep enough to prevent washing out under the rig while drilling the surface hole. In most of these cases, the first casing head is attached to the surface casing once it is in place and cemented so the conductor serves no further purpose. For many shallow wells with hard surface soils, the conductor may be set at depths of 50 ft or so, some times 100 ft. On the other hand, in areas where the surface soils (or ocean bottom) are extremely soft, it may be necessary to set the conductor 200–500 ft below the surface (or ocean bottom) just to drill the hole for the surface casing. In some situations, the surface formations are so incompetent or problematic that two strings of conductor casing may be required. In other cases, the conductor casing is also a support structure for the well and must support a small platform attached to the wellhead and some minimal amount of production equipment; this is not as uncommon as many might think, hundreds of these wells exist in shallow waters. While conductor pipe usually is considered the simplest of the casing strings we will run in our well, it often is the most complicated in terms of both setting depth and design. The setting depth of conductor, in many cases, must be determined by soil bearing tests and coring. This gets more into the realm of the civil engineer than the petroleum engineer's domain. Most companies have their own specifications or they rely on the standard practice in the area that has already proven successful.

Unfortunately, no handy formulas are available for determining the setting depth of conductor casing. Just too many variables and complexities must be considered here. That probably sounds like an avoidance of the issue, and it is. About the only guide we can offer in the absence of soil bearing tests similar to those performed for foundations of bridges, tall buildings, and similar structures is to use what has proven successful in the area. And as much as we hate to say it, that brings us to a rule of thumb.

In the absence of soil mechanics data and analysis, the only way to reliably select the depth of conductor casing is to use the depth already proven successful in the area. In other words, do what everyone else does. The main thing is that, if you do not have data to support your choice, do not attempt to set your conductor casing at a lesser depth than is standard in an area. If it is a critical well and there is nothing in the area, then get soil data.

3.2.3 Surface Casing Depth

A number of factors affect the setting depth of surface casing:

- Pore pressures.
- Fracture pressures.
- Depth of freshwater bearing zones.
- Legal regulations and requirements.

Which of these do we choose? Which are the most important? The answer is almost always the one that requires the deepest casing string. Strictly from a design point of view, the first two are the most important: they are related and are our basis for maintaining bore-hole integrity. We intend that to include well safety. The last two may also be related. Protecting surface freshwater sands is of extreme importance in populated areas, and in truth, it should be everywhere. Regulations require this in most areas now. However, it sometimes is possible to obtain a variance from the regulations if the freshwater sands will be protected by the next string of casing. Damaging a freshwater aquifer is not only a bad thing to do; in some parts of the world, it could put your company out of business!

The question of regulations, as already mentioned, usually is a matter of protecting freshwater aquifers, but in many cases, regulations also address safety aspects of setting sufficient surface casing. Unfortunately, regulations do not always take specific situations into account, and they may require more casing than is really needed and sometimes less than what is needed. In those cases, it is best to consult with regulatory agencies as to what exceptions and variations from the regulations might be possible.

Aside from the regulations, the surface casing must allow us to drill to the next (or final) casing point with the mud density required to contain the formation pressures encountered, so as not to cause fracture failure of the exposed formations near the upper part of the hole. If more than one additional string of casing (an intermediate casing string) is required, then the two become interdependent as to setting depths.

3.2.4 Intermediate Casing Depth

The most common cause for needing intermediate casing is that the bore hole below the surface string may require a mud density too high (or sometimes too low) for the formations between the final drilling depth and the surface casing depth. A high mud density may fracture exposed weak zones or a mud density too low may allow higher-pressured zones to flow into the well bore. Additional reasons for running intermediate casing include the presence of unstable zones and corrosive zones. Instability in some zones, usually shale, may make it impossible to drill to total depth without isolating these zones. The presence of corrosive zones may require isolation to protect the production string.

3.2.5 Setting Depths Using Pore and Fracture Pressures

Aside from regulations and known problem zones, casing depths typically are selected using formation pore pressures and formation fracture pressures, and that is what we address now. The best way to understand how these two parameters are used is to make a plot of pore pressure and fracture pressure versus depth. Figure 3–6 is a plot of the two parameters for a simple well.

The figure shows a plot of the formation pore pressure versus depth on the left and the fracture pressure on the right. Note that the pressure is given in terms of equivalent mud density (specific gravity here) to make the plot more easily used by drilling personnel. Drillers use plots like this to determine mud densities required at various depths for drilling the well. The mud density must be slightly higher than the formation pressure to prevent formation fluids from entering the well bore; at the same time, the density must be less than the fracture pressure so that the drilling fluid does not fracture and enter the formations. The lines shown in the chart do not include any safety margins. Drillers typically drill with the density slightly higher than that required to balance the formation pressures. This allows some safety margin, especially when making trips because the action of pulling the pipe tends to cause a negative pressure surge or a reduction in the hydrostatic pressure while the pipe is in motion.

Likewise drillers like to keep the maximum density slightly lower than the fracture pressure because running the drill string back into the hole causes a positive surge pressure, but more important, the maximum also is considered a "kick margin," so that during a well control event, the formation is not fractured in the process of killing the well. Different companies have their own policies on the amount of safety margin required, and it may vary with type and location of individual wells. In Figure 3–7, we use a margin of 0.06 specific gravity (~0.5 lb/gal or 60 kg/m^3).

The figure shows the addition of these two safety margins. We can see that the mud density required to contain the pore pressure plus the safety margin at 12,000 ft is 1.4 sg, but above 1700 ft, that mud density begins to exceed the kick margin. In other words, we cannot drill safely to 12,000 ft in the well unless the hole is cased down to 1,700 ft or more, because the

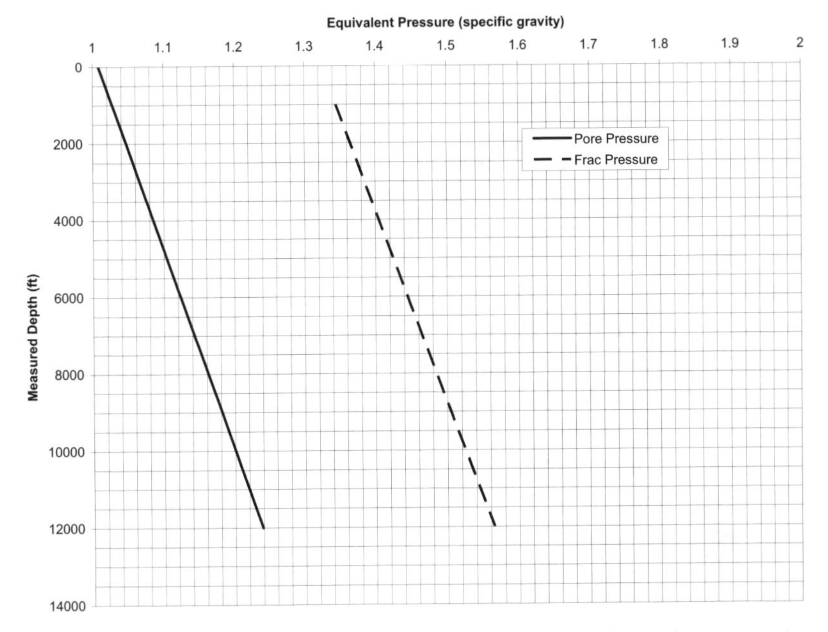

Figure 3–6 *Pore pressure and fracture pressure plot used in selecting casing setting depths.*

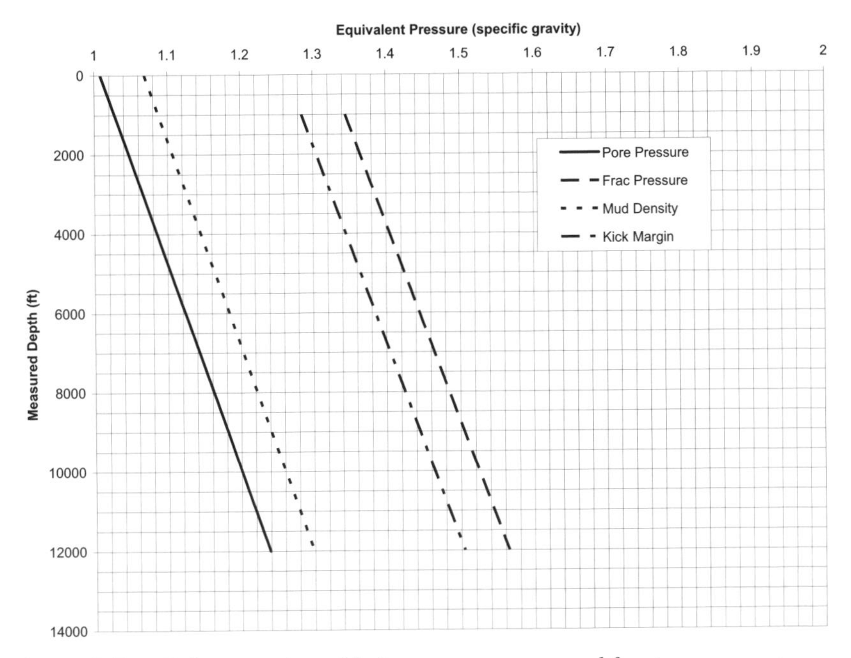

Figure 3–7 *Safety margins added to pore pressure and fracture pressure.*

mud density required to contain the pore pressure at bottom is greater than the fracture pressures at the surface (including the safety margins). That is exactly how we determine the setting depth of the surface casing in this well (Figure 3–8).

If we start at the mud density at 12,000 ft (point *a*) and draw a line vertically until it intersects the kick margin line (point *b*) then horizontally to the vertical axis (point *c*), we can read the setting depth of the surface casing, which in this case is about 1700 ft.

This particular well requires only a surface casing string at 1,700 ft and a production string at 12,000 ft. If the surface casing depth of 1700 ft meets the regulatory requirements for this well, then our setting depth selection is complete. If the regulations require more casing, say, 2500 ft, we simply move our surface casing depth to 2500 ft and it will give us more safety margin in our mud densities as far as a kick is concerned.

This is a relatively simple well. But, before we dismiss it as trivial, that is exactly the circumstance for the vast majority of all wells drilled in the world. That said, we now look at an example in which an intermediate string is required.

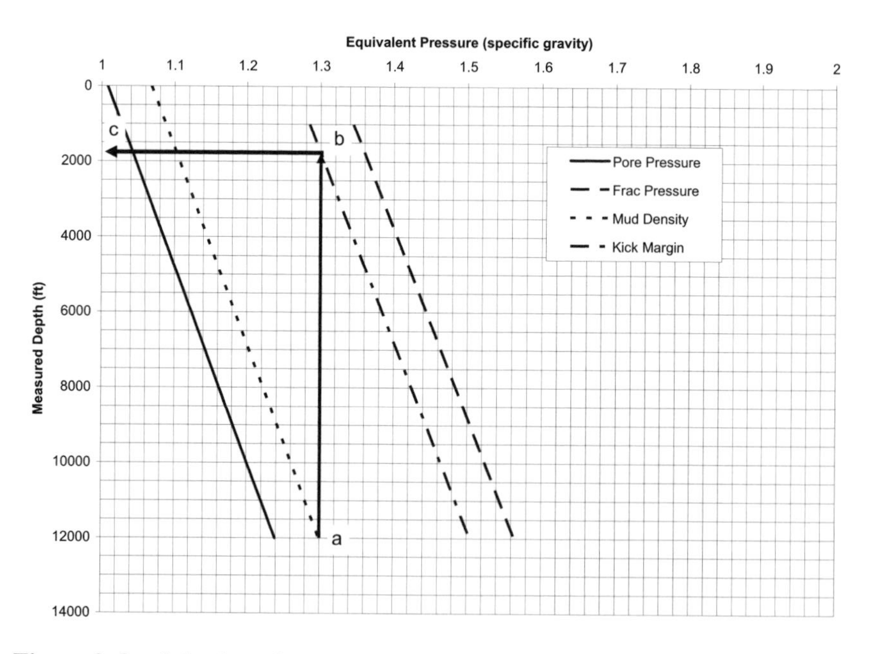

Figure 3–8 *Selection of casing setting depths.*

Example 3–1 Casing Depths

In this example (see Figure 3–9), we see that the mud density of 1.83 sg required at 14,000 ft exceeds the kick margin at all depths above 10,500 ft. So, we must set a string of casing at that depth. Moving horizontally to the left, we see that the mud density required at 10,500 ft is 1.42 sg. This mud density exceeds the kick margin at all depths above 3000 ft. So 3000 ft becomes the surface casing depth.

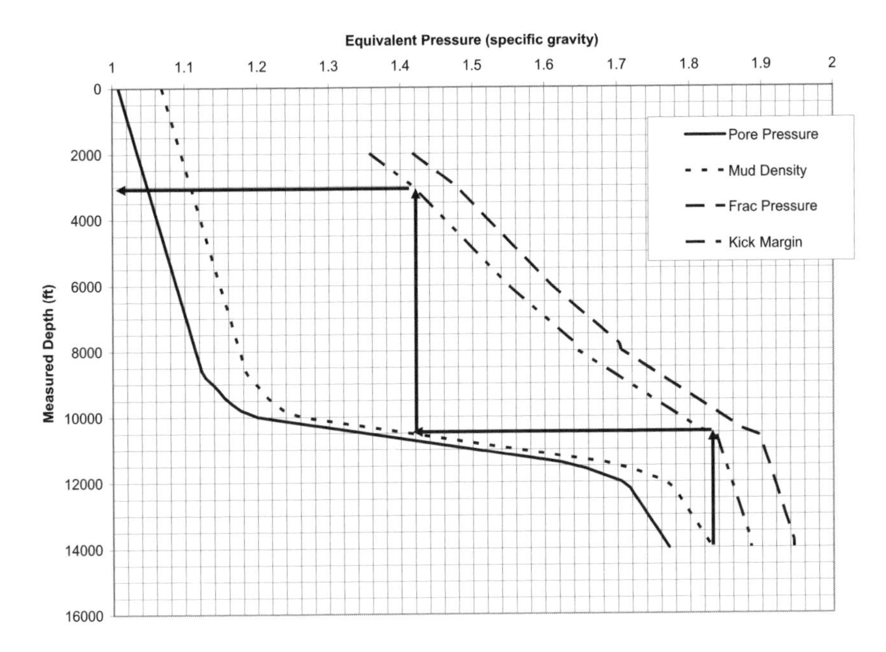

Figure 3–9 *Casing depth selection for example well.*

This is a straight forward procedure, but sometimes it can be complicated by depleted zones that have lowered pore pressure and fracture pressure but are located among normally pressured zones. Some situations may require more than one intermediate casing string, in which case, we typically install a liner (usually called a *drilling liner*) before reaching total depth rather than a second intermediate string. There are many possibilities, but that is the basic procedure.

Next Step

After determining the number of casing strings required and the setting depths, the next step in the design procedure is to select the sizes of casing required. What size casing and what size bits do we require?

3.2.6 Casing Size Selection

Once the setting depths have been determined the next step is obviously to select the sizes of the casing strings to be set. The sizes depend on a number of things.

Two important things to know about selection of casing size:

- Hole size determines casing size.
- Hole size at any point in the well except the surface is determined by the previous string of casing.

This means that, in selecting casing size, we usually start with the casing size at the bottom of the hole and work to the top.

The size of the last string of casing run in a well generally is determined by the type of completion that will be employed. That decision usually is the function of an interdisciplinary team of reservoir, production, and drilling personnel. This decision is based on numerous criteria, so we assume, for our purposes, that the size of the last string is predetermined and proceed from that point. From the standpoint of drilling, our input into that process is to assess the risks and allow for alternatives. For example, if we know there are serious hole stability problems in an area and our drilling experience in the area is limited, we may be well advised to recommend a final size that is large enough for us to set an extra string of casing or liner and still reach the objective with a usable size of hole for a good completion. This point unfortunately too often is overlooked in the desire to keep well costs low.

Once we know the diameter of the final string of casing or liner, the process proceeds like this:

- Determine the hole size (bit size) for the final string of casing.
- Determine what diameter casing allows that size bit to pass through it. That is the size of the next string of casing.
- Repeat the procedure until all the hole sizes and casing sizes have been determined.

Many times in actual practice, casing sizes are determined by what is readily available in the company's, partner's, or vendor's inventory and delivery times. The cost of leaving surplus pipe in inventory or excessive delivery times often supersede any optimum design based strictly on engineering calculations.

Precaution: After the casing strings have been designed be sure to check the drift diameters to be certain that the desired bit sizes can be accommodated.

3.2.7 Well Bore Size Selection

What is the proper bore-hole size for various sizes of casing? What do we require of the borehole size?

- A bore hole must be large enough for the casing to pass freely with little chance of getting stuck.
- There should be enough clearance around the casing to allow for a good cement job.
- In general, the bigger the bore hole, the more costly it is to drill.
- There are no formulas for determining the ideal bore-hole size.

Selecting the bore-hole size is based primarily on current practices in the area or areas with similar lithologies. There are a number of charts and tables in the literature, some good for some areas, but greatly lacking for other areas. The best advice we can offer is to use what is common practice in the area, unless there is good reason to do otherwise. No matter what specific charts we suggest here, they going to be wrong for some particular locale or application. That notwithstanding, Figure 3–10 and Figure 3–11 illustrate some typical choices. One chart is for hard rock and the other is for unconsolidated rock.

Figure 3–10 starts with the last string of casing or liner and works downward to the first casing string of the well. You can see on this chart there are many options even for those situations where the same size liner or casing is to be run. In general, hard rock offers us more choices, and clearance between the casing and bore-hole wall can be less than for unconsolidated wells. Figure 3–11 is a similar chart for unconsolidated formations.

Note, in Figure 3–11, there are still some options, but not as many. A few may not be available even though shown on the chart. For instance, on the fourth row from the top it shows that either an 8½ in. or 8¾ in. bit may be used from 9⅝ in. casing. That may be true in some cases, but if

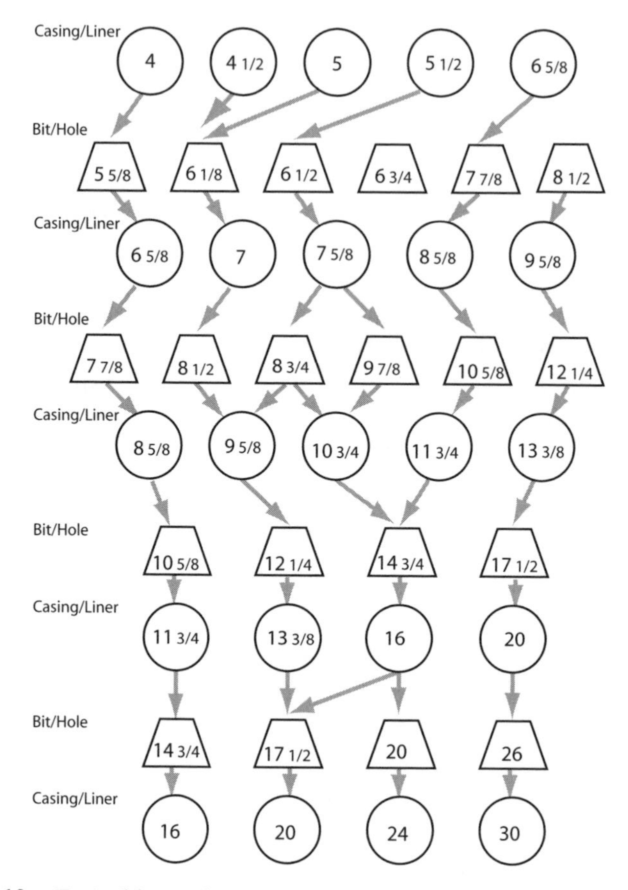

Figure 3–10 *Typical bit and casing sizes for hard rock formations.*

the 9⅝ in. casing string contains any 40 lb/ft or heavier pipe, then the 8¾ in. bit cannot be used. What is common practice in one area may not work in another, because formation pressures may require a heavier pipe.

Example 3–2 Casing Size Selection

Continuing with the same example we just looked at, assume that we have determined the following casing depths:

Surface casing = 3,000 ft

Intermediate casing = 10,500 ft

Production casing = 14,000 ft

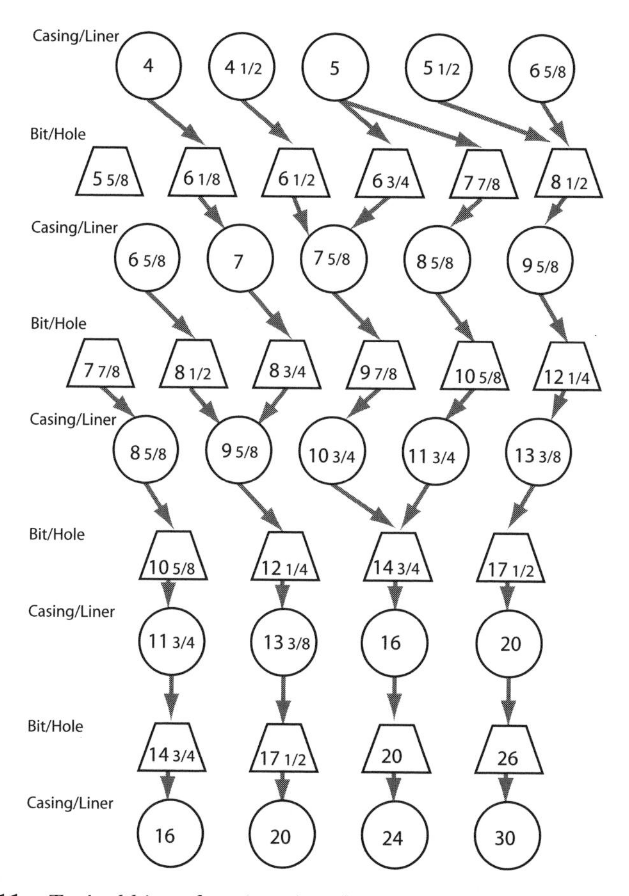

Figure 3–11 *Typical bit and casing sizes for unconsolidated formations.*

The production engineers tell us they require a production casing diameter of 7 in., so the production casing size is determined. Assume that the well is in an area of unconsolidated formations. Use the soft formation chart to determine the intermediate casing size, the surface casing size, and the conductor casing size:

Intermediate casing = 9⅝ in.

Surface casing = 13⅜ in.

Conductor casing = 20 in.

Although not shown in the chart as a possible path, some operators in areas where bore-hole stability is a serious problem elect an alternative for 7 in. casing as follows:

Intermediate casing = 10¾ in.

Surface casing = 16 in.

Conductor casing = 24 in.

That choice would be a case where experience in a particular area might influence the decision in order to allow more margin for the effects of anticipated problems.

3.2.8 Bit Choices

Obviously, from the preceding charts, we select the hole size for our particular casing and that automatically sets our bit size, too. While that is true, another aspect to the bit sizes should be mentioned. Those charts are based on the most commonly available bit sizes. There are special cases where it will be necessary to use an unusually thick wall casing, and you find that the common bit used in that casing will not work—the bit is too large. Other diameters of bits are available for special applications that are not shown in these charts. In general, they tend to cost more, but the biggest problem often is that there is a limited choice of types when it comes to uncommon bit sizes. For instance, for one common size, we may have a choice of 25 tooth and hardness characteristics, just from a single manufacturer, and maybe 50–100 choices if we include all manufacturers. However, with some odd-size bit, we may be constrained to a limited range of tooth and hardness choices and possibly only one manufacturer. That may be acceptable for some special cases, so it always should be considered.

Actual Bit Clearance

To determine the bit clearance, we look at the casing tables for the internal diameter and make sure it is larger than the diameter of the bit. In the tables we see two diameters listed. One is the internal diameter and the other is the internal drift diameter, which is slightly smaller than the internal diameter. The internal diameter is the diameter to which the tube is supposedly manufactured. Once it has gone through the milling process, it is inspected for final diameter by passing through it a mandrel of the diameter listed as the internal drift diameter. So, its internal diameter might be the same as the specified internal diameter or it might be slightly smaller (or larger), but we know that it is at least as large as the drift diameter (assuming the manufacturer does its job). We normally assume that the drift diameter is the maximum bit diameter we can be assured will pass through the casing. But, in many cases, bits greater than the drift diameter have been used. In that case, you must drift the casing with a mandrel the size of the bit first and cull out those joints that are undersized. Some steel mills actually do this for customers (usually for extra cost).

3.2.9 Alternative Approaches

There are additional approaches to allow for more clearance for the casing. One method is to underream the open hole below the current casing string. This allows additional clearance and is a proven method where the expense of the extra time and reaming can be justified. A similar result can be obtained with a bicentered bit for drilling below the current string of casing. Such a bit drills a hole larger than its nominal diameter. This technique can eliminate the extra expense of underreaming and accomplish the same result.

Another option is the use of expandable casing. This is a relatively new technology and has proven successful in a number of applications. The hole typically is drilled with a bicenter bit or is underreamed to give more clearance. The casing itself is run just like a conventional liner and is expanded after it is in place. There are obvious advantages to this expandable casing technology; however, there are also some disadvantages. Expandable casing is discussed in Chapter 10.

3.3 Closure

In this chapter, we examined the procedures for selecting casing setting depths and casing sizes. We used a plot of formation pore pressures and fracture pressures to select the setting depths. This straightforward method may appear deceptively simple. The truth is that the procedure is not complex but the data for use in the plot often is not readily available nor totally reliable. When this type of procedure first came into use, many operators looked at it as a way to save money by reducing the number of casing strings traditionally run in some areas. It appeared that, in many cases, it was possible to run surface casing a bit deeper and eliminate an intermediate casing string. When it worked, it did save costs; but when it did not work it not only added an intermediate string that had been "eliminated" but often an additional string as well and resulted in a very small hole size at bottom and significant additional costs. The problem in these situations was that the data proved unreliable in some cases and the margins were too close for operating personnel to adhere to in others. So, in all cases, the data used in the depth selection process must be scrutinized with care. A prudent philosophy might be stated like this:

- Exploratory wells or critical wells. Data are possibly scarce or unreliable, so allow for the unexpected with contingencies in casing size and depths. Usually, this means allowing for the possibility of running one more casing string than the plan calls for. These are not the wells where we try to save money on casing.
- Development wells. Data reliability and risks are well known. These are the wells where casing costs can be minimized and smaller margins can be used.

No matter what method you use to determine the casing setting depth, always keep in mind that it is one of the most critical steps in assuring a well's success. Do not be caught in the trap of compromising the chances of success by trying to save money by unnecessarily minimizing casing depths and sizes.

3.4 References

Fertl, W. H. (1976). *Abnormal Formation Pressures.* Amsterdam, Oxford, New York: Elsevier Scientific Publishing Co.

Fjær, E., R. M. Holt, P. Horsrud, A. M. Raaen, and R. Risnes. (1992). *Petroleum Related Rock Mechanics.* Amsterdam, Oxford, New York: Elsevier Scientific Publishing Co.

Terzaghi, K., and R. B. Peck. (1948). Theoretical Soil Mechanics in Engineering Practice. New York: John Wiley and Sons.

CHAPTER 4

Casing Load Determination

4.1 Introduction

In this chapter, we discuss the collapse and burst loading used for the design of casing strings. To illustrate the process, we continue with the example used in the previous chapter, where we determined the setting depths and casing sizes. We use several types of loading for illustrational purposes and again emphasize that operating companies have a variety of views on the use of the various types of loading.

4.2 Casing Loads

To determine what strength of casing we need, we must next consider the types and magnitudes of the loads the casing must safely bear. A number of different considerations and possibilities accompany each string of casing run in a well. Some simple load situations suffice for most casing strings, but often special conditions may apply to a specific well or type of well. We look at the types of loads commonly used for design for each type of casing string. Three basic types of loads commonly are encountered:

1. *Collapse loads.* These are differential pressure loads where the external pressure exceeds the internal pressure, tending to cause the casing to collapse.
2. *Burst loads.* These are differential pressure loads in which the internal pressure exceeds the external pressure, tending to cause the casing to rupture or burst.
3. *Axial loads.* These are tension or compression loads caused by gravitational and frictional forces on the pipe.

The first two are dictated by well conditions and anticipated operations in the well. Those are the two we cover in this chapter. The third type of load, axial, is a function of the casing selection process itself and is discussed in the next chapter. The first two are functions of pore pressures, fracture pressures, and drilling fluid (or cement) pressures.

Magnitude of Loading

The magnitudes of the loads that a particular casing string actually experiences in service for the most part are unknown. Certainly, we can calculate the loads we are likely to encounter if all operations are perfectly successful, but the problem is that often there are imperfections in our cementing results or problems in our drilling operations. We almost always are able to determine loads if all is perfect, and we can almost always determine the type of loading that would take place if things go totally awry. But between those two situations is a great unknown. Hence, our most logical approach is to assume the worst case that can happen, within reason, and that is the one we typically use for our casing design. We assume that we can reasonably predict the nature of the worst case loading and calculate its values. For now, we do not concern ourselves with the probability of such loading occurring.

The process then consists of determining the following:

- Collapse loading.
    - Minimum internal pressures.
    - Maximum external pressures.
- Burst loading.
    - Maximum internal pressures.
    - Minimum external pressures.

Some typical sources of these pressures are

- Formation fluids.
    - Water (fresh or salty).
    - Oil.
    - Gas.
- Drilling fluids.
    - Whole mud.
    - Mud filtrate.

- Unset cement.
  - Whole cement.
  - Cement filtrate.
- Stimulation fluids.
- Ocean or surface water.
- Atmosphere.

These are the pressure sources that contribute in various combinations to the pressure loading of casing in wells.

4.3 Collapse Loading

In the case of collapse loading, our task is to determine the least amount of pressure the casing will have inside and the maximum amount of pressure the casing will have on the outside. For collapse loading, we typically consider some static fluid gradient on the inside and outside of the casing string. The internal loads are the loads that resist collapse. We would like these loads to be high, but in the design process, we try to determine the minimum internal load that might reasonably occur.

Internal Loads, Collapse

- Empty casing (atmospheric pressure).
- Gas.
- Oil.
- Freshwater.
- Field saltwater or stimulation fluids.
- Drilling or workover fluids.
- Combinations and partial columns of these.

In most wells, the most serious internal collapse loading is atmospheric pressure (an empty well bore). It can happen in surface casing and even intermediate casing, if severe lost circulation is encountered while drilling below the casing. It can occur during underbalanced drilling operations, where air or gas is the drilling fluid. It can even happen in some production casing strings with gas lift or pumps or if the casing is "blown dry" after a stimulation. (For the skeptic, this does happen and more often than one might think.)

At the surface, the external load usually is taken to be zero or atmospheric pressure, except in the case of subsea completions, where the external pressure is the seawater pressure at the subsea wellhead. Below the wellhead lie a number of possibilities for external loads.

External Loads, Collapse

- Freshwater.
- Saltwater.
- Formation pressure.
- Drilling fluid.
- Cement (unset).

Generally, the worst case here might be the possibility of a poor cement job and the external pressure is equal to the mud pressure in which the casing is run (which is higher than the formation pressures of the formations covered by the casing). While the density of the cement usually is higher than the mud, there is always the displacement fluid inside the casing when the cement is placed. However, in some cases, the differential pressure between the cement outside and the displacement fluid on the inside represents a more severe collapse situation than any other that may be encountered in a particular well, so that should be considered. It is important to note that we never consider that the hardened cement gives us any protection from external loads. It might well do that, but we cannot assume it will.

In many other wells, it is not possible that the casing would ever be empty. In intermediate strings that were set because of abnormally high pressures below, the likelihood of the casing being empty usually is remote. However, in the presence of partially depleted zones and an underground blowout, it could happen.

4.4 Burst Loading

In low pressure wells or wells that will not flow, burst is seldom considered. However, one should keep in mind the possibility of a future fracture job or high rate stimulation that might be pumped down the production casing. In burst loading, the external pressure is the resisting load, and the external loading in a burst situation normally is taken to be the lowest possible pressure externally. At the surface of the string, that pressure is taken to be zero or atmospheric pressure. In a subsea casing string, it would be the seawater pressure at the wellhead. The external pressures other than at the surface could be from a number of sources.

External Loads, Burst

- Atmospheric pressure (at surface of string).
- Seawater pressure (at surface of string).
- Freshwater.
- Saltwater.
- Formation pressure.
- Drilling fluid.

It is never acceptable to assume that hardened cement will give us support in burst, even though it will. The problem with cement is that we have to design our string before the well is cemented. If our cement job is near perfect, then we have additional support in those sections covered by cement. However, if there is even a small interval where the cement is poor, then we have no support at that interval, and there is nothing we can reasonably do the change that. Hence, we can never safely assume that the hardened cement gives us any benefit when we are in the design stage.

Internal Loads, Burst

- Gas.
- Oil.
- Water.
- Combinations of gas and liquids.
- Stimulation fluids in combination with pumping pressures.

In most wells where burst is considered, the internal loading is a gas pressure, but that is not always the case. In many wells, the internal loading is oil pressure or even water. It is relatively easy to design for burst if the internal pressure is due to a liquid, but several complications arise when the fluid is gas.

In cases where gas is present, additional factors usually are taken into consideration. If drilling will take place below the casing shoe, then there is a possibility that the gas pressure will exceed the fracture pressure at or below the casing shoe. In such a case, there is no point in designing the casing string for full pressure of the gas all the way to the surface, if it is not possible for that pressure to ever be reached, since the gas pressure will not exceed the fracture pressure at the shoe. The same thing can be said for oil and water. Typically, this condition is considered in designing surface and intermediate strings only, since the condition does not arise in production strings. There also is another approach where we consider a combination of liquid and gas in the fluid column, fracture at the shoe, and a limited working pressure on the blowout preventer (BOP). These various loading conditions are demonstrated in the following sections, which are specific to the type of string being designed.

4.5 Surface Casing

In this section, we examine the particular loads as they apply to surface casing. Typically, the loads in surface casing are relatively low compared to other casing strings in the well, but many casing failures occur because of underdesigned surface casing.

4.5.1 Surface Casing Collapse Loads

The collapse load for surface casing depends on the worst-case scenario anticipated, in which the pressure outside the casing exceeds the internal pressure. There are a number of possibilities, but the most commonly accepted situation assumes that the surface casing is empty inside (possibly due to lost circulation while drilling somewhere below) and has mud pressure on the outside of the same magnitude as when the casing was run. We can modify the internal pressure, if we have some knowledge of the worst case of lost circulation that could be encountered and how far the drilling fluid would drop in the surface casing should that occur. But, in the absence of such knowledge, we should assume the lost circulation situation could be severe enough to empty the surface casing. On the outside of the surface casing, we know the pressure when the casing is run; it is the hydrostatic pressure of the mud column. If the cement is of greater density than the mud (and it usually is) we easily can calculate the pressure due to the cement. The question is, what is the pressure after the cement hardens? We can be fairly certain that it will not be as high as the cement pressure before it hardened, but the actual pressure depends on the integrity of the cement job, that is, whether there are channels in the cement or some formations are not cemented properly. Typically, a safe assumption is that the highest pressure outside the casing after cementing is the mud pressure before cementing. It may be less, but it is unlikely to be greater.

Typical Surface Casing Collapse Design Load

- Internal pressure—atmospheric pressure or zero.
- External pressure—mud pressure when run.

That is the collapse design load we use here, but be aware that there are other possibilities.

4.5.2 Surface Casing Burst Loads

The burst load of the surface casing is based on the maximum anticipated internal pressure and the minimum anticipated external pressure. Let us look at the external pressure first. In collapse, we looked for the maximum external pressure, now we are interested in the minimum. The minimum external pressure is likely to occur some time after cementing. It is believed that, when cement hardens, fluid in the spaces where the cement has channeled or is absent often is similar in density to freshwater. For that reason, many assume that the minimum external pressure is equivalent to a freshwater gradient. Some believe that a freshwater gradient is not really likely and they use the mud pressure on the outside, just as we did in collapse. That is also a valid external load but not the most critical that could occur.

The internal pressure for burst is a little more complicated. If we drill a well some distance below the surface casing, encounter a gas kick, and get a large volume of gas in the casing, then the pressures could get quite high. However, if the pressures get very high, the formations at the bottom of the surface casing will fracture and flow will go into those formations. That being the case, it does not make sense to design a surface casing string to withstand say 6000 lbf/in.^{2} internal pressure if the formation below the surface casing fractures at 3500 lbf/in.^{2}. The typical procedure for determining burst load is to assume that the maximum internal pressure is equivalent to the fracture pressure beneath the casing shoe and gas from there to the surface. In cases where gas is known to not be present, we could use oil or water as the internal fluid; however, gas gives us the most critical load and should always be used unless there is absolute certainty that no gas is present in the zones between the surface casing and the next string of casing.

Typical Surface Casing Burst Design Loads

- Internal pressure—equivalent of gas kick that fractures and flows into formation(s) below the casing shoe.
- External pressure—freshwater gradient.

Again, we must emphasize there many possibilities and different companies have a variety of approaches. These, however, are simple and should be safe in most cases.

4.5.3 Surface Casing Load Curves

One of the easiest ways to work with casing loads is to construct a set of design load curves. The anticipated loads, such as collapse pressures and bust pressures, are plotted graphically as pressure versus depth. This makes it very easy to visualize the loading, rather than relying on a lot of formulas. (We still need formulas and calculations to construct the load curves, but they require very few calculations.)

Example 4–1 Surface Casing Example

Possibly the best way to understand the construction of the load curves is with an example. We use the depth selection curve used in Chapter 3. Assume that the bottom-hole temperature is 326° F, the average surface temperature is 74° F, and that the temperature gradient is linear from the bottom to the top. The curve is shown in Figure 4–1.

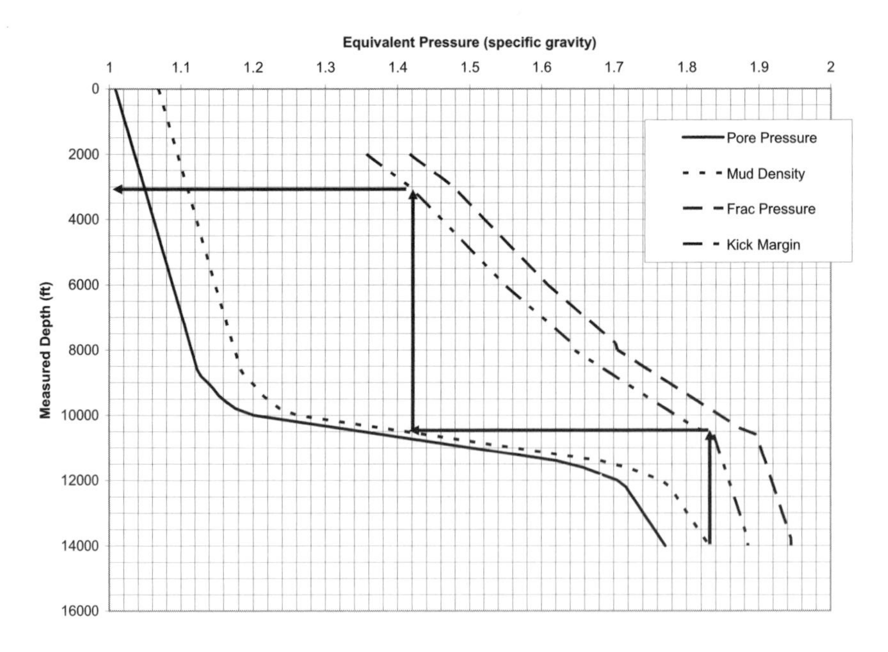

Figure 4–1 *Depth selection chart for example well.*

In this example, we are going to set surface casing at 3000 ft, the mud specific gravity is 1.10, and the fracture gradient is equivalent to 1.45 sg.

Collapse Load

First, we plot a collapse curve. Assume the internal casing pressure is 0 lbf/in.^{2} and the external pressure at 3000 ft is due to the mud pressure. The collapse load at 3000 ft is

$$\Delta p_{\text{shoe}} = \gamma_{\text{mud}} h - 0 = 0.052(1.11)(8.33)(3000) - 0 \approx 1440 \text{ lbf/in}^2 \quad (4.1)$$

Note that we are going to round off to the nearest 10 lbf/in.^{2} to keep these calculations simple—after all, we are not doing rocket science here. Refreshing your memory from Chapter 2, γ is the specific weight of the mud and h is the depth. We are not going to show examples in SI units again, but just to be sure you are clear on the calculation, in SI units, it would be written as

$$\Delta p_{\text{shoe}} = \gamma_{\text{mud}} h - 0 = 9.81(1.11)(998)(914) - 0 \approx 9930 \text{ kPa} \quad (4.2)$$

If you are not perfectly clear on this, then you probably did not read Chapter 2, Section 2.4. Go back and read it now, otherwise you will be lost from here on.

The collapse load at the surface is zero, since there is no external pressure nor any internal pressure.

$$\Delta p_{\text{surf}} = 0 \text{ lbf/in}^2 \quad (4.3)$$

We plot this collapse load curve in Figure 4–2.

Burst Load

Next, we examine the burst load. At the shoe, the burst load is the fracture pressure of the formation below the casing less the external pressure at the casing shoe, which we said is equivalent to a freshwater gradient. Some companies add some amount of additional pressure to account for extra frictional pressure for flow into the formation—most do not. We do not do that here. Next, we calculate the burst load at the shoe at 3000 ft, which is the fracture gradient at the shoe:

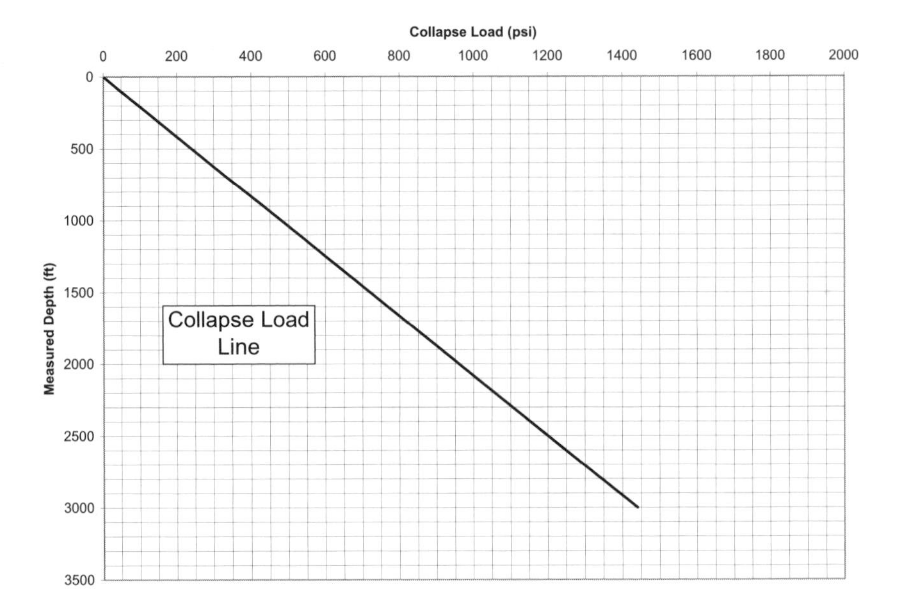

Figure 4–2 *Surface casing collapse load for the example well.*

$$\Delta p_{\text{shoe}} = p_{\text{frac}} - p_{\text{wtr}} = (\gamma_{\text{frac}} - \gamma_{\text{wtr}})h = 0.052(1.48 - 1.00)(8.33)(3000)$$
$$\approx 620 \text{ lbf/in}^2 \tag{4.4}$$

You can see that the burst load at the shoe is quite low, and that is usually the case with most casing strings. An exception to that might occur at some known depleted formation near the shoe.

Next, we need to calculate the burst load at the surface. The worst-case scenario here is to have the surface casing full of gas, all the way from the shoe to the top, this gives us the maximum pressure at the surface, and such pressure is quite possible in a kick situation. We use equation (2.45) (from Chapter 2) to calculate the gas pressures, assuming pure *methane*.

Calculate the pressure at the shoe at 3000 ft, which is the fracture pressure of the formation:

$$p_{\text{frac}} = 0.052(1.48)(8.33)(3000) \approx 1920 \text{ lbf/in}^2 \tag{4.5}$$

Assuming the temperature gradient is linear, we can calculate the temperature at 3000 ft, knowing that it is 74° F at the surface and 326° F at 14,000 ft:

$$T = 74 + \frac{3000}{14000}(326 - 74) = 128\ ^{\circ}\mathrm{F} \tag{4.6}$$

The average temperature is

$$T_{avg} = \frac{74 + 128}{2} + 460 = 561\ ^{\circ}\mathrm{R} \tag{4.7}$$

Then the surface gas pressure is

$$p_{surf} = 1920\, e^{\left[\frac{16(0-3000)}{1544(561)}\right]} \tag{4.8}$$

$$p_{surf} \approx 1820\ \mathrm{lbf/in^2}$$

Since there is no external pressure at the surface, then that value is also the burst load at the surface. We now plot the surface casing burst load in Figure 4–3. That completes the load curves for the surface casing example.

4.6 Intermediate Casing

The intermediate casing loading often is straightforward, like the surface casing, except that the magnitude of the loads generally is greater. For many designs the procedure is exactly the same as our surface casing example.

4.6.1 Intermediate Casing Collapse Loads

Collapse loading in intermediate casing is not often critical, but it can be. Many companies use a mud gradient outside the intermediate casing and no pressure on the inside. This almost always is the case if the intermediate casing later will become part of the production string after a production liner is set. If the intermediate casing later will be covered by the

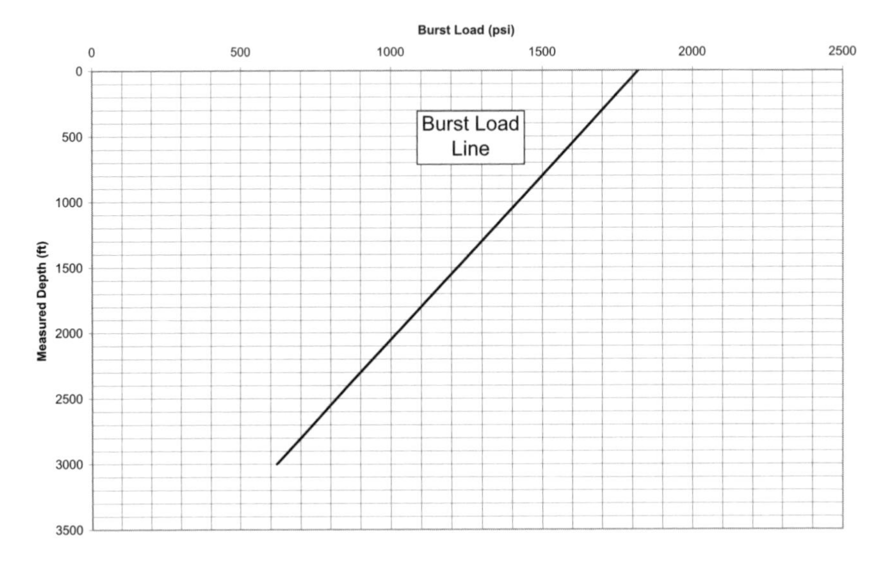

Figure 4–3 *Surface casing burst load for the example well.*

production casing or a tieback string, then the issue of collapse load may change. Since there is very little likelihood that the intermediate string will ever be empty, some other internal load often is chosen. Typically, freshwater is selected. This is a likely scenario in case of an underground blowout below the casing shoe, where the mud level falls and the casing must be kept full of liquid to prevent gas from reaching the surface. In those cases, it is not uncommon that the casing might be filled with freshwater when there is not enough weighted drilling fluid for the purpose. However, in some cases, intermediate casing is set to allow the drilling of low-pressure formations below the casing. In those cases, it is possible that the casing could be empty or nearly so.

4.6.2 Intermediate Casing Burst Loads

The typical burst load on intermediate casing usually is assumed to be one similar to the surface casing. That is a case of a freshwater gradient or a drilling mud gradient on the outside, using the mud density at the time the casing was cemented, and gas pressure on the inside with the pressure at bottom equal to the fracture pressure at the shoe. An alternate approach sometimes is used when the surface BOP and wellhead selection limits the burst rating at the surface. In those cases, the reason the BOP does not withstand full well pressure is that the formations below the shoe fracture before the maximum pressure is reached at the surface, so it is common practice in many areas to use a BOP stack that contains the kick pressure to the extent that the formation below the shoe fractures before the BOP fails. This often is a cost-saving measure, but in other cases, it is an issue of availability. We do not argue the merits of such a choice here, other than to say that it is not an uncommon practice. If that type of well plan is chosen, then the surface pressure is fixed at the maximum service pressure rating (MSP) of the BOP and wellhead, and the pressure at the bottom of the intermediate casing is fixed at the formation fracture pressure. (In some cases, where high rates of gas flow might be anticipated, it is common practice to add some incremental amount of pressure to the fracture pressure to account for the high injection rate into the formation.) Given those two pressures, one must determine the configuration of the mud and gas column that would impose the highest burst loads on the casing. It seems intuitive that the highest load on the casing occurs with gas at the surface and mud below, but that is not the case. Prentice (1970) showed that the maximum burst load actually occurs with a mud column on top and gas beneath. We illustrate the use of this procedure in our example well in Figure 4–4.

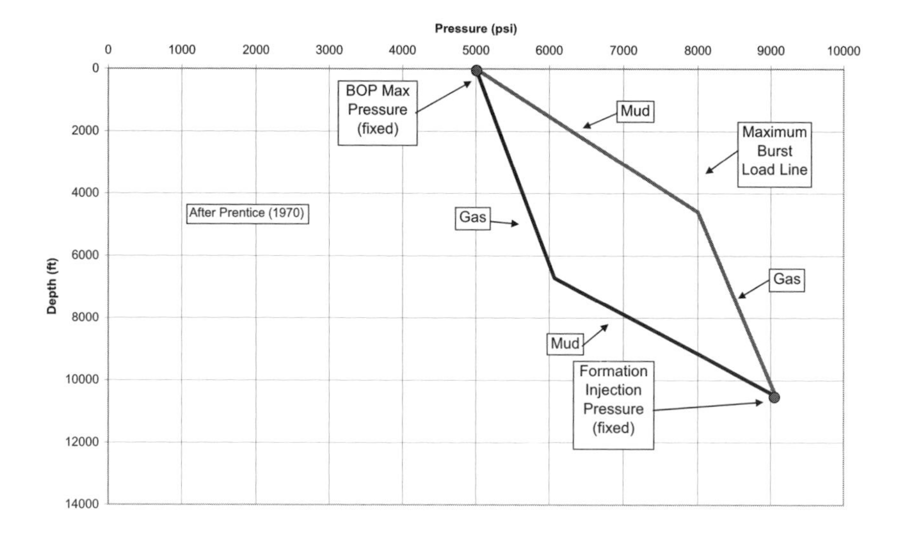

Figure 4–4 *Maximum burst load method (after Prentice, 1970).*

Example 4–2 Intermediate Casing Example

Continuing with our example from the previous chapter (and here), we now determine the collapse and load curves for the 9⅝ in. intermediate casing to be set at 10,500 ft. At 10,500 ft, the pore pressure is equivalent to a mud column with a specific gravity of 1.36, and the fracture pressure is equivalent to a 1.88 sg mud. The mud density when the casing is set is 1.42 sg. The bore-hole temperature at the shoe is 263° F and the surface temperature is 74° F.

Collapse Load

The load curve for collapse is similar to a surface casing load curve. However, there is very little chance the intermediate casing could ever be empty of all fluids like the surface casing, and if we design an intermediate string for collapse with no fluid inside, then we likely will have a greatly overdesigned and expensive string of casing, whose function in the well is only temporary. We consider that the worst possible case for collapse of our example intermediate casing is one in which there is freshwater on the inside and the mud it was run in on the outside. Note that some companies would not find this assumption acceptable and would apply a more severe collapse load criterion. That always depends on the specific well, and the design generally should be for the worst case even if it means a very expensive string of casing.

For our case, then, the net collapse load at the bottom is

$$\Delta p_{\text{shoe}} = (\gamma_{\text{mud}} - \gamma_{\text{wtr}})h = 0.052(1.42 - 1.00)(8.33)(10500) \approx 1910 \text{ lbf/in}^2 \tag{4.9}$$

This is a very low value for collapse, and it will probably not even affect the design. The collapse load at the surface is zero, see Figure 4–5.

Burst Load for Gas

As for the burst load, there are many possibilities, as already mentioned. The most common method is exactly like we did with the surface casing. We illustrate that method first, then show two other approaches. The gas loading example is as follows:

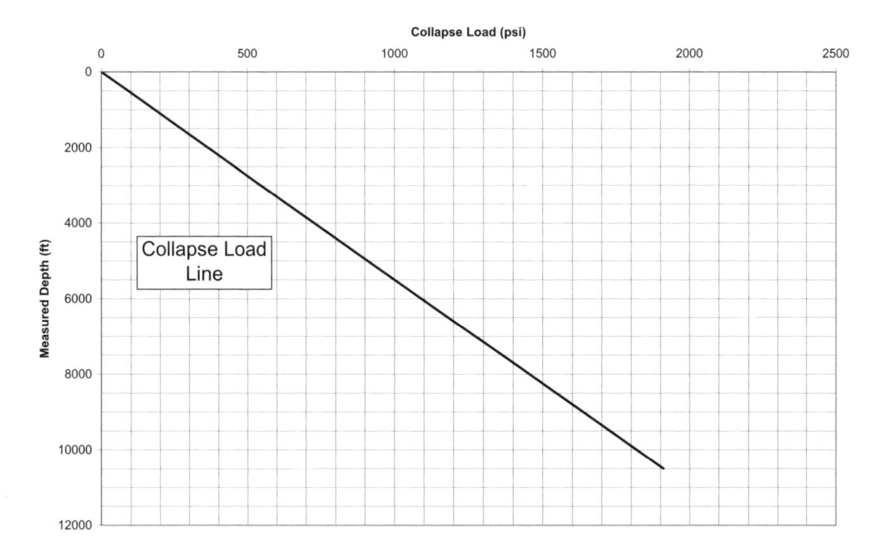

Figure 4–5 *Intermediate casing collapse load for the example well.*

- Internal pressures.
  - Fracture pressure at the shoe.
  - Gas inside with gas pressure at the surface.
- External pressures.
  - Freshwater gradient behind the casing.

Then the net burst pressure at the bottom is

$$\Delta p_{\text{shoe}} = (\gamma_{\text{frac}} - \gamma_{\text{wtr}})h = 0.052(1.88 - 1.00)(8.33)(10500) \approx 4000 \text{ lbf/in}^2 \tag{4.10}$$

The net bust pressure at the surface is due to a column of gas whose pressure at the casing shoe is equal to the fracture pressure at the shoe. We calculate the fracture pressure first:

$$p_{\text{frac}} = \gamma_{\text{frac}}\, h = 0.052(1.88)(8.33)(10500) \approx 8550 \text{ lbf/in}^2 \tag{4.11}$$

Next, we calculate the gas pressure at the surface:

$$T_{avg} = \frac{74 + 263}{2} + 460 = 628.5 \ \ ^{o}R \tag{4.12}$$

$$p_{surf} = 8550\, e^{\left[\frac{16(0-10500)}{1544(628.5)}\right]} \approx 7190 \ \ \text{lbf/in}^2 \tag{4.13}$$

This gas pressure is the burst load at the surface, since there is no pressure outside the wellhead. This load curve is plotted in Figure 4–6.

Maximum Burst Load Method

We will now examine the method of Charlie Prentice (1970). This method of burst design assumes that the minimum rating of the wellhead equipment is one limiting factor and the injection pressure of gas or liquids into the formation just below the casing shoe is the other. Between the surface and the casing shoe is some combination of gas and mud. As previously mentioned, it would seem that the worst case would be gas at the surface and mud from some point in the casing down to the shoe, such that the surface pressure of the gas is equal to the working pressure of the BOP, and the combined column is such that the pressure at the shoe is equal to the injection pressure of the formation at the shoe (or just below). But, as stated earlier, Prentice showed that the worst case is exactly the opposite, with the mud on top and the gas below.

We assume for this example that the maximum service pressure of our BOP is 5000 lbf/in.^{2}. Clearly, if we allow the gas to get all the way to the surface, the pressure will exceed that of the BOP working pressure by almost 2200 lbf/in.^{2}. For a real well, we should seriously question such a choice, but for this example, we assume it is not our choice to make: furthermore, we want to illustrate how the method works. We design a string for that limitation.

By knowing the working pressure of the BOP, the fracture pressure of the formation below the shoe, and the density of the mud, we can calculate the length of the column of mud and gas from the following equations of Prentice (1970):

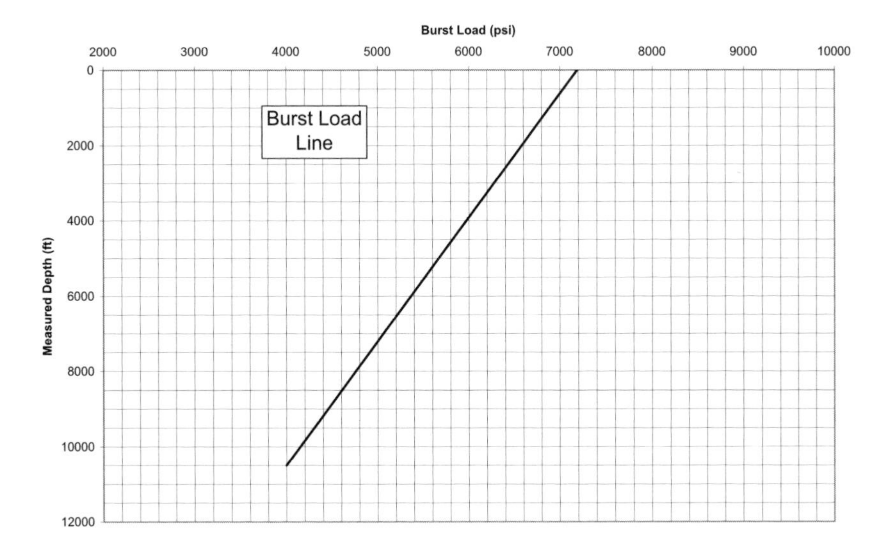

Figure 4–6 *Intermediate casing burst load for the example well with a full gas column.*

$$h_{mud} + h_{gas} = h_{csg}$$

$$\gamma_{mud} h_{mud} + \gamma_{gas} h_{gas} = p_{frac} - p_{surf} \tag{4.14}$$

where

h_{csg} = vertical length of the casing
h_{mud} = vertical length of the mud column
h_{gas} = vertical length of the gas column
p_{frac} = fracture pressure at the shoe
p_{surf} = surface equipment pressure rating
γ_{mud} = specific weight of mud
γ_{gas} = specific weight of gas

Please note that these equations are not unit specific and will work with either SI or oil-field units, as we previously defined specific weights in Chapter 2. Also, I made a variation in nomenclature here from Prentice's paper, the use of the symbol h for the vertical length (or vertical displacement), so that these formulas are valid for either vertical or directional wells. This is a system of linear equations we might set up and solve as follows:

$$\begin{bmatrix} 1 & 1 \\ \gamma_{\text{mud}} & \gamma_{\text{gas}} \end{bmatrix} \begin{Bmatrix} h_{\text{mud}} \\ h_{\text{gas}} \end{Bmatrix} = \begin{Bmatrix} h_{\text{csg}} \\ p_{\text{frac}} - p_{\text{surf}} \end{Bmatrix} \tag{4.15}$$

Solving this simple system we get equations for the vertical lengths of the mud column and the gas column:

$$h_{\text{mud}} = \frac{p_{\text{frac}} - p_{\text{surf}} - \gamma_{\text{gas}} h_{\text{csg}}}{\gamma_{\text{mud}} - \gamma_{\text{gas}}}$$

$$h_{\text{gas}} = \frac{p_{\text{frac}} - p_{\text{surf}} - \gamma_{\text{mud}} h_{\text{csg}}}{\gamma_{\text{gas}} - \gamma_{\text{mud}}} \tag{4.16}$$

We need not calculate both. If we calculate the vertical length of the mud column, we can subtract it from the vertical length of the casing to get the vertical length of the gas column. The difficulty with this sort of formula is the specific weight of the gas. A gas gradient is not constant; it varies with depth and temperature. Typically, in casing design, an average value is used. We use an average value of a pressure at the shoe equal to the fracture pressure and a pressure at the surface equal to a full column of gas. This is not precise, but then neither is casing design. At least, it gives us a reasonable approximation. Actually, we already did these calculations. The fracture pressure was calculated in equation (4.11), and the gas pressure at the surface was calculated in equation (4.13). We take the difference between those two pressures and divide by the depth of the casing to give the average specific weight of the gas:

$$\gamma_{\text{gas}} = \frac{8550 - 7190}{10500} = 0.130 \ \frac{\text{lbf}}{\text{in}^2 \cdot \text{ft}} \tag{4.17}$$

This is the average specific weight of the gas. (Note that, if you are using SI units, the pressures are in Pa, the depth in m, and the result is in Pa/m, so as long as we are working with specific weights, no conversion factors are needed.)

Now, a question arises. What is the density of the mud when the kick occurs? Is it the mud density at the time the casing was set or at some future time while we are drilling below the shoe? The difficulty here is that we do not know what the mud density will be when the well kicks. We could choose either the maximum weight we expect to use in drilling to total depth or the minimum, which is the density used when the shoe is drilled. But which one actually gives us the worst loading case? As it turns out, the worst case occurs with the highest mud density. The higher the mud density, the closer the gas can get to the surface without exceeding the pressure limitations of the surface equipment. So, we calculate the values using the maximum mud density for the well, which is 1.82 sg:

$$\gamma_{\text{mud}} = 0.052(1.82)(8.33) = 0.788 \ \frac{\text{lbf}}{\text{in}^2 \cdot \text{ft}} \tag{4.18}$$

$$h_{\text{mud}} = \frac{p_{\text{frac}} - p_{\text{surf}} - \gamma_{\text{gas}} h_{\text{csg}}}{\gamma_{\text{mud}} - \gamma_{\text{gas}}} = \frac{8550 - 5000 - (0.130)(10500)}{0.788 - 0.130} \approx 3320 \text{ ft} \tag{4.19}$$

$$p_{\text{int}} = p_{\text{surf}} + \gamma_{\text{mud}} h_{\text{mud}} = 5000 + 0.788(3320) \approx 7620 \text{ lbf/in}^2 \tag{4.20}$$

$$\Delta p_{\text{surf}} = p_{\text{surf}} - 0 = 5000 - 0 = 5000 \text{ lbf/in}^2$$

$$\Delta p_{\text{int}} = p_{\text{int}} - \gamma_{\text{wtr}} h_{\text{mud}} = 7620 - 0.052(1.00)(8.33)(3320) \approx 6180 \text{ lbf/in}^2$$

$$\Delta p_{\text{shoe}} = p_{\text{frac}} - \gamma_{\text{wtr}} h_{\text{csg}} = 8550 - 0.052(1.00)(8.33)(10500) \approx 4000 \text{ lbf/in}^2 \tag{4.21}$$

We can see that using a mud with a specific gravity of 1.82 gives us a net burst pressure of 6180 lbf/in.^{2} at a vertical depth of 3320 ft. Had we used a lower mud density, the net burst pressure would have been less and at greater depth. In fact, if we used the same gas density, it would plot on the same line between the maximum net burst we calculated previously and the net burst pressure at the shoe, see Figure 4–7.

Burst Load for Oil

Many engineers work in areas where there is essentially no gas, only oil. And, no doubt, they are weary of seeing all examples of intermediate casing design based on gas—in fact, some have told me so. We almost never see examples of intermediate casing designed for oil, although that type of loading often occurs in many parts of the world. Here is an example of the same well using oil as an internal loading as opposed to gas.

Assume an oil gradient instead of gas and assume that the oil in our well is 35 API gravity at 60° F. Since the density of oil is temperature dependent, we make a temperature correction first, using an average temperature:

$$T_{avg} = \frac{74 + 263}{2} = 168.5\ ^{\circ}\text{F} \tag{4.22}$$

Using API tables, we find that the API gravity at 168.5° F is approximately 43. Then, using the API formula relating specific gravity and API gravity,

$$\text{Specific Gravity} = \frac{141.5}{\text{API Gravity} + 131.5} = \frac{141.5}{43 + 131.5} = 0.811 \tag{4.23}$$

Please note that this is an approximation; oil also is compressible. This should be close enough for most casing design, but there are more accurate methods for calculating pressures from compressible oil at varying temperatures. Now, assuming a pure column of oil inside the casing and a fracture pressure of 8550 lbf/in² at the shoe, we calculate the surface pressure:

$$\Delta p_{surf} = p_{frac} - \gamma_{oil} h = 8550 - 0.052(0.811)(8.33)(10500) \approx 4860\ \text{lbf/in}^2 \tag{4.24}$$

If we were using a 5000 lbf/in.² maximum service pressure BOP on this well, it would be adequate. This is the net burst load at the surface, and the net burst load at the shoe is exactly as previously calculated. The burst load at the shoe is the same as before, and the burst load curve would look like that in Figure 4–8.

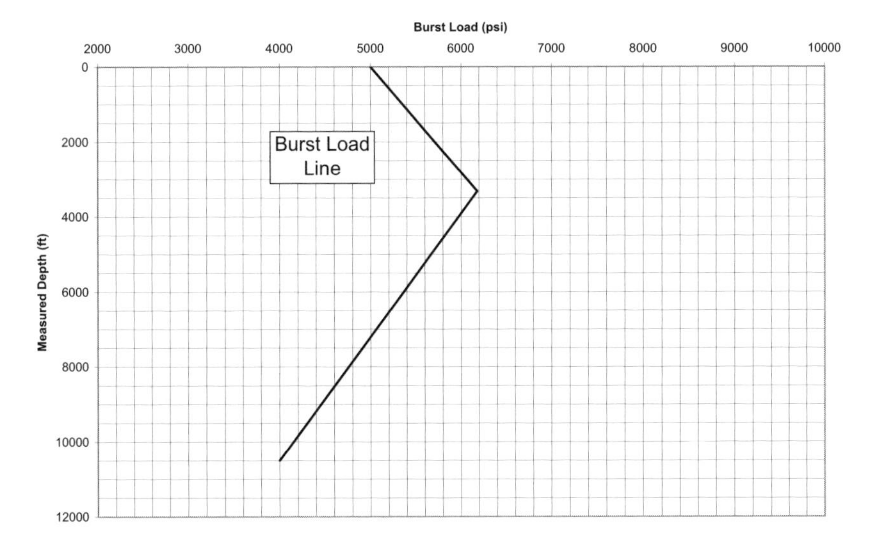

Figure 4–7 *Intermediate casing burst load for example well using maximum load method of Prentice (1970).*

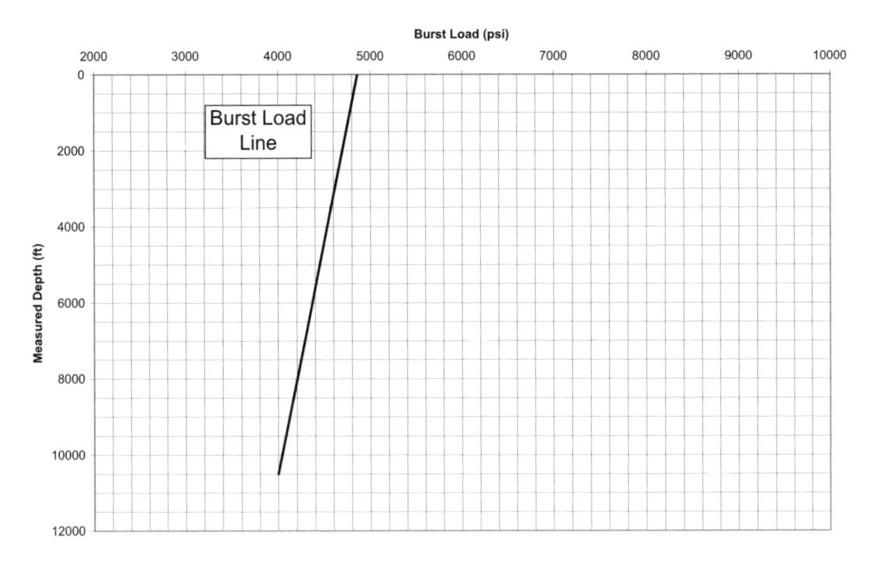

Figure 4–8 *Intermediate casing burst load for example well with 35 API gravity oil instead of gas.*

4.7 Production Casing

As one might imagine, there are a number of different ways to consider the loads in the production casing. One purpose of the production casing is as a backup string for the production tubing, so that it can support the same loads as a tubing string. However, there are some major differences in the maximum loads that a production string might encounter as compared to a tubing string. In some cases, the collapse load might be higher, and in other cases, the burst load might be significantly higher.

4.7.1 Production Casing Collapse Loads

The most common approach to collapse loading in a production string is to assume that the worst collapse scenario is one in which the casing is empty and open to the atmosphere. This is not common, but it does happen. Another possibility is to assume that the well always has some amount of liquid or pressure inside it, equal to the formation pressure at the time (usually taken to be the depletion pressure). The situation for each well may be different and can be complex. One should always keep in mind that what may actually occur in the future may be difficult to foresee. Casing often collapses during the producing life of some wells, because later in the life of the well someone attempted some operation that was not foreseen when the casing string was designed.

4.7.2 Production Casing Burst Loads

As far as burst is concerned, the most common procedure is to assume that the casing must withstand the maximum shutin formation pressure in the form of a gas column for a gas well (or oil for an oil well) from the perforations all the way to the surface. In other words, the production casing is a backup for the tubing as far as burst pressure is concerned. And, there are many ways in which a situation such as that can occur. However, one other situation can be much worse, especially with a gas well. Suppose the tubing is set in a packer and a leak develops in the tubing near the surface. There is no problem with casing burst at the surface, because it was designed for that pressure. But, what happens downhole because of the gas pressure on top of the packer fluid? The burst load is much higher in a situation like this than with a pure gas column in the casing. Designing for a case like this can lead to a very expensive casing string. Although this particular scenario is often ignored in production casing design, it is not at all an uncommon situation in the producing life of many gas wells. A lot of wells in the world have pressure relief valves on the tubing or casing annulus because of this very situation; the alternative is a downhole rupture of the production casing and an underground blowout.

Example 4–3 Production Casing Example

Looking again at our example, we see that the production casing is set from the surface to 14,000 ft. We assume a gas well. The bottom-hole pressure is equivalent to a 1.76 sg mud. We need not be concerned with the fracture pressures in the production casing loading. The collapse loading we consider is that the casing could be empty and the pressure on the outside is equivalent to the mud it was run in, 1.82 sg.

Collapse Load

The collapse loading is

- Empty on the inside.
- Mud on the outside, 1.82 sg.

For collapse, the net load at the surface is 0, and at 14,000 ft, the net collapse pressure at the bottom of the production casing is due to the 1.82 sg mud on the outside:

$$\Delta p_{\text{surf}} = p_{\text{o}} - p_{\text{i}} = 0$$

$$\Delta p_{\text{shoe}} = p_{\text{o}} - p_{\text{i}} = \gamma_{\text{mud}} h - 0 = 0.052(1.82)(8.33)14000 \approx 11040 \text{ lbf/in}^2 \tag{4.25}$$

The collapse load is shown in Figure 4–9.

Burst Load

For burst, we again assume that the pressure on the outside is equivalent to freshwater (although many use the mud weight it was run in), and on the inside, we consider that the packer might fail during production, so that the packer fluid is produced with the gas, resulting in a full column of gas in the annulus between the tubing and production casing.

The formation pressure is equivalent to 1.76 sg, and from that, we calculate the pressure at the bottom of the casing:

$$p_{\text{shoe}} = \gamma_{\text{eq}} h = 0.052(1.76)(8.33)(14000) = 10670 \text{ lbf/in}^2 \tag{4.26}$$

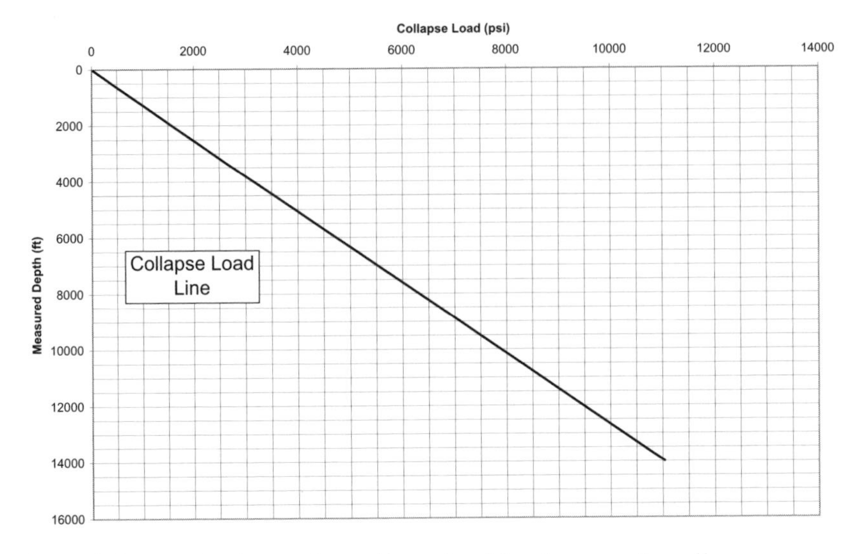

Figure 4–9 *Production casing collapse load for the example well.*

Then, the gas pressure at the surface is calculated using methane:

$$T_{\text{avg}} = \frac{74+326}{2} + 460 = 660 \text{ °R} \tag{4.27}$$

$$p_{\text{surf}} = 10670\, e^{\left[\frac{16(0-14000)}{1544(660)}\right]} = 8560 \text{ lbf/in}^2 \tag{4.28}$$

The net burst loads are then calculated.

$$\Delta p_{\text{surf}} = 8560 - 0 = 8560 \text{ lbf/in}^2$$

$$\Delta p_{\text{shoe}} = 10670 - 0.052(8.33)(14000) \approx 4610 \text{ lbf/in}^2 \tag{4.29}$$

Now, we are ready to plot the load curves for the production casing, see Figure 4–10.

In this simplified approach, we calculated a net burst pressure at the shoe and at the surface. We connected them with a straight line. Implicit in this is the assumption that both the liquid outside the casing and the gas or liquid inside the casing have constant densities from the shoe to the surface. For the most part, that is reasonable for the liquids, but we know that the gas density varies with depth, since it is temperature and pressure dependent. For the purposes of basic casing design, this nonlinearity of the gas density usually is ignored. That is our choice here also. However, for critical wells, it may be considered better to calculate the gas pressure at several points in the casing to account for the nonlinearity of the gas density.

Before we leave the production casing, we should look at the possible burst load that would occur if a tubing leak developed near the surface. We already calculated the gas pressure at the surface, which still constitutes the net burst load at the surface. Now, we add that pressure to the hydrostatic pressure of the 1.82 sg packer fluid less the hydrostatic pressure of freshwater (or some other fluid) outside the casing to get the net burst load at the bottom of the production casing:

$$\Delta p_{\text{shoe}} = 8560 + 0.052(1.82 - 1.00)(8.33)(14000) \approx 13530 \text{ lbf/in}^2 \quad (4.30)$$

We can see that this almost triples the burst load at the bottom of the casing. This might require a very expensive casing string, but the reality is that it is not uncommon to develop a tubing leak at or near the surface. Near-surface tubing corrosion from freshwater condensation mixed with CO_2 to form carbonic acid is quite common in many gas wells. Could we expect to rely on the cement to resist such a burst load? Some operators do, but it is not a good idea. This is a point where we really have to seriously question the use of a freshwater gradient outside the pipe. For a production string, as in our example, for an overpressured interval at 14,000 ft and below the intermediate casing, it is a stretch of the imagination to visualize a freshwater gradient outside the production casing. In this case, it is much more reasonable to assume something closer to formation pressures rather than a freshwater gradient. Let us look at a gradient equivalent to the formation pressure, 1.76 sg, behind the casing, which is slightly less than the mud the casing was run in:

$$\Delta p_{\text{shoe}} = 8560 + 0.052(1.82 - 1.76)(8.33)(14000) \approx 8920 \text{ lbf/in}^2 \quad (4.31)$$

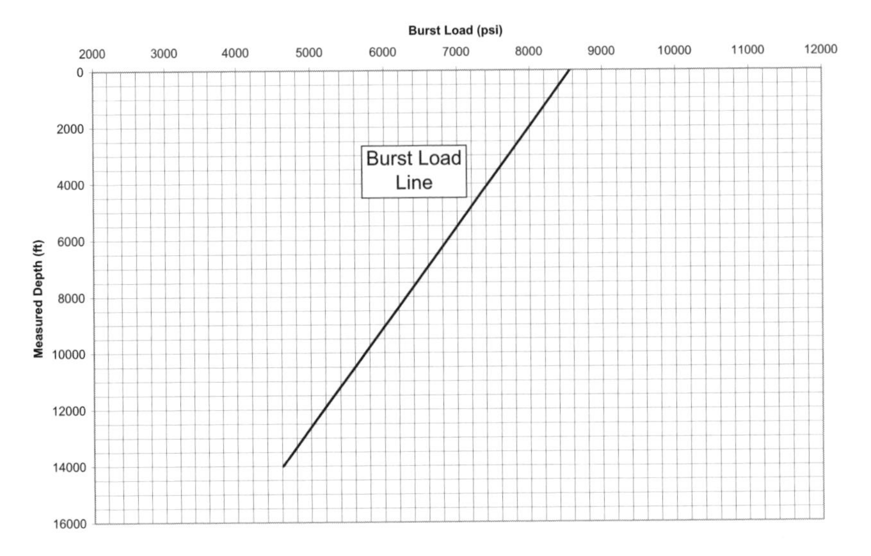

Figure 4–10 *Production casing burst load for the example well.*

While this value is still quite high, it is about the minimum burst load we can reasonably expect if our well develops a tubing leak at or near the surface in the early life of the well, before the zone depletes (see Figure 4–11). Another fact that should be mentioned is that we used methane in our gas calculations, which is a worst-case scenario. In an actual design for the production casing, it would be much more useful to use the actual gas that will be produced rather than methane. That would result in lower surface pressures than what we calculated. For our example well, we use the loads we calculated here, but the point is that, when designing a production casing string for a gas well, we should consider the best data we have rather than rely on the simplifications we use for designing other strings in the well.

4.8 Liners and Tieback Strings

Liners and tieback strings are special situations; however, the approach is very similar to that of either the intermediate or production casing. The thing that is different in the load curve for a liner or a tieback is that the load curve is not just for the liner or tieback but for the casing in which it hangs if it is a liner or the liner and tieback combination. Sometimes, liners must meet the requirements of two functions (see Figure 4–12). In other words, a liner or a tieback is never designed by itself but as a contiguous part of another string of casing. The only thing that really differs as

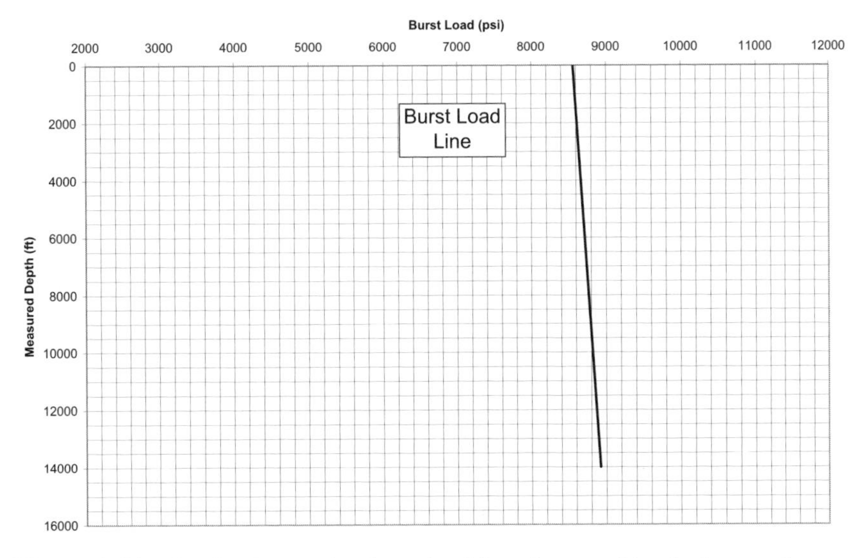

Figure 4–11 *Production casing burst load based on a tubing leak near the surface.*

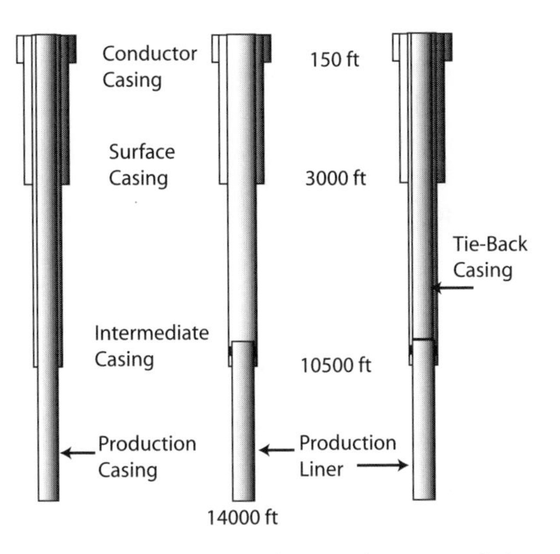

Figure 4–12 *A well with a production liner and two completion options.*

far as the load is concerned is the tension load, since it is a separate part of a longer string.

In the figure, we see a well with a production liner and two possibilities for final completion. On the left, the well could be completed as is, with the production liner and the intermediate casing forming the final production string. In this case, the intermediate string is designed to function as both the intermediate string and the upper portion of the production string. In the second case, where a tieback is run, the intermediate casing serves only as an intermediate string and the liner and the tieback together serve as the production casing. Figure 4–13 shows another common liner situation.

In this case there are two liners, a drilling liner and a production liner. On the left, the intermediate casing serves its normal purpose, but it also serves as a portion of a second intermediate string in conjunction with the liner, so both have to be designed as one string and the string has to satisfy both functions. On the right, the drilling liner is tied back to the surface and a production liner run below it. In a case like this, the design depends on when the tieback is run. If the tieback is run immediately after running the drilling liner, the intermediate casing serves as intermediate only until the tieback is run, then the drilling liner and tieback serve as a second intermediate string, and finally, in conjunction with the production liner, they serve as a production string. If the tieback is run after the production liner is run, then the intermediate casing has to be designed to perform its first function as well as a second intermediate string with the drilling liner. Finally, like before, the tieback, the drilling liner, and the production liner all function as the final production string. It may sound more complex than it actually is, but the only thing to keep straight is to be sure all strings are designed to meet all the required loads to which they will be subjected in their various roles during drilling and production.

4.9 Closure

It should be pointed out with emphasis that the loads used in this chapter are more or less typical, but still they represent a number of simplifying assumptions. One always should evaluate the possibilities in each individual well rather than rely on common practice. More and more often, new wells are drilled in fields with depleted reservoirs present. This may change significantly the load curves for a particular well, and one should always be wary of using common "recipes" for the loads in these types of wells.

Our example load curves for collapse and burst are complete. We have not mentioned load curves for axial tension yet. That is because the well itself does not impose the axial load (discounting bore-hole friction and curvature for now). The axial load is not determined until we make our preliminary selection of pipe for the well, because it is a function of the weight of the specific pipe and the density of the drilling fluid. We

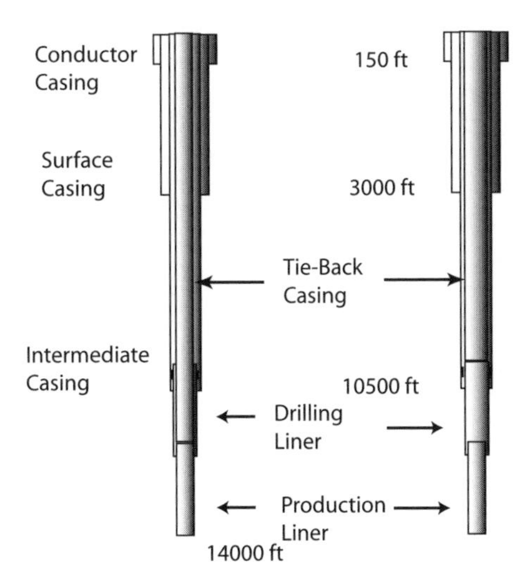

Figure 4–13 *The example well with a drilling liner and a production liner.*

address the axial load in the next chapter, where we use these curves to arrive at basic casing designs for all three of our example casing strings.

4.10 Reference

Prentice, Charles M. (1970). *Maximum Load Casing Design.* SPE 2560. Richardson, TX: Society of Petroleum Engineers.

CHAPTER 5

Design Loads and Casing Selection

5.1 Introduction

In the previous two chapters, we covered the first three steps of basic casing design:

1. Selection of casing depths.
2. Selection of casing sizes.
3. Development of load curves for collapse and burst.

In this chapter, we continue the process to make our initial casing design then refine it to account for combined loads:

- Development of design loads for collapse and burst.
- Initial casing selection for collapse and burst.
- Development of axial load curves.
- Development of axial design curves.
- Selection of casing for axial loads.
- Refinement of basic design selection for combined loads.

Casing selection is primarily a two-step procedure when done manually. Just like writers make a first draft then revise it to make it better, we make a preliminary casing selection based on published strength properties of the tube then refine it, if necessary, to account for the effects of combined loads. It is very easy to use the published values to get a preliminary design; and when used with appropriate design factors, many of these preliminary designs become a final design with no need for further refinement. However, the currently published values for collapse, burst, and tension are based on tests and formulas that assume no other loads are on the casing. In other words, the collapse rating you see in the tables is the collapse rating with no tension in the tube; the collapse rating is lower if the tube is in tension. We begin with the initial selection process then discuss ways to refine it for combined loading.

5.2 Design Factors

Everything we do not know about a problem we lump into something called a design factor. That may seem a bit flippant in tone, but essentially that is the process. So, design factors represent the unknown. That does not necessarily mean we are covering up our ignorance, but rather employing a means of accounting for things we cannot reasonably measure. Essentially, we use design factors to account for uncertainties in the properties of materials, uncertainties in the dimensional tolerances of casing, and uncertainties in the casing loads. In the absence of exact data, that is the common approach used in structural engineering for centuries. We used to call these factors safety factors, but with the ever-growing malignancy of litigation that term has fallen out of fashion because of the connotation of the word, safety. Now, the fashionable term is design factor, and that is the term we use.

The issue of the magnitudes of casing design factors is a difficult one. There was a time when some industry recommended standards appeared in various publications. Most companies seemed to generally accept those values, even though almost everyone deviated from them on occasion. Now almost no two companies use exactly the same design factors. Table 5–1 presents a range of the commonly used design factors.

Table 5–1 Common Casing Design Factors

Collapse	Burst	Tension
1.0–1.125	1.0–1.25	1.6–2.0

We reiterate, these are not industry standards nor are they necessarily recommended. They are merely some industry values that have been used for over 50 years and in hundreds of thousands of applications. Most companies likely have their own design factors, and usually the design factors vary depending on the type of well and possibly its proximity to populated areas. Also, it is not necessary, nor always advisable, to use the same design factors for each string of casing in a well.

In Alberta, Canada, where many wells have high H_2S or CO_2 concentrations in the produced fluids, specific minimum design factors are required by regulation. While these may be subject to change, they are listed in Table 5–2 for added information. There are provisions for using reduced design factors in Alberta under some circumstances, where the casing to be used has met certain test requirements.

Table 5–2 Minimum Design Factors for Alberta (EUB Dir. 010, 2004 draft)

	H_2S pp		H_2S pp	CO_2 pp
	<0.34 kPa	≥0.34 kPa	>500kPa	>2000kPa
Collapse	1.0[a]	1.0[a]	1.0[a]	1.0[a]
Burst	1.0	1.25	1.35	1.35
Tension	1.6[b]	1.6[b]	1.6[b]	1.6[b]

[a] Casing evacuated internally.
[b] No allowance for buoyancy.

5.3 Design Curves for Collapse and Burst

To make a preliminary selection of specific casing for a casing string, it is first necessary to apply design factors to the collapse and burst load curves. The result is called a design curve. The best way to illustrate the process is to continue with the examples for the surface, intermediate, and production strings we began in previous chapters. The selection of design factors for these examples are arbitrary, with the primary intent being that of illustrating different possibilities. In practice, one would weigh the choice of design factors in light of company standards, experience in an area, perceived risk, and so forth. Those are decisions we cannot teach here.

Example 5–1 Surface Casing Example

From previous chapters, we determined that the surface casing string for our well would be 13⅜ in. casing set at 3000 ft. For this example, we select the following design factors:

Collapse = 1.125

Burst = 1.125

We then apply these design factors to the load curves developed in the previous chapter. Since the collapse and burst curves for the surface casing are linear, we merely multiply the design factors by the upper and lower values of each curve and plot those on the load curve (or we could plot them on a separate graph to avoid any possible confusion). In the surface casing collapse load, the value at 3000 ft is 1440 lbf/in.^{2}, which we multiply by the design factor of 1.125 to give 1620 lbf/in.^{2}. That is the design load at the surface casing shoe at 3000 ft. At the surface, the collapse load is 0 lbf/in.^{2} so our design load is also zero at that point. We plot those points on our graph, connect them with a line, and that becomes our design curve, Figure 5–1.

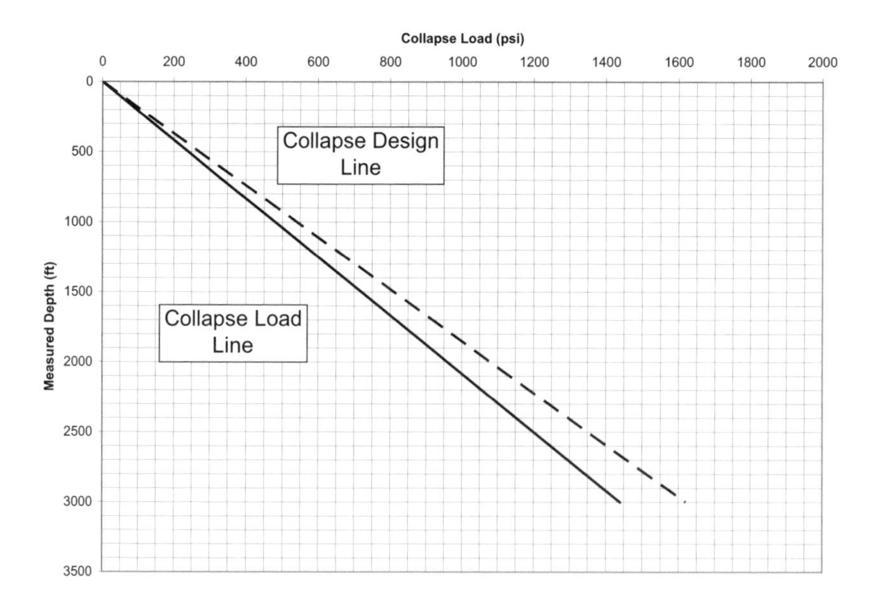

Figure 5–1 *Design load for the surface casing collapse example.*

Next, we do essentially the same thing for the burst load. The design factor we chose for burst also is 1.125, and we multiply that by the burst load at the top and bottom and plot that line.

$$\Delta p_{\text{surf}} = 1.125(1820) \approx 2050 \text{ lbf/in}^2$$

$$\Delta p_{\text{shoe}} = 1.125(620) \approx 700 \text{ lbf/in}^2$$

We plot these two values on the burst load curve to give us our design curve for burst, Figure 5–2.

That is a pretty straightforward procedure and takes only a few minutes to do. The only point of difficulty in the procedure is deciding on the magnitude of the design factors to use. The preliminary design process for collapse and burst of the surface casing is complete. The only steps that remain are to select the casing that meets or exceeds the design requirements in collapse and burst. We do that later; for now we construct the design curves for collapse and burst of the intermediate and production strings.

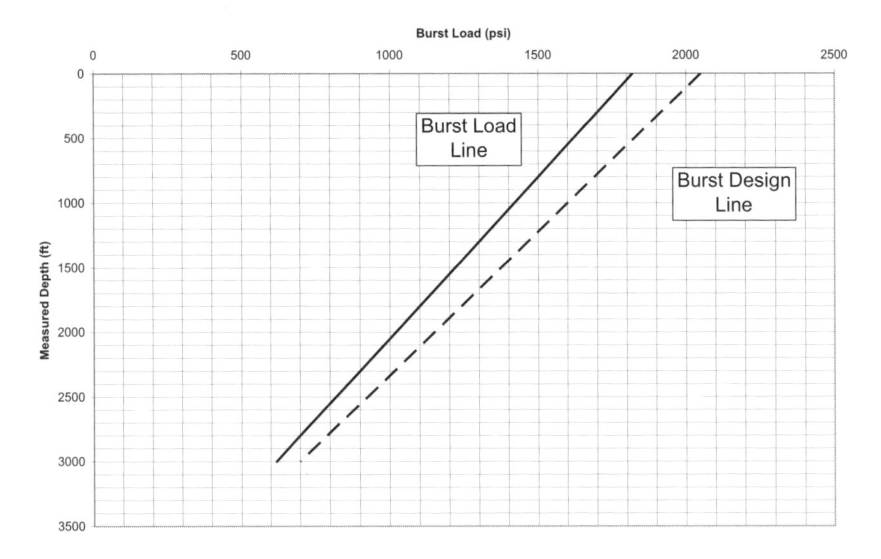

Figure 5–2 *Design load for the surface casing burst example.*

Example 5–2 Intermediate Casing Example

For the 9⅝-in. intermediate casing string to be set at 10,500 ft, we use the collapse load curve and the maximum load curve for burst developed in the previous chapter. The design factors we use for the intermediate string are:

Collapse = 1.125

Burst = 1.2

We used the same design factor for collapse as we did for the surface casing, but we used a higher design factor for burst. Typical reasons for this might be that we have the possibility of a gas kick in this string and we might also have a reduction in wall thickness due to wear while we are drilling to total depth after the string is in place. For the collapse design, we have to determine the design load only at the shoe as the design load at the surface is zero:

$$\Delta p_{\text{shoe}} = 1.125(1910) \approx 2150 \text{ lbf/in}^2$$

We plot this on the same chart as the collapse load curve to give us our collapse design curve, Figure 5–3.

The burst load curve we generated using the maximum load method of Prentice (1970) in the previous chapter has three points at which we must apply the design factor—the surface, the point of the interface between the mud column above and the gas column below, and the shoe:

$$\Delta p_{\text{surf}} = 1.2(5000) = 6000 \text{ lbf/in}^2$$

$$\Delta p_{\text{int}} = 1.2(6180) \approx 7420 \text{ lbf/in}^2$$

$$\Delta p_{\text{shoe}} = 1.2(4000) = 4800 \text{ lbf/in}^2$$

We plot these points on the load curve (Figure 5–4), and we now have the design curve for the intermediate casing in burst.

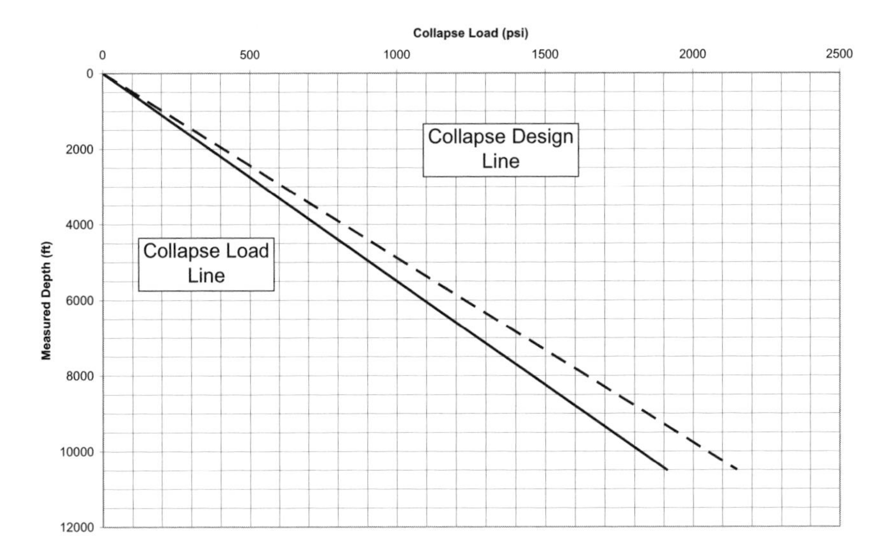

Figure 5–3 *Design load for the intermediate casing collapse example.*

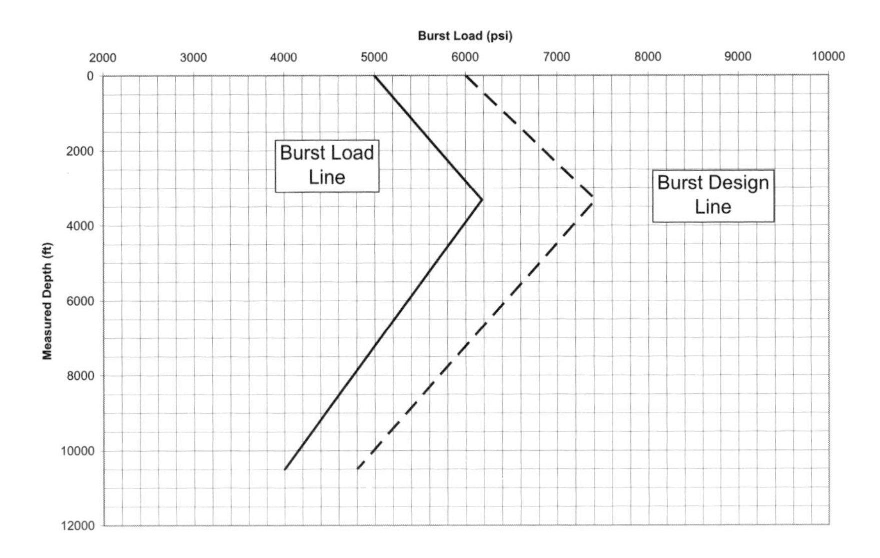

Figure 5–4 *Design load for the intermediate casing burst example.*

Example 5–3 Production Casing Example

Finally, we come to the production casing, which is the last casing string in our example. We expect that it should be capable of containing full well pressure throughout the producing life of the well. It should not collapse if the well becomes depleted significantly or during any operations conducted in the well bore during work overs or stimulations. For the 7 in. production casing at 14,000 ft, we use the following design factors:

Collapse = 1.125

Burst = 1.2

We apply the collapse design factor to our collapse load to get the collapse design and plot the results on our load curve, Figure 5–5.

$$\Delta p_{\text{shoe}} = 1.125(11040) \approx 12420 \text{ lbf/in}^2$$

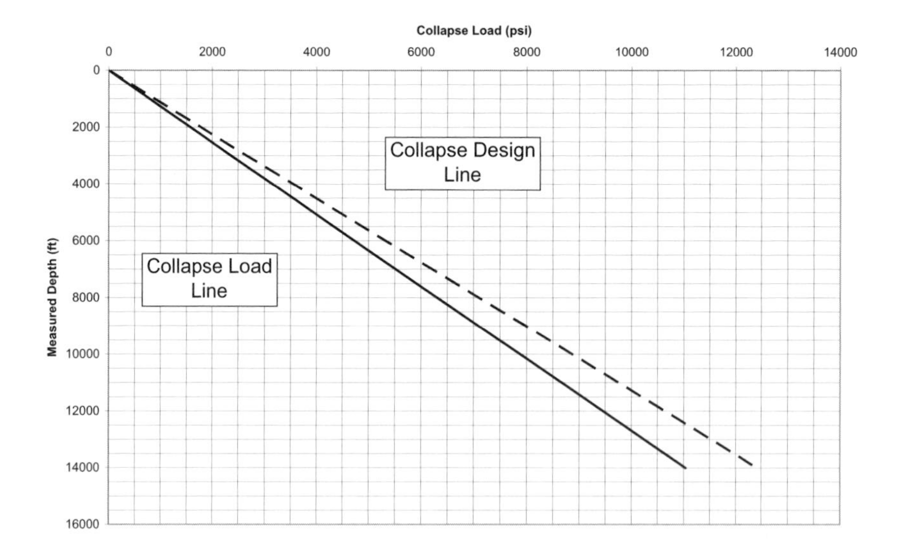

Figure 5–5 *Design load for the production casing collapse example.*

Then the burst design, plotted in Figure 5–6:

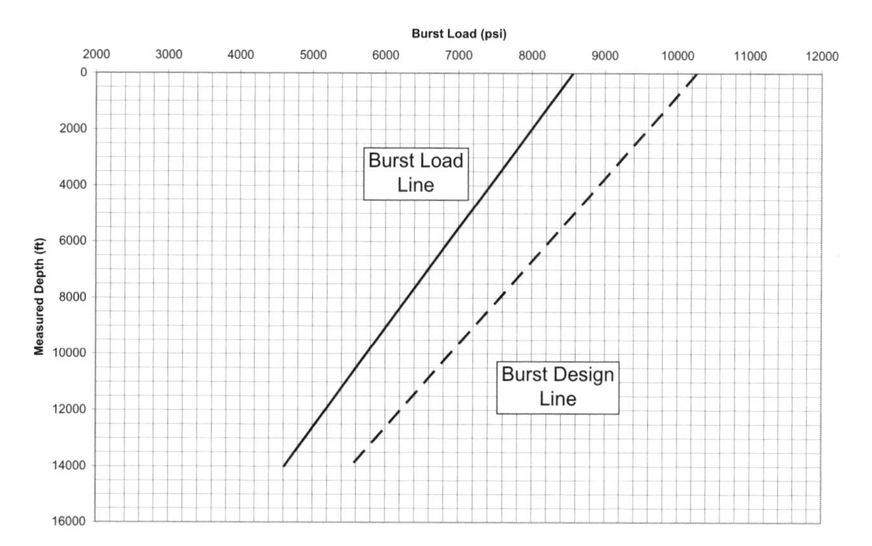

Figure 5–6 *Design load for production casing burst example.*

$$\Delta p_{\text{surf}} = 1.2(8560) \approx 10270 \text{ lbf/in}^2$$

$$\Delta p_{\text{shoe}} = 1.2(4610) \approx 5530 \text{ lbf/in}^2$$

Note that we did not use the burst load curve we generated assuming a near surface tubing leak. In practice, we might have chosen that load curve, even though it is not common practice in most designs. We did not choose it here, because some points in this example would not be illustrated had we chosen that option, and the purpose of this example is to illustrate as many points as possible. Now that we generated the design curves for collapse and burst it is time to begin selecting specific casing for the example well.

5.4 Preliminary Casing Selection Process

We are now ready to select the casing that will meet our design requirements in collapse and burst. We constructed design curves that show the specific collapse and burst requirements, but many choices of casing meet those requirements. So what are some of the things we need to consider at this point?

We have so many variables and choices that we could spend a great deal of time discussing all the possibilities. The most prevalent consideration is cost, or perhaps we should say minimum cost. Minimum cost, however, can be misleading, because that does not necessarily translate to the market price of the casing. Minimum cost also includes logistics, transportation, availability, current inventory, rig costs, and so forth. Many considerations go into the selection process. In the examples used here, we stay with some rather simple choices, but there are additional considerations.

5.4.1 Selection Considerations

Weight and Grade

In selecting the casing for our string, we often are presented with a choice of a particular weight and grade of pipe versus a different weight or grade, both of which might satisfy our design. For example, we might have a choice between 7 in. 23 lb/ft N-80 or 7 in. 26 lb/ft K-55, either of which would work in our string. The most obvious selection criterion might be cost or availability, as previously mentioned, but what else might enter into the decision? A thicker wall pipe might offer better corrosion or wear life; hence, we might choose the thicker wall 26 lb/ft K-55. But, if it is a directional well where the pipe is below the critical inclination angle (the angle below which nothing will move due to its own weight, 70° ±), then the heavier it is, the greater the force required to push it in the hole, so in that case the 23 lb/ft N-80 might be a better choice. Also the preferred or available bit size or completion equipment dimensions can enter into the selection process, so that one might favor a specific limit on the internal diameter reduction of a thicker wall pipe. The choice of grade of pipe also is significantly affected by the presence of corrosive fluids or hydrogen sulfide.

Connections

In the process of selecting casing to meet our load requirements, we are confronted with many different types of connections. What type do we need? For most normal pressure applications, we can use standard API ST&C or LT&C couplings; but for higher pressures and temperatures, bending in curved well bores, rotating torque, high-tensile loads, gas containment, and so forth, integral and proprietary connections may be necessary. In those cases, one must refer to the proprietary manufacturer's specifications and recommendations. We can comment on a couple of things though. If there is considerable bore-hole friction or problems with unconsolidated formations, then one should consider the use of beveled couplings or integral joint connections to reduce sliding friction. These can significantly reduce frictional drag on the casing. Another consideration is clearance problems, both in the open-hole section and inside of existing casing strings. In some cases, flush joint casing might be the choice because of clearance problems, and in other cases, special clearance couplings might be the choice. There are just too many variables to write out a decision chart for all the different possibilities.

Design Strengths

In selecting casing that meets our design requirements, we rely on published values of strengths for the various sizes and types of casing. The source of these design strengths is API Bulletin 5C2, which essentially is a collection of tables listing the dimensions and strengths of the various sizes and grades of API casing. The source of the strength values for these tables is the collection of formulas published in API Bulletin 5C3. These formulas have been used for many years with good success.

It is also necessary to specify what we mean by collapse strength and burst strength as used in this text for basic casing design. What we call the collapse strength is listed in API Bulletin 5C2 as "Collapse Resistance." It is the minimum external pressure at which the pipe collapses, as calculated from formulas in API Bulletin 5C3. It assumes no internal pressure or axial load on the pipe. The burst strength as we use it here is listed in API Bulletin 5C2 as "Internal Yield Pressure." It is the internal pressure at which the inner wall of the pipe or coupling yields, as calculated from the formulas in API Bulletin 5C3. It is not the pressure at which the pipe actually ruptures or bursts, but we use it as the limiting pressure. It assumes no external pressure or axial load on the pipe.

In recent times, certain limitations have been recognized with some of the formulas of API Bulletin 5C3, and an effort is well underway to revise these in light of modern manufacturing processes and casing requirements. Currently, in the adoption procedure for the ISO 10400 standards, the new formulas are recognized by most to be an improvement over previous formulas, however, they are not yet officially adopted at the time of this writing. For now, we take all strength values from the current API

Bulletin 5C2 and formulas from API Bulletin 5C3. When these documents are revised, they will not affect this basic design procedure, other than to change the published strength values slightly. We discuss the new formulas and approaches in Chapter 8, but for now, we just mention it to inform you that at least some of the strength values we use are likely to change in the near future, but those changes do not affect the basic casing design procedures we use.

Simplicity—the Key to Success

One thing to always keep in mind about different weights and grades of pipe, as well as connections, is that the fewer different types you have in a single string, the better. The more different types you have, the easier it is for a mistake to occur while running it in the hole. For every point in our design where we change from one type connection to another, we require a crossover sub or joint. In fact, if we are prudent operators, we require two crossovers on location for each of those points, in case one is damaged while running the casing. Not many things can be worse than running a string of 13⅜ in. casing to 10,000 ft and damage a crossover joint by cross-threading it when the casing is 2000 ft from bottom. Yes, it does happen! If you do not have a spare, then you have to pull 8000 ft of 13⅜ in. casing out of the hole, laying it down in singles as you pull it. Another thing you will learn if you ever have to pull casing is that often the mill end of the coupling will back out instead of the field end, so you need backup tong jaws that fit the coupling as opposed to the pipe body, as used when running the casing. You also will discover that casing made up to the maximum recommended makeup torque often has galled threads when it is backed out and requires a good number of the joints to be replaced. These are things we must avoid. The casing running process with most rigs of the world is an intense and continuous operation. To stop or interrupt the process, even momentarily, in many areas, will cause the string to stick off bottom. Pipe is rolled off pipe racks or off-loaded from barges as it is being run into the well bore and usually part, if not most, of this operation occurs at night. Hence, the simpler the design, the less is the likelihood of a costly mistake. Today, in many cases, the rig costs are so high compared to the casing costs that there is no cost benefit to having more than one type of casing in a particular string. This is especially true for shallower strings. In those cases, we might choose just one weight, grade, and connection that meets all the load criteria and disregard any cost savings of multiple weights or grades as inconsequential compared to the rig cost. This is common on many offshore wells and remote wells. However, that does not apply for most wells drilled in the world, and we would not learn much about casing design if we were to adopt that philosophy here.

Another point about simplicity is that the best way to let people in the field know that you are inexperienced and have never run any casing yourself is to send a casing design and casing string to the field that has several short sections of various weights, grades, or connections that might require crossover joints. Most operators seldom run a section of different type of casing that is less than 1000 ft in length and some set 2000 ft as a minimum. There is almost never any justification for running a section less than 500 ft in length, except in short strings of conductor and surface casing. Another exception might be made in the case where a few thick wall joints are placed at the top of a string for wear, wellhead support, or gauge control, where that same pipe is also in the string further down hole.

Example 5–4 Surface Casing Example—Preliminary Selection

For this example, we assume that we have the casing described in Table 5–3 available in our inventory for this particular well.

Table 5–3 Available 13⅜ in. Surface Casing for the Example

Wt. lb/ft	Grade	Connection	ID in.	Collapse Pressure lbf/in.²	Internal Yield lbf/in.²	Joint Strength 1000 lbf
54.5	K-55	ST&C	12.615	1130	2730	547
61	K-55	ST&C	12.515	1540	3090	633
68	K-55	ST&C	12.415	1950	3450	718
68	N-80	ST&C	12.415	2260	5020	963
72	N-80	ST&C	12.347	2670	5380	1040

We can begin with either the burst or collapse, and it is really immaterial where we start. For most surface casing strings, collapse usually is more critical than burst, and the initial selection for collapse often satisfies the requirements for burst, too. That said, we start with collapse using the design curve we previously constructed (Figure 5–1). We typically start at the top of the design curve and the lowest collapse strength pipe we have and see where the collapse rating of that section intersects our design line. In this case, the pipe with the lowest collapse rating, the 54.5 lb/ft K-55 which has a collapse rating of 1130 lbf/in.^{2}, can be run to a depth of 2100 ft before its collapse rating is exceeded on our design curve (Figure 5–7). At that point, we go to the casing with the next higher collapse rating; it is 61 lb/ft K-55 with a collapse rating of 1540 lbf/in.^{2}. We see that rating is exceeded by our design curve at a depth of about 2850 ft. Then, we select the casing with the next higher collapse rating, which is 68 lb/ft K-55 with a collapse rating of 1950 lbf/in.^{2}, which exceeds our maximum collapse design load of 1620 lbf/in.^{2} at bottom.

So our collapse design is satisfied by a string of 13⅜ in. K-55 casing with 54.5 lb/ft from 0 to 2100 ft, 61 lb/ft from 2100 to 2850 ft, and 68 lb/ft from 2850 to 3000 ft. This string will work and probably is the least costly string we could run using our available inventory. However, remember what we said about simplicity. We have a 150 ft section of casing on bottom. What should we do about this? There is nothing wrong with it as far as our design is concerned, but do we really want to send three different sections of pipe to the location, one of which is only 150 ft in length (five joints)? We are going to opt for simplicity and say we do not. We have two obvious choices. One, we could just run the 61 lb/ft casing all the way to bottom and say that the chances are slim that it would ever experience the worst-case collapse load. But there is a problem with that approach. Suppose a joint of that casing is defective and it did collapse during the drilling of this well, not in the last 150 ft but somewhere above that point. It clearly is not the fault of the design but rather a defective joint. Now suppose that one of the investors in this well is a lawyer who knows nothing about the oil field but decides he wants his money back. The pipe supplier may claim that you cannot prove the joint was defective, since it is not available for inspection. The lawyer does not care whose fault it is; he just wants out of his obligation to pay for his share of the working interest cost of a disastrous well that had to be abandoned before reaching its objective. From his point of view, it is not his fault; it is yours, because you designed a casing string and then did not follow your own design parameters. When you are on the witness stand, what is going to be your answer as to why you ignored the design factor you specified? This is not some preposterous scenario; this is the way it works in real life. The point here is that, if you select a design factor of 1.125 in collapse, you must stay with it. At this point, you may elect to change your design factor to say 1.0, citing the remote possibility of a worst-case collapse load actually occurring. That might be a reasonable choice, although it might appear you were more influenced by cost than

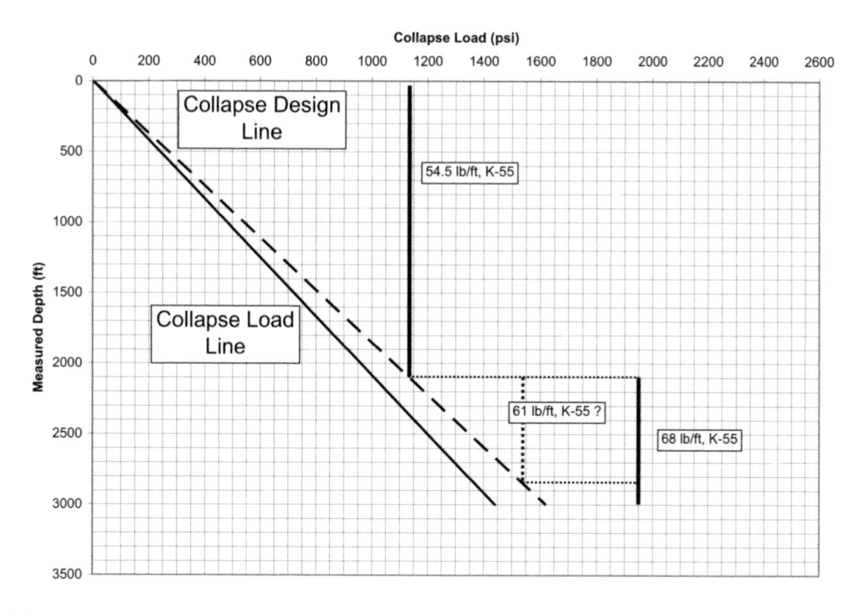

Figure 5–7 *Collapse selection for the surface casing example.*

engineering judgment. If you make that choice, you should make a new design curve and not just mark up the original one. (And, you should probably destroy the first one. This is not to be construed as legal advice, just common sense.) Now, I am not going to advise such a step here, because we are going to stay with our original design factor. To simplify our design, we elect to eliminate the 61 lb/ft section and run 68 lb/ft pipe from 2100 to 3000 ft. Many operators would have just elected to run 68 lb/ft all the way from surface to 3000 ft. That is fine, too, if the additional cost is not a consequence and you are not trying, as we are here, to learn about casing design.

Now that we have made our selection based on collapse design load, we check that selection for burst and adjust it as necessary. To do this, we plot the burst ratings of our selected casing string on the burst load chart (see Figure 5–8). We see that the burst rating of the casing string exceeds the burst design at all points. It is typical that a surface casing sting that meets the collapse requirement also meets or exceeds the burst requirements without necessity of modification, but sometimes it does not. In any case, we always check it to be sure.

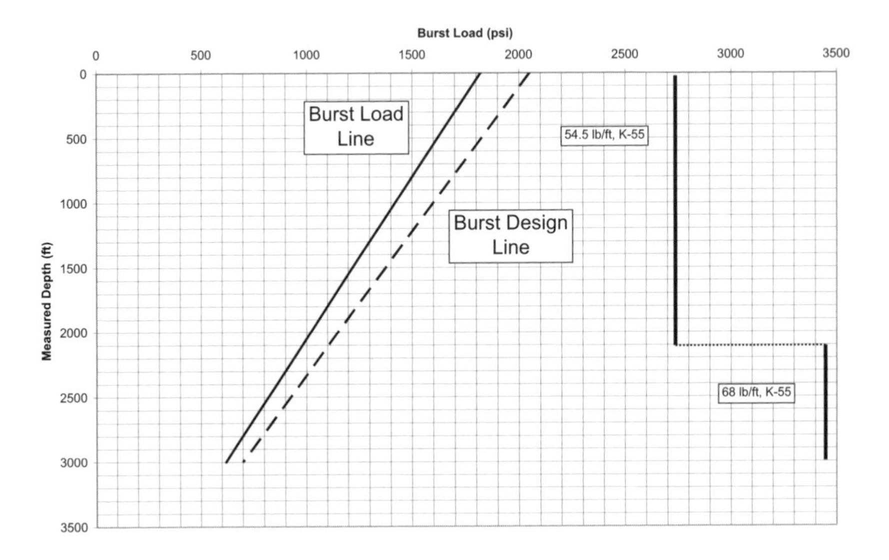

Figure 5–8 *Burst selection for the surface casing example.*

This concludes the preliminary selection process for the surface casing. Once we conclude this selection process, we can determine the weight of the string and the axial loads. We proceed with the selection preliminary selection process for the remaining strings before we consider the axial loads in a later section.

Example 5–5 Intermediate Casing Example—Preliminary Selection

We assume that we have available the 9⅝ in. casing of Table 5–4 in our inventory for use in this well.

Note that, since we elected to drill an 8½ in. hole from the bottom of the intermediate casing to total depth, we may have a problem with some of the casing in this inventory. If we use any 53.5 lb/ft casing in the intermediate string, it must be specially drifted for an 8½ in. bit. The 58.4 lb/ft casing cannot be used at all unless we use a smaller bit, and we do not consider this an option for our well.

In a precursory examination of the available pipe and the loads, we can see almost immediately that the collapse loading is very small and the weakest pipe in our inventory easily could sustain the maximum collapse load. We also note that the burst load is relatively high and the first three items in our inventory will not sustain the burst load at the bottom of the

Table 5–4 Available 9⅝ in. Intermediate Casing for the Example Well

Wt. lb/ft	Grade	Connection	ID in.	Collapse Pressure lbf/in.^{2}	Internal Yield lbf/in.^{2}	Joint Strength 1000 lbf
36	K-55	ST&C	8.921	2020	3520	423
40	K-55	ST&C	8.835	2570	3950	486
40	K-55	LT&C	8.835	2570	3950	561
40	N-80	LT&C	8.835	3090	5750	737
43.5	N-80	LT&C	8.755	3810	6330	825
47	N-80	LT&C	8.681	4750	6870	905
53.5	N-80	LT&C	8.535*	6620	7930	1062
58.4	N-80	LT&C	8.435*	7890	8650	1167
43.5	P-110	LT&C	8.755	4420	8700	1105
47	P-110	LT&C	8.681	5300	9440	1213
53.5	P-110	LT&C	8.535*	7950	10900	1422

string, where the burst load is the lowest. Therefore, it looks like the best place to start the selection is with the burst design, and that is fairly typical of intermediate strings run to protect lower-pressured formations from higher pressures below. In the cases where the intermediate casing is run to protect lower-pressured formations below, we would probably start with the collapse selection first. Again, it really makes no difference whether we start with the collapse or burst selection, but if we start with the most critical one first it results in less revision.

We start our burst selection at the top of the string and plot the various sections as to their burst ratings onto the design curve, see Figure 5–9.

In this design, we selected some 40 lb/ft, 43.5 lb/ft, 47 lb/ft, and 53.5 lb/ft N-80 casing in the string. This might constitute an optimum design from an engineering point of view, but do we really want to run something like this in our well? First of all, two things about this string are questionable. First is that the top section is only about 750 ft in length, and we said we would not run a section of less that 1000 ft in length. So we can eliminate that section by running the 47 lb/ft pipe from 2000 ft to

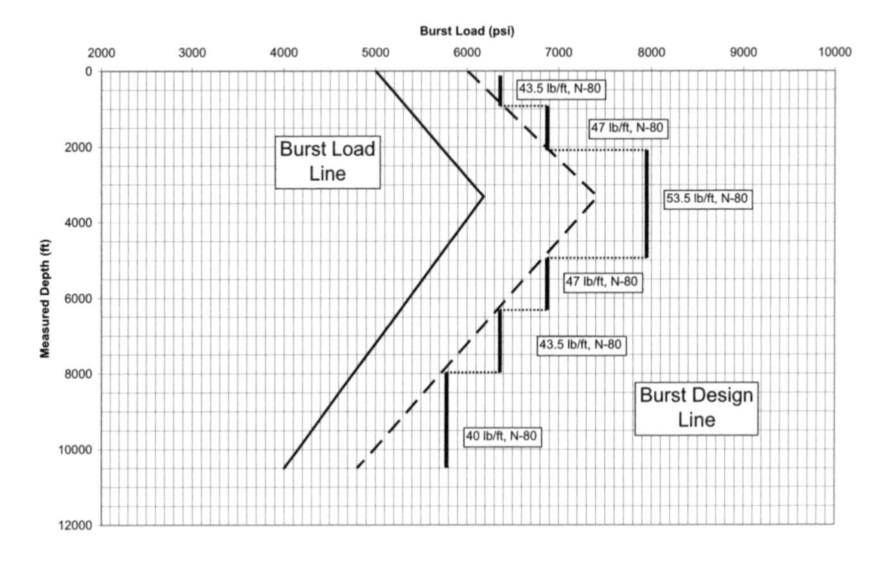

Figure 5–9 *Tentative burst selection for the intermediate casing example.*

the top. The second problem we face is that the 53.5 lb/ft pipe must be specially drifted to be sure that an 8½ in. bit will pass through it. What if it will not? We would then go to one of the P-110 grades, say, the 43.5 lb/ft. And that brings up another point. If this well is in a hard rock area, where we will be drilling for a long time below the intermediate casing, do we think that the 43.5 lb/ft pipe has enough wall thickness to sustain the wear caused by the rotating tool joints and still maintain sufficient burst resistance. This is not an easy question to answer, but is quite typical, because wear often is a problem with intermediate strings and the worst wear often occurs nearer the surface than further down hole. We discuss wear later in this text and bring it up at this point only because this is the kind of question that often arises with intermediate casing.

For our purposes, we assume that we drifted the 53.5 lb/ft N-80 casing for an 8½ in. bit, and it will pass freely through the pipe. This design probably still has more sections in it than most operators would choose to run, but we are going to stay with it to illustrate the design process. Figure 5–10 shows our selection.

A quick check shows that all the casing we selected easily exceeds the collapse requirements. In a case like this, we seldom bother to plot it, but we will do so in Figure 5–11 for illustration.

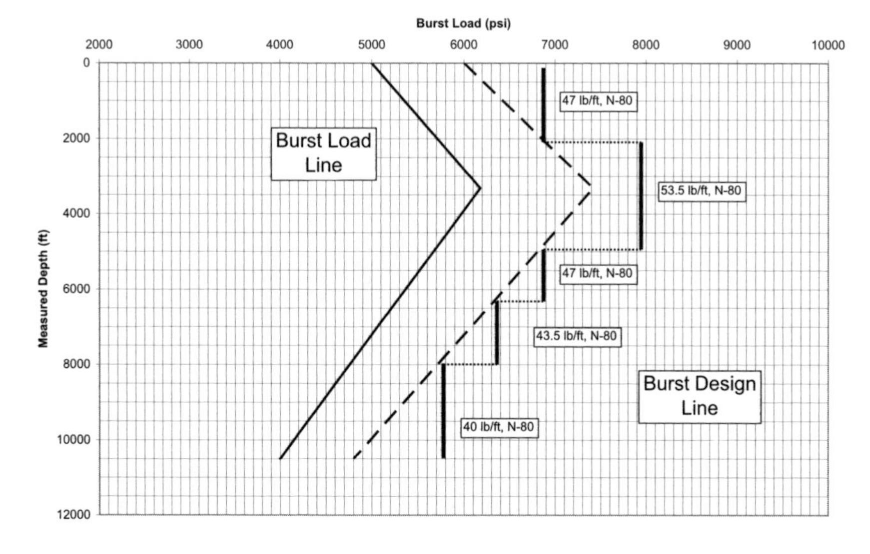

Figure 5–10 *Burst selection for the intermediate casing example.*

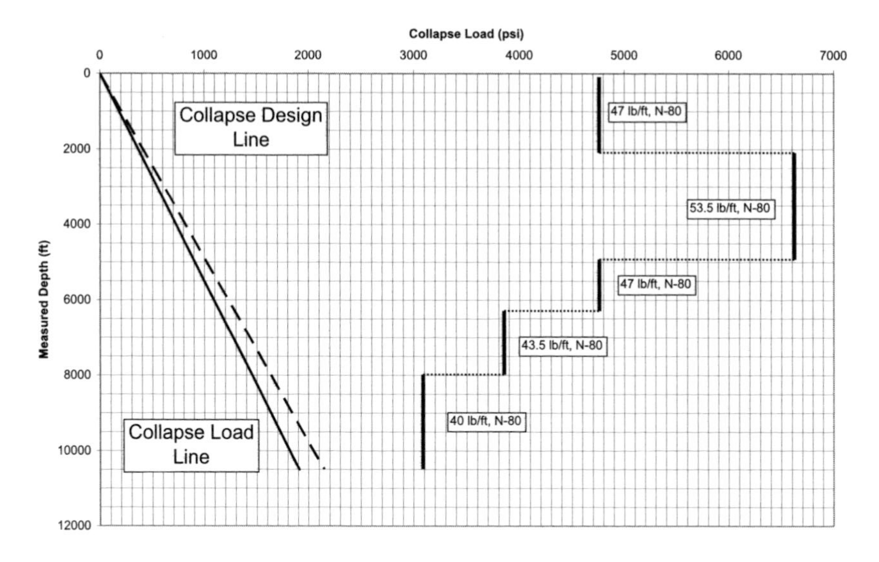

Figure 5–11 *Collapse selection for intermediate casing example.*

Example 5–6 Production Casing Example—Preliminary Selection

Table 5–5 shows the 7 in. casing available to us for this production string.

As to whether to start with the collapse or burst design curves for the selection of the production casing, it makes little difference, because most often we have to adjust the string for burst if we start with collapse and vice versa. For the example, we start at the bottom of the string with collapse, Figure 5–12. Note that we did not select casing all the way to the top based on collapse. That is because the burst load requires us to change it anyway. Now, using the burst design curve, we plot the casing we selected for collapse on the chart (Figure 5–13) and add or modify the upper portion for burst.

We can see on Figure 5–13 that the 29 lb/ft N-80 will not work at all, so we actually went one step too far in the collapse selection. We started at the top on the burst design and worked down to the collapse selection.

This completes our preliminary selection for collapse and burst on all three casing strings in our example well. Next, we look at the axial loads.

Table 5–5 Available 7 in. Production Casing for the Example Well

Wt. lb/ft	Grade	Connection	ID in.	Collapse Pressure lbf/in^2	Internal Yield lbf/in^2	Joint Strength 1000 lbf
26	N-80	LT&C	6.276	5,410	7,240	519
29	N-80	LT&C	6.184	7,030	8,160	597
32	N-80	LT&C	6.094	8,600	9,060	672
26	P-110	LT&C	6.276	6,230	9,960	693
29	P-110	LT&C	6.184	8,530	11,220	797
32	P-110	LT&C	6.094	10,780	12,460	897
35	P-110	LT&C	6.004	13,030	12,700	996

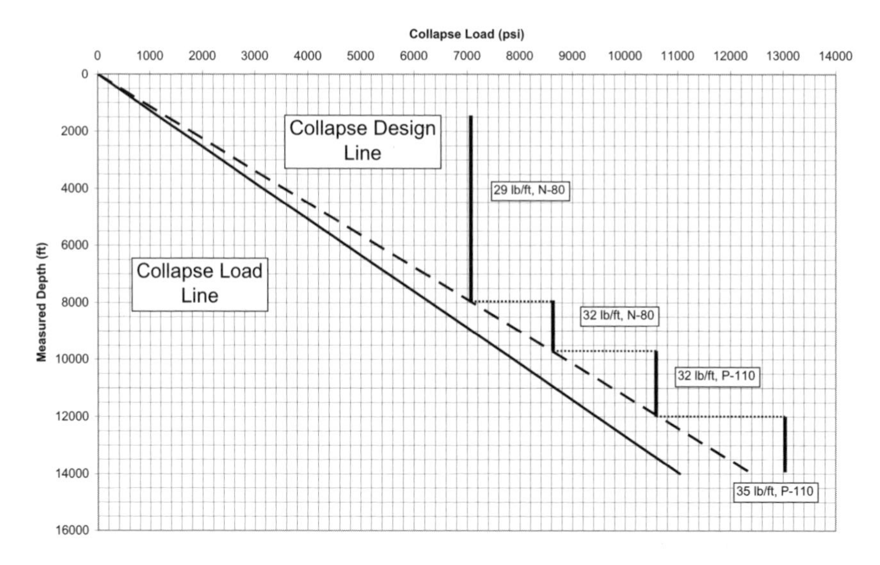

Figure 5–12 *Collapse selection for the production casing example.*

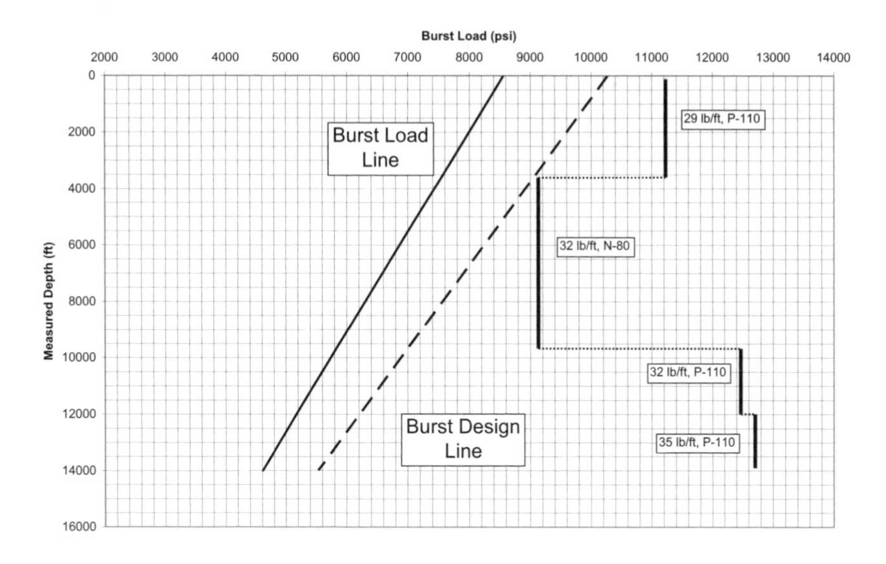

Figure 5–13 *Burst selection for the production casing example.*

5.5 Axial Loads

We did not consider axial loads at the time we made our collapse and burst load curves for the simple reason that we cannot know the axial loads until we know the weight of the casing. Therefore, we selected casing that would satisfy our design parameters for both collapse and burst, and having made that preliminary selection, we now determine the axial loads and possibly adjust our selection if the axial loads are too great for the casing we selected for collapse and burst.

There are four sources of axial load (tension or compression) in a casing string:

1. Gravitational forces (weight and buoyancy).
2. Bore-hole friction.
3. Bending.
4. Temperature changes.

The axial load in a casing string at any point due to gravity or weight is a function of the buoyancy of the drilling fluid and the inclination of the well bore. The bore-hole friction is a function of gravity, buoyancy, well bore inclination, and curvature, also the axial load in the pipe. In the case of a curved well bore, the axial load is a function of the friction, but the friction itself is also a function of the axial load; in other words, they are not independent of each other. We are not going to consider directional wells or bore-hole friction at this time but discuss them in Chapter 9.

There are a number of considerations when it comes to determining the design criteria for axially loaded casing. Here are a few questions we might have:

- Weight of casing—in air or buoyed weight?
- Bore-hole friction—how much?
- Design factors—an overpull margin or a design factor?

We discuss these in the following sections.

5.5.1 Axial Load Considerations

Weight of Casing

When we work with casing in a well bore, we must consider its weight and the amount of tension in the string due to that weight. What measure do we use for the weight? Do we use the weight of the casing in air or the buoyed weight of the casing in the drilling fluid in the hole? As hard as it may be to believe, this question has no universally accepted answer in oil-field practice today. Many use the weight in air, claiming that it gives an extra margin of safety. Others say the buoyed weight is more realistic. We illustrate both methods. The calculation procedures were shown in Chapter 2, but we give examples in the following sections.

Bore-Hole Friction

We know that there is friction in a well bore, and as we move the pipe, the friction increases or decreases the axial load in the casing, depending on whether the pipe motion is down or up. In directional wells, we have software that can predict the friction with reasonable accuracy while we are in the design process. For "vertical" wells, we know there is some amount of friction, but we have no means of calculating it, unless we assume some well-bore path and use the software as we would for a directional well. We can measure the pickup and slack-off weight while drilling the well, whether it is a vertical well or a directional well. The problem with this is that we usually have to design and purchase the casing string far in advance of the point where we can measure the actual friction in a particular well. Also, the friction load we measure with the drill string is not the same as the friction load the casing string experiences. For most near-vertical wells, we do not consider the friction specifically, but we allow for it with a design factor. That is one reason the design factor for the axial load usually is much larger than the design factor for collapse or burst. We have a much better chance of predicting the worst-case loading for collapse and burst than in tension. At least, we have a much better chance when we are sitting in an office several months before the well actually is drilled. We discuss bore-hole friction in much more detail in Chapter 9, but for now, we assume we can avoid estimating it if we select an appropriate design factor.

Design Factor

When it comes to the tensile design of casing, there are two schools of thought. One is to use a design factor, say 1.6, and the other is to use a specified amount of overpull, say 100,000 lbf. It is quite common to use both and say that the design should incorporate whichever one leads to the strongest design. In cases where the design factor results in the higher value, that usually is the case only near the surface and the over-pull is greater near the bottom.

The significance of the design factor or over-pull is especially critical in casing design, because of the bore-hole friction and the fact that its magnitude generally is not known when the casing string is designed. Friction force opposes the motion of the pipe, so we might think that it is of little significance in the design, since it reduces the axial tension only as the casing is run into the well. While that is true, there are two other considerations. One is that, if we intend to reciprocate the casing during cementing (as is desirable for a good primary cement job), then the friction increases the axial tensile load when the pipe is in an upward motion. The second, and extremely important, consideration is that, if a problem is encountered in running the casing, the casing string may have to be pulled out of the hole before reaching bottom. While this is rare, it does happen. So the design factor or over-pull must account for the fact that the casing might be subjected to the full amount of friction in an upward motion. That is one reason for the popularity of an over-pull margin rather than a typical design factor. It is easier for the driller if he knows that he can safely pull a certain amount, say, 100,000 lbf, above the weight of the casing string.

5.5.2 Types of Axial Loads

If we chose to use the weight of the casing in air, the design process is quite simple. The drawback to the approach is that it often leads to an overdesign of the string, since the casing never actually is suspended in air. While the weight in air approach was quite common at one time, it is less favored by most operators today.

The buoyed weight of the casing in the drilling fluid generally is assumed as the most common approach for designing casing to withstand axial loads. There are two ways to go about this. One way is to use the true axial load and the other is to use the effective axial load. We discussed these in Chapter 2 when we discussed hydrostatics.

- True axial load comes from the actual hydrostatic forces acting on the tube and is valid for all bodies.
- Effective axial load comes from Archimedes' principle, which does not give us information about the axial load within the casing body.

Figure 5–14 shows a plot of the axial load of our example surface casing string in air (unbuoyed), the true axial load in 1.1 sg mud, and the effective axial load, also in 1.1 sg mud.

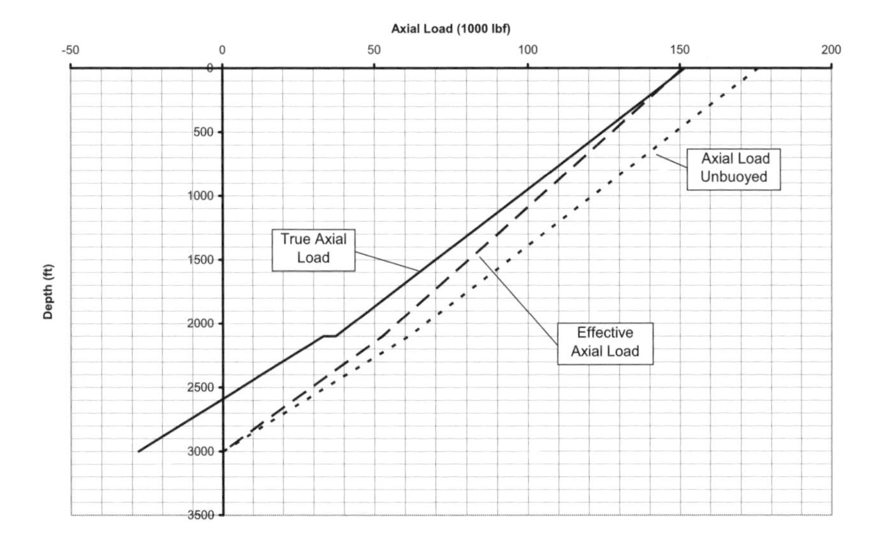

Figure 5–14 *Distributed axial load in the surface casing example.*

Note the true axial load curve on the left. It actually is in compression at the bottom, because of the hydrostatic pressure on the cross-sectional area of the tube at the bottom. Note also that, as you move up the curve from bottom to 2100 feet, the curve shifts slightly. That is due to the difference in cross-sectional area of the 54.5 lb/ft and the 68 lb/ft casing at that point. The tension increases, meaning that the net hydrostatic force is acting downward, because the internal diameter of the 68 lb/ft casing below is smaller than the internal diameter of the 54.5 lb/ft pipe above. (Had the heavier pipe been on top, the curve would have shifted in the opposite direction.)

Another thing to note about these curves is that the unbuoyed load curve essentially parallels the true load curve, except for the change in cross-sectional area. It is much easier to calculate manually, since there are no differences in cross-sectional areas and hydrostatic pressures to consider. That is why, in the past, many used this as the basis for their design (along with an appropriate design factor). And, as stated previously, many still use it, especially when doing manual calculations for calculating the axial load. Some also justify its use by stating that, since we do not know the magnitude of bore-hole friction in the well when we are designing the casing string, the axial load in air is a better approach.

True Axial Load

We showed in Chapter 2 how to calculate the true axial load, using formulas that contain the pressure and cross-sectional areas of the pipe at various depths. We also saw how we could make the calculation using the effective axial load and correcting it to true axial load a key points by subtracting the pressure/area effects.

Effective Axial Load

Also in Chapter 2, we showed how the effective axial load is calculated using a buoyancy factor and the specific weight of the various sections of casing. If you are using torque and drag software to determine the axial load, then you should be aware that the axial load most models give are the effective axial load plus friction. You easily can determine if your torque and drag software is giving you the effective axial load by looking at the load at the bottom of the casing string. If it is zero, then it is the effective axial load. The effective axial load can easily be converted to the true axial load by subtracting the pressure/area term from the effective axial load at key points where the cross-sectional area changes.

Example 5–7 Surface Casing Example—Axial Loads

Now that we have discussed the various methods of calculating the axial load curve, let us proceed with the axial load design of the surface casing string. For this example, we use the unbuoyed axial load and a design factor of 1.6 and 100,000 lbf overpull, whichever is greater (see Figure 5–15).

In this case, the design factor of 1.6 is less than the 100,000 lbf overpull at all points, so we use the overpull line as the design line. In Figure 5–16, we plot the casing we already selected to meet the collapse and burst requirements, and we find that it easily exceeds the tension requirements also.

This is fairly typical of many surface strings, but the tensile design should always be checked to be certain. The design summary of the surface casing is shown in Table 5–6 on page 179.

Example 5–8 Intermediate Casing Example—Axial Loads

Now we look at the axial load for the intermediate casing string. The casing is run in 1.4 sg mud, and for this example, we use the true axial load of the casing. For a design factor, we use 1.6. Rather than use the

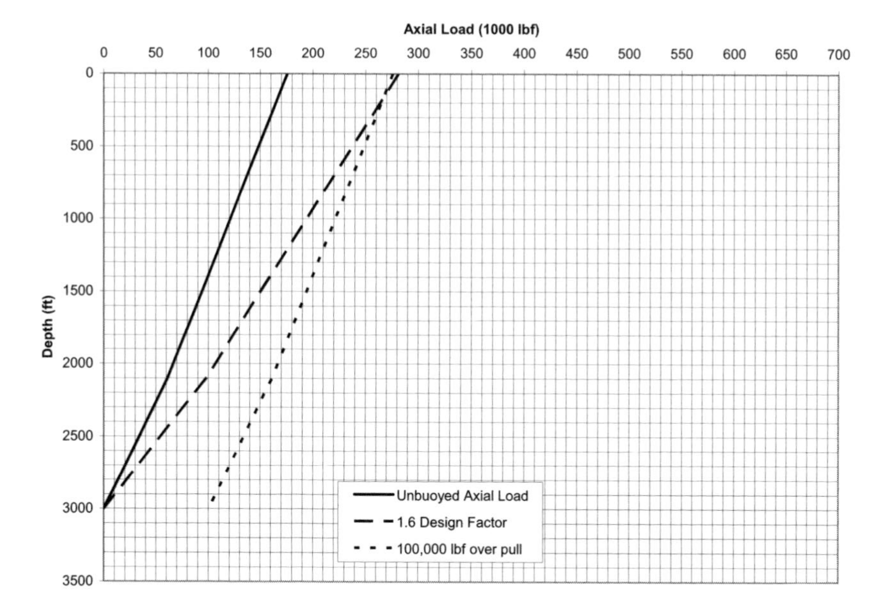

Figure 5–15 *Surface casing axial design load example.*

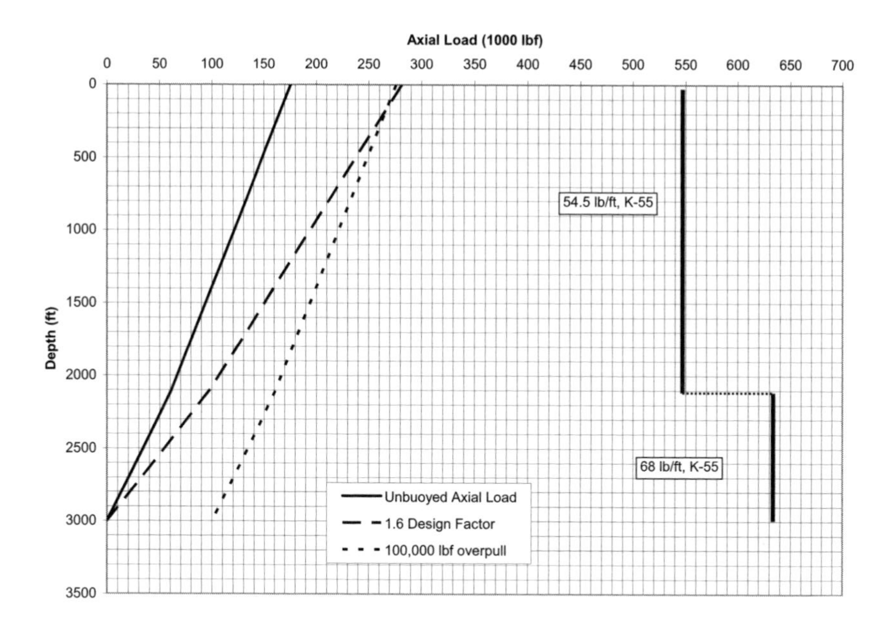

Figure 5–16 *Surface casing design load with axial strengths of preliminary collapse and burst selection.*

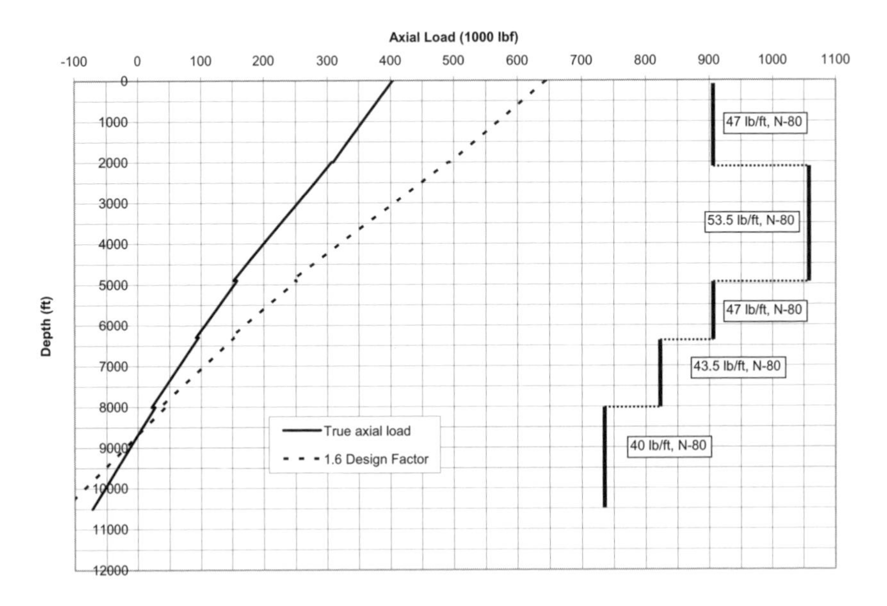

Figure 5–17 *True axial load for the 9 ⅝ in. intermediate casing example.*

direct formulas for the true axial load, we calculate the effective load first and then correct it to the true axial load. We show the calculation results in Table 5–8 on page 180. The plot of Figure 5–17 shows us that all of the casing is well above the tensile limit.

The casing selection that satisfies burst and collapse also satisfies our tension load and design factors. If these had not been satisfied, at this point, we would adjust the casing string so that it would satisfy the tension requirements. This example is summarized in Table 5–9 on page 180.

Example 5–9 Production Casing Example—Axial Loads

For the production casing example, we use the true axial load in 1.82 sg mud and a design factor of 1.6 or 100,000 lbf overpull. Then we plot these on Figure 5–18 for designing the tension.

As can be seen, we got lucky, because the string we designed for collapse and burst meets the design load for tension, too. While this is often the case with higher pressures, the general case is that deeper wells with lower pressures require adjustment of the preliminary selection for ten-

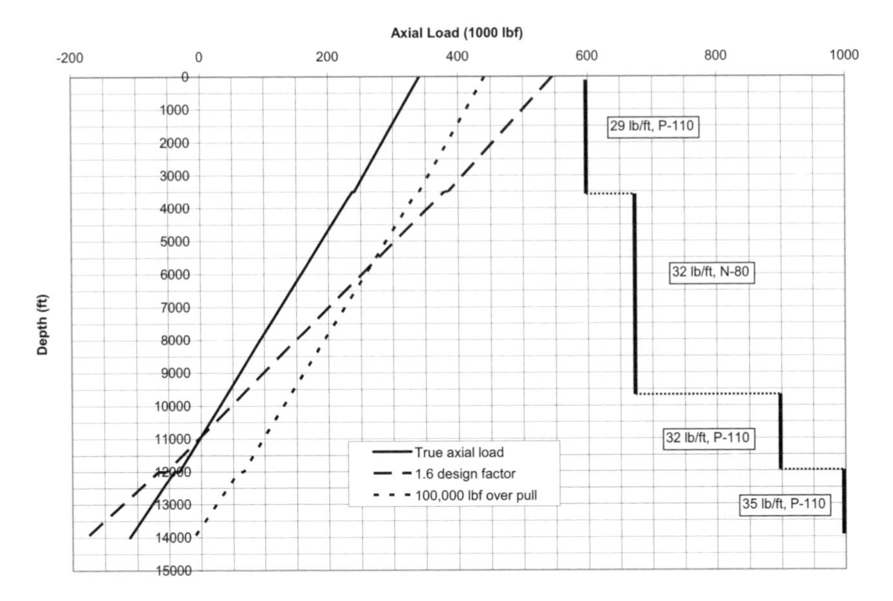

Figure 5–18 *Axial design load for the production casing example.*

sion. This completes our production casing design, which is summarized in Table 5–7 on page 179.

5.6 Collapse with Combined Loads

All casing is loaded with a combination of loads, such as tension, internal pressure, and external pressure. The significance of this is that, in the case of combined loads, the table values we used for collapse, burst, and tension no longer are valid. For instance, the collapse value of the 13⅜ in. 54.5 lb/ft K-55 casing was listed as 1130 lbf/in.^{2}. But that value is valid only if there is no axial tension or compression in the casing. In the presence of tension, the collapse value is lower, and in the presence of compression, the collapse value may be higher. So we look at combined loads and determine whether or not we need to refine our casing design.

5.6.1 Combined Loads

Casing designs for many wells ignore altogether the effects of combined loads, and many operators have never suffered any consequences for having done so. The reasons are that they have used large-enough design factors in collapse, so that combined loading effects are never seen, the actual loads on their casing strings have been lower than the worst-case design loads, or some combination of those two. However, casing failures due to combined loading do occur. And, when they happen, the consequences are serious to the extent that the well may be lost.

The subject of combined loading is a bit complicated and we delay a discussion of the mechanics until Chapter 8, but we need some tools we can use for basic casing design. We present some methods in this section without discussion as to the background, which we will leave for later chapters.

5.6.2 Simplified Method

There is a very simple method for adjusting casing to account for combined loads. It has been used by a number of operating companies for many years, and when combined with design factors, it has proven workable for most normal vertical wells. It has its basis in a more theoretical context but has been simplified for easy use. Some would say it is oversimplified and that is an accurate statement, because it departs from theory and the results tend to be somewhat conservative. It is based on the graph shown in Figure 5–19.

The way this chart works is that we take the tension load of the casing at some point. We divide it by the joint tensile strength of the casing to get a decimal fraction. Then, we then locate that point on the horizontal tension/compression axis. From that point, we go down vertically to the point of intersection with the ellipse. From that intersection, we go horizontally to the vertical burst/collapse axis and read the collapse fraction. We multiply that fraction times the collapse rating of the pipe: that gives us the reduced collapse rating of the casing under that amount of tensile load.

Example 5–10 Example of the Simplified Method

Joint tensile strength = 547,000 lbf

Joint collapse rating = 1,130 lbf/in.^{2}

Tensile load = 61,200 lbf

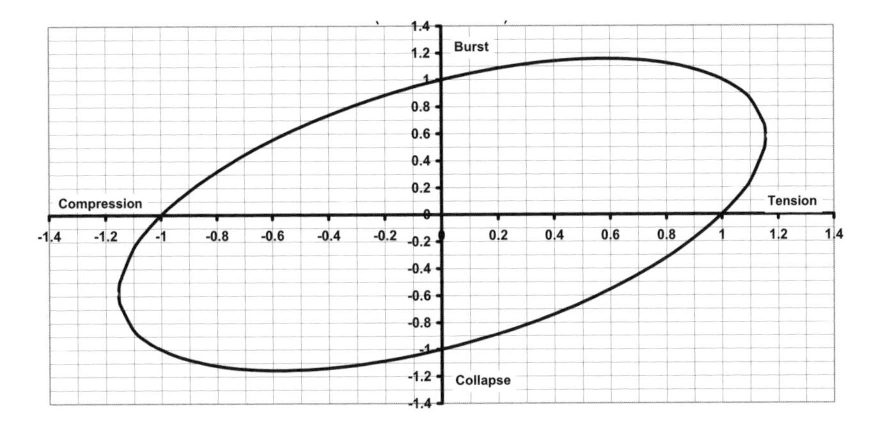

Figure 5–19 *Chart for simplified combined loading corrections (use with caution).*

We determine the fraction of tensile load to the tensile strength:

$$\frac{61200}{547000} \approx 0.1$$

We go to the positive 0.1 point on the horizontal axis, down to the intersection with the ellipse, then horizontally to the vertical axis, where we read a value of 0.94. We multiply this fraction times the collapse rating of the pipe to give us a reduced collapse rating:

$$0.94(1130) \approx 1060 \text{ lbf/in}^2$$

This value of 1060 lbf/in.^{2} is taken to be the collapse rating at that point. If our design curve has a higher value, then we must adjust our casing string accordingly.

The chart in this method is based on a yield criterion for steel known as the von Mises yield criterion, which we discuss in more detail in Chapter 7. There have been a number of similar uses of the chart in Figure 5–19. A formula can be derived from the curve for the quadrant concerning tension and collapse:

$$f_{\text{clps}} = \sqrt{1 - 0.75 f_{\text{tens}}^2} - 0.5 f_{\text{tens}} \tag{5.1}$$

where

f_{clps} = fraction of collapse rating, i.e. $\tilde{p}_{\text{clps}} / p_{\text{clps}}$
(reduced collapse pressure/collapse pressure rating)

f_{tens} = fraction of tensile rating, i.e. F / F_{tens}
(tensile load/joint tensile strength)

To calculate the reduced collapse pressure of casing with axial tension, the factor calculated in equation (5.1) is multiplied by the published collapse pressure (without tension):

$$\tilde{p}_{\text{clps}} = f_{\text{clps}}\, p_{\text{clps}} \tag{5.2}$$

This method should be used with caution: perhaps I should not even include it here, but it has been used for many years by many people for noncritical wells with success. It is not the API method for dealing with combined loads, nor is it the improved method up for adoption by the ISO. It might be called a quick-and-dirty method that proved successful in many normal pressured wells all over the world before the days of electronic calculators and computers. While that approach is simple in its graphical form, it is not as well accepted if one is actually going to do the calculations. The reason for this is that the graph is based on stresses rather than loads and joint strength. The joint strength of a tube in tension is based on the connection strength rather than the cross-sectional area of the tube itself.

You may also note that the chart in this procedure shows a reduction of burst strength in axial compression. Additionally, it shows an increase in collapse strength in axial compression and an increase in burst strength in axial tension. While this is true, almost no one uses a simple chart like this for those cases in practice. Increases in burst or collapse due to axial loads are seldom considered in basic casing design. Likewise, a case of reduced burst, which almost always results from axial compression due to thermal expansion, is not considered using such a simple method. These types of combined loads are generally considered only in more-advanced designs and more-sophisticated methods are used rather than reading simple values from a chart like this. We discuss those issues in Chapter 8.

5.6.3 Better Simplified Method

A more consistent way of expressing the reduced collapse fraction based on stresses is as follows:

$$f_{clps} = \sqrt{1 - 0.75\left(\frac{\sigma_{axial}}{Y}\right)^2} - 0.5\frac{\sigma_{axial}}{Y}$$

or

$$f_{clps} = \sqrt{1 - 0.75\left(\frac{F}{A_t Y}\right)^2} - 0.5\frac{F}{A_t Y} \tag{5.3}$$

where

σ_{axial} = axial stress, lbf/in^2 or kPa

Y = yield strength, lbf/in^2 or kPa

F = axial load, lbf or N

A_t = cross sectional area of tube body, in^2 or m^2

and the reduced collapse rating is calculated with equation (5.2). (Note that the difference between the two versions is that the first one uses axial stress, which is the axial load divided by the cross-sectional area of the tube that appears in the second version.) A similar approach was proposed by Wescott, Dunlop, and Kimler (1940) that attempts to account for the difference in the thickness of the tube body and the area under the threads so that one may use the axial loads rather than calculating the axial stress. Their formula is

$$f_{clps} = \sqrt{1 - 0.932 f_{tens}^2} - 0.26 f_{tens} \tag{5.4}$$

where, as before,

f_{clps} = fraction of collapse rating, i.e. $\tilde{p}_{\text{clps}}/p_{\text{clps}}$
(reduced collapse pressure/published collapse pressure)

f_{tens} = fraction of tensile rating, i.e. F/F_{tens}
(tensile load/joint tensile strength)

Using equation (5.1) in our example, we get the same conservative collapse value we got from the curve in Figure 5–19 (assuming we can read the graph accurately):

$$f_{\text{clps}} = \sqrt{1-0.75f_{\text{tens}}^2} - 0.5f_{\text{tens}} = \sqrt{1-0.75\left(\frac{61200}{547000}\right)^2} - 0.5\left(\frac{61200}{547000}\right)$$
$$= 0.939$$

$$\tilde{p}_{\text{cpls}} = f_{\text{clps}}p_{\text{clps}} = 0.939(1130) \approx 1060 \text{ lbf/in}^2$$

If, instead, we use equation (5.3) for this example, we get a slightly higher value for the reduced collapse:

$$f_{\text{clps}} = \sqrt{1-0.75\left(\frac{61200}{\frac{\pi}{4}\left(13.375^2-12.615^2\right)(55000)}\right)^2}$$

$$-0.5\left(\frac{61200}{\frac{\pi}{4}\left(13.375^2-12.615^2\right)(55000)}\right) = 0.962$$

$$\tilde{p}_{\text{clps}} = f_{\text{clps}}\, p_{\text{clps}} = 0.962(1130) \approx 1090 \text{ lbf/in}^2$$

Using equation (5.4) we get

$$f_{\text{clps}} = \sqrt{1 - 0.932\left(\frac{61200}{547000}\right)^2} - 0.26\left(\frac{61200}{547000}\right) = 0.965$$

$$\tilde{p}_{\text{clps}} = 0.965(1130) \approx 1090 \text{ lbf/in}^2$$

The difference in these last two equations is that equation (5.3) is based on the von Mises ellipse:

$$x^2 - xy + y^2 = 1$$

and equation (5.4) is a modified version that uses the axial load and the joint strength as developed by Wescott et al. (1940), based on the ellipse:

$$x^2 - 0.52xy + y^2 = 1$$

The axial stress formula, equation (5.3) has been used for many years with success and generally is preferable to the conservative method of equation (5.1). It also appears in the minimum casing design requirements of the Alberta Energy and Utilities Board in Canada as the preferred formula (EUB 010, 2004, draft). The second formula, equation (5.4), currently appears in the catalog of a major casing manufacturer and has seen many years of use. It might appear that the difference between these formulas is a bit trivial, but since they appear in various sources, usually without explanation, our purpose here has been to explain the differences.

5.6.4 Historical API Method

The API has a method for calculating the effects of combined loads, and it is published in API Bulletin 5C3. The reason it is referred to as the historical API method is because it may be phased out with the adoption of the new ISO 10400 standards. But, as of this writing, the new ISO standard is not yet official and the ISO standard currently in effect is just a copy of the current API standard. The API method for dealing with combined loads utilizes the same ellipse used previously but with different labels on the axes. We explain the derivation in Chapter 8, but for now, we do not worry about that. The axes in this case (see Figure 5–20) include the

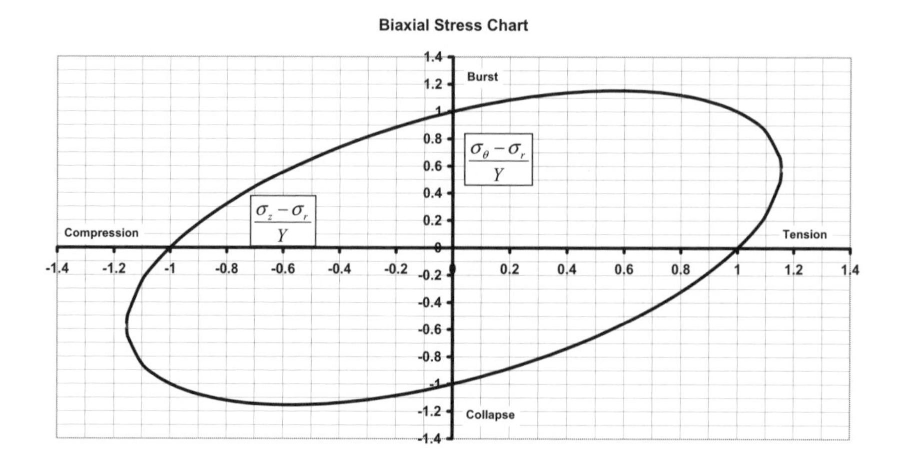

Figure 5–20 *The API ellipse for combined loading.*

internal pressure and are in terms of axial stress, which is the axial load divided by the cross-sectional area of the tube.

The way the API method works is that we determine the axial tension in the pipe divide it by the cross-sectional area of the tube then divide that by the yield strength of the pipe. The value is a decimal fraction, which we locate on the horizontal axis exactly as we did before. The value on the vertical axis, we take to be a fraction of original yield strength of the tube in collapse (instead of the reduced collapse rating). We multiply the original yield of the pipe by this fraction to give us a reduced yield value. This reduced yield value is substituted into one of the four API collapse formulas to calculate the reduced collapse rating of the pipe. Although this method is not especially good engineering, it is preferable to the previous simplified method. Let us look at an example that requires a little more data than the previous example but with the same load.

Example 5–11 Example of API Method

Joint tensile strength = 547,000 lbf

Collapse rating = 1,130 lbf/in.^{2}

Yield strength, Y = 55,000 lbf/in.^{2} (K-55)

Diameter (outside), d_o = 13.375 in.

Diameter (internal), $d_i = 12.615$ in.

Internal pressure, $p_i = 0$ lbf/in.^{2}

Axial load = 61,200 lbf

We first determine the axial stress plus the internal pressure divided by the yield strength, which is the positive horizontal axis of Figure 5–20 (remember, the axial stress, σ_z, is the axial load divided by the cross-sectional area of the tube):

$$\frac{\sigma_z + p_i}{Y} = \frac{\dfrac{61200}{\dfrac{\pi}{4}\left(13.375^2 - 12.615^2\right)} + 0}{55000} = 0.072$$

We find 0.072 on the positive horizontal axis, go down to intersect the ellipse, then left horizontally to the vertical axis, and read the fraction there, which is about 0.96. Or we could calculate it directly using equation (5.3), in which case it would be 0.962:

$$f_{\text{clps}} = \sqrt{1 - 0.75\left(\frac{61200}{\dfrac{\pi}{4}\left(13.375^2 - 12.615^2\right)(55000)}\right)^2}$$

$$-0.5\left(\frac{61200}{\dfrac{\pi}{4}\left(13.375^2 - 12.615^2\right)(55000)}\right) = 0.962$$

We multiply this times the yield strength to give us a reduced yield strength, $\tilde{Y}$:

$$\tilde{Y} = 0.962Y = 0.962(55{,}000) \approx 52900 \text{ lbf/in}^2$$

So far, this is exactly what we did numerically with equation (5.3), except that, instead of multiplying the factor by the collapse pressure as before, we multiply it by the yield strength to get a reduced yield strength value. We plug this reduced yield value into the appropriate API collapse formula to calculate the reduced collapse rating:

$$\tilde{p}_{\text{clps}} = \tilde{Y}\left[\frac{F}{(d_o/t)} - G\right] = 52900\left[\frac{1.99403}{\dfrac{13.375}{0.5(13.375-12.615)}} - 0.035422\right]$$

$$\approx 1120 \text{ lbf/in}^2$$

where

$\tilde{p}_{\text{clps}}$ = reduced collapse pressure
d_o = outside diameter
t = wall thickness
$\tilde{Y}$ = reduced yield strength
F = API constant (see Chapter 8)
G = API constant (see Chapter 8)

We see that this gives us a different answer and our simplified method is a bit conservative. That is probably one reason why it was used so often. You are probably wondering about the API formula we used, since we have not referred to it previously. Four API formulas are used for calculating collapse:

- Yield strength collapse formula.
- Plastic collapse formula.
- Transition collapse formula.
- Elastic collapse formula.

The first three are dependent on the yield strength and the ratio of the outside diameter to the wall thickness. The formula we just used is the transition collapse formula. There is a range of values for the d_o/t ratios and yield strengths for which the individual collapse equations are valid. Our particular yield and d_o/t ratio fell within the valid range of that particular equation, the transition collapse formula. The constants F and G in the formula are dependent on the yield strength, and there are formulas for calculating them also. In the case of standard yield strengths, like 55,000 lbf/in.^{2}, for instance, there are tables specifying the values of the constants, but for our case of a yield of 52,800 lbf/in.^{2}, a nonstandard yield value, all the constants must be calculated. There are five constants in all, though only two appear in the transition collapse formula. In this case, all five had to be calculated to determine the correct formula to use. All these formulas appear in API 5C3 and ISO 10400.

It may appear that we are avoiding a lot of detail here, and we are. This topic is covered in Chapter 8 in its entirety, but for now, all I am trying to show you is that it is a tedious process to do manually using the historical API method. I am not trying to make a case for the simplified method, however. One can easily program the API method to a spreadsheet and avoid the tedium and inherent errors in doing the calculations manually.

The API Method with Tables

One need not do the preceding calculations to use the historical API method. Tables published in API Bulletin 5C2 allow one to look up the reduced collapse value directly. For instance, in the example, we could use Table 4 in 5C2. It is in terms of axial stress rather than axial load, but it also gives the cross-sectional area of the tube to make the axial stress calculation easier:

$$\sigma_z = \frac{61200}{15.514} = 3945 \text{ lbf/in}^2$$

The table gives the collapse pressure corresponding to axial stress in 5000 lbf/in.^{2} increments, so for zero stress. the collapse pressure is 1130 lbf/in,^{2}, as we already knew, and at 5000 lbf/in.^{2} axial stress, the collapse value is 1120 lbf/in.^{2}. Since collapse values are rounded to the nearest 10 lbf/in.^{2}, our value of axial stress of 3945 lbf/in.^{2} gives a collapse value of 1120 lbf/in.^{2}, the same as our calculation. If you do not have a spreadsheet programmed for the API method, it is much easier to use Table 4 in API Bulletin 5C2.

5.6.5 New Combined Loads Formula

There is a newer way to calculate the effects of combined tension and collapse loading than the current API method just shown. New and improved formulas have been proposed, and as already mentioned, these formulas may be adopted by the ISO in the near future. We discuss those in Chapter 8.

Example 5–12 Combined Loads for the Production Casing Example

If we look at the summary of the production casing, we see that there might be a combined tension and collapse problem at the bottom of sections 2 and 3, where the actual collapse design factors are 1.14, without tension. At the bottom of section 2, at 12,000 ft, we see that the casing actually is in compression, so there is no problem there. The bottom of section 3 is at 9600 ft, and there is about 42,300 lbf tension at that point. The casing in section 3 is 32 lb/ft N-80 with a collapse pressure of 8600 lbf/in.^{2}. If we use equation (5.3) and calculate the reduced collapse, we get

$$f_{\text{clps}} = \sqrt{1 - 0.75\left(\frac{42300}{\frac{\pi}{4}\left(7.000^2 - 6.094^2\right)(80000)}\right)^2}$$

$$-0.5\left(\frac{42300}{\frac{\pi}{4}\left(7.000^2 - 6.094^2\right)(80000)}\right) = 0.970$$

$$\tilde{p}_{\text{clps}} = f_{\text{clps}}\, p_{\text{clps}} = 0.970(8600) \approx 8340 \text{ lbf/in}^2$$

The collapse load at 8600 ft is

$$\Delta p_{\text{clps}} = 0.052(1.82)(8.33)(9600) = 7568 \text{ lbf/in}^2$$

And actual the design factor with the reduced collapse is

$$f_s = \frac{8340}{7568} = 1.102$$

We said that our minimum design factor in collapse is 1.125, so we must modify our design. We move the bottom of section 3 up to 9500 ft. The collapse load is less, but there also is greater tension at that point. The tension is about 45,500 lbf at 9,500 ft. The reduced collapse is

$$f_{\text{clps}} = \sqrt{1 - 0.75\left(\frac{45500}{\frac{\pi}{4}\left(7.000^2 - 6.094^2\right)(80000)}\right)^2}$$

$$-0.5\left(\frac{455000}{\frac{\pi}{4}\left(7.000^2 - 6.094^2\right)(80000)}\right) = 0.968$$

$$\tilde{p}_{\text{clps}} = f_{\text{clps}}\, p_{\text{clps}} = 0.968(8600) \approx 8325 \text{ lbf/in}^2$$

And the collapse load at that point is

$$\Delta p_{\text{clps}} = 0.052(1.82)(8.33)(9500) = 7489 \text{ lbf/in}^2$$

And the actual collapse design factor is

$$f_s = \frac{8325}{7489} = 1.112$$

We are closer but not there yet. Now, we could go on with this trial and error, which would be easy if we had it programmed on a spreadsheet, but we can be smarter than that. We can turn this into an iterative technique and do it on a computer or graphically. Here is how a graphical method works. Suppose we combine all the processes we did already and call the combined process a function of depth, *h*. If we plug in the correct depth, it will give us the correct design factor of 1.125. We could write that like this:

$$f(h) = 1.125$$

or

$$y = f(h) - 1.125 = 0$$

In other words, if we plug in the correct value of the depth, *h*, then y = 0, otherwise y ≠ 0. We can plot our points in a graph of *y* and *h*. The point where y = 0 should give us the correct value of the depth, *h*. We already calculated two points for the graph:

$$y_1 = f(h_1) - 1.125 = f(9600) - 1.125 = 1.102 - 1.125 = -0.023$$

$$y_2 = f(h_2) - 1.125 = f(9500) - 1.125 = 1.112 - 1.125 = -0.013$$

We can plot these two points. Figure 5–21 shows the two points, and a line drawn through them shows that, at $y = 0$, $h = 9380$.

We calculate the design factor at that point to be sure. The tension at 9370 ft is 49,700 lbf:

$$f_{\text{clps}} = \sqrt{1 - 0.75\left(\frac{49700}{\frac{\pi}{4}\left(7.000^2 - 6.094^2\right)(80000)}\right)^2}$$

$$-0.5\left(\frac{497000}{\frac{\pi}{4}\left(7.000^2 - 6.094^2\right)(80000)}\right) = 0.966$$

$$\tilde{p}_{\text{clps}} = f_{\text{clps}}\, p_{\text{clps}} = 0.966(8600) \approx 8308 \text{ lbf/in}^2$$

$$\Delta p_{\text{clps}} = 0.052(1.82)(8.33)(9370) = 7386 \text{ lbf/in}^2$$

$$f_s = \frac{8308}{7386} = 1.125$$

That is accurate to within our round off, but for practical purposes, we probably just picked a depth of 9300. So our final 7 in. production casing corrected for tension and collapse is shown in the Table 5–10 on page 181. This completes our basic casing design example.

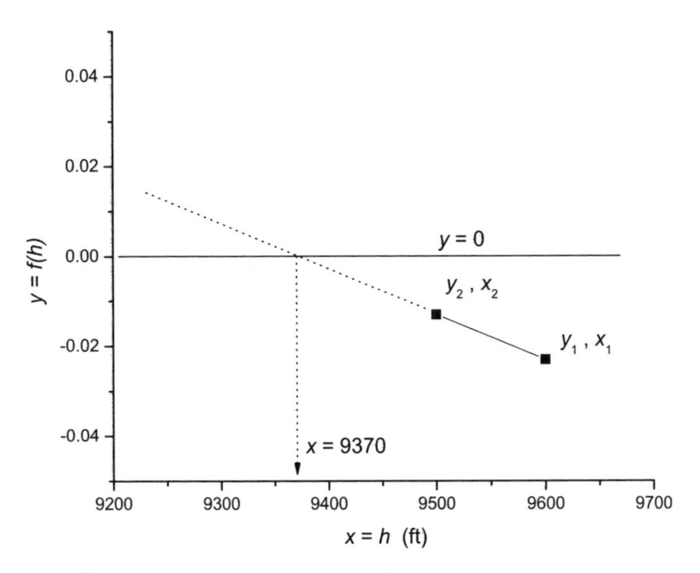

Figure 5–21 *Collapse/tension interpolation.*

5.7 Additional Consideration—Cost

Generally, cost is the overriding factor in deciding which type of casing to select when several types of casing satisfy the load requirements of the design. Obviously, we could select a string of some weight of P-110 grade pipe that might meet all our design criteria easily. However, the cost of such a string would far exceed that of a string made up of several weights of N-80, K-55, and even some P-110, if required. In the designs in this chapter, our basic premise was to try to select the lowest grade first, then the lowest weight, because that is how costs tend to run. We also tended to stay away from the heaviest weight in any grade, since that usually is a special item, not readily available and often with too small an internal diameter to use common bit and tool sizes. Market costs vary considerably, and we do not attempt to put casing costs into our examples here, but in general, the lower the grade, the lower is the cost. The other thing that complicates the cost picture is the inventory status within a company and the availability of certain weights and grades. It may be more costly to purchase some K-55 casing than to use some N-80 already owned by the company or the company's partners in some venture. These considerations may override what we tend to call an optimum design based on a common scale of prices.

5.8 Closure

In the basic designs for surface casing, intermediate casing, and production casing that we just examined, we used a variety of design factors. In two cases, we considered the buoyed weight of the casing in the tension design; in the other, we did not. We did that primarily to illustrate the different approaches. Typically, a company has a set of design criteria for a specific area or field or even one used companywide and stays with those criteria for all designs.

Another point we should make is that we selected from our inventory of pipe without explanation as to why we chose one as opposed to another. Many possible combinations would work just as well if not better than the selections we made. In general, the choice between two different types of casing for a particular section is based on

- Cost.
- Availability.
- Simplicity of design.
- Minimum number of crossovers.
- Wear considerations.

The basic casing design process we considered so far in this text is adequate for the vast majority of all wells drilled in the world every year. What we covered was more or less a method for basic casing design. We briefly covered some aspects of combined loading with little explanation. Again, the reason was to give the reader who has made it thus far through this text the ability to do basic casing design. We could have gone a bit further and also included some simple formulas for curved well bores, but at some point, we have to stop and say that we have covered an adequate amount for basic casing design and some topics require a better understanding of the underlying principles.

Before leaving this chapter, here is something to think about when you decide that casing design for shallow wells is just a routine. We might classify this in the "Just when I thought I knew what I was doing department." A production string of 4½ in. H-40 grade casing was run in a shallow well. It was to be a pumping well, and this was the same type casing that had been run in hundreds of wells in the same field. The cement was being displaced, and just as the top wiper plug landed on the float collar, circulation was lost. The drilling fluid level and cement in the annulus fell rapidly. And it fell, and it continued to fall. Then suddenly, the casing parted at the first connection below the elevator. A big surprise. A check of its rated joint strength showed that it certainly had enough tensile strength to be suspended in air all the way to total depth without failure, but the trouble was that it did not have enough strength to be suspended in air and full of water that could not fall out the bottom because of the wiper plug. An unusual case to be sure, but also it is a reminder that casing design should not become routine even for shallow wells.

5.9 References

API Bulletin 5C2. (October 1999). *Bulletin on the Performance Properties of Casing, Tubing, and Drill Pipe.* Washington, DC: American Petroleum Institute.

API Bulletin 5C3. (October 1994). *Bulletin on Formulas and Calculations for Casing, Tubing, Drill Pipe, and Line Pipe Properties.* Washington, DC: American Petroleum Institute, Washington, D.C.

EUB Directive 010. (2004, draft). *Minimum Casing Design Requirements.* Calgary, Alberta, Canada: Alberta Energy and Utilities Board.

ISO/DIS 10400. (2004, draft). *Petroleum and Natural Gas Industries—Formulae and Calculations for Casing, Tubing, Drill Pipe, and Line Pipe Properties.* Geneva: International Organization for Standardization.

Prentice, Charles M. (1970). *Maximum Load Casing Design.* SPE Paper no. 2560. Richardson, TX: Society of Petroleum Engineers.

Wescott, Blaine B., C. A. Dunlop, and E. N. Kimler. (1940). *Setting Depths for Casing, Drilling and Production Practice 1940.* Washington, DC: American Petroleum Institute.

Table 5–6 Summary of 13⅜ in. Surface Casing Design[a]

						Section			Cumulative		Axial Load		Actual Design Factors		
Section	OD	ID	Weight	Grade	Conn.	Bottom Depth	Length	Weight	Weight Air	Buoyed Weight	Bottom	Top	Collapse	Burst	Tension
2	13.375	12.615	54.5	K-55	ST&C	2100	2100	114	176	151	37	151	1.13	1.5	3.11
1	13.375	12.415	68	K-55	ST&C	3000	900	61	61	53	–28	33	1.37	3.52	10.38
Total							3000	176							

[a] Minimum design factors: Collapse = 1.125 empty; Burst = 1.125 gas; Tension = 1.6/1000 air

Table 5–7 Summary of the 7 in. Production Casing Design[a]

						Section			Cumulative		Axial Load		Actual Design Factors		
Section	OD	ID	Weight	Grade	Conn.	Btm Depth	Length	Weight	Weight Air	Buoyed Weight	Bottom	Top	Collapse	Burst	Tension
4	7	6.184	29	P-110	LT&C	3500	3500	102	444	341	240	341	2.55	1.31	2.34
3	7	6.094	32	N-80	LT&C	9600	6100	195	342	263	42	237	1.14	1.2	2.84
2	7	6.094	32	P-110	LT&C	12000	2400	77	240	113	–34	42	1.14	2.13	High
1	7	6.004	35	P-110	LT&C	14000	2000	70	174	54	–112	–42	1.72	2.45	High
Total							14000	444							

[a] Minimum design factors: Collapse = 1.125 empty; Burst = 1.2 gas; Tension = 1.6 true

Table 5–8 True Axial Load, 9 5/8 in. Intermediate Casing

						Section			Cumulative		Axial Load	
Section	OD	ID	Weight	Grade	Conn.	Bottom Depth	Length	Weight	Weight Air	Buoyed Weight	Bottom	Top
5	9.625	8.681	47	N-80	LT&C	2000	2000	94	489	402	308	402
4	9.625	8.535	53.5	N-80	LT&C	4900	2900	155	395	325	151	306
3	9.625	8.681	47	N-80	LT&C	6300	1400	66	240	197	91	157
2	9.625	8.755	43.5	N-80	LT&C	8000	1700	74	174	143	21	95
1	9.625	8.835	40	N-80	LT&C	10500	2500	100	100	82	–73	27

Table 5–9 Summary of 9 5/8 in. Intermediate Casing Design.[a]

						Section			Cumulative		Axial Load		Actual Design Factors		
Section	OD	ID	Weight	Grade	Conn.	Bottom Depth	Length	Weight	Weight Air	Buoyed Weight	Bottom	Top	Collapse	Burst	Tension
5	9.625	8.681	47	N-80	LT&C	2000	2000	94	489	402	308	402	High	1.2	2.25
4	9.625	8.535	53.5	N-80	LT&C	4900	2900	155	395	325	151	306	High	1.28	3.47
3	9.625	8.681	47	N-80	LT&C	6300	1400	66	240	197	91	157	High	1.21	High
2	9.625	8.755	43.5	N-80	LT&C	8000	1700	74	174	143	21	95	2.62	1.2	High
1	9.625	8.835	40	N-80	LT&C	10500	2500	100	100	62	–73	27	1.35	1.21	High
Total							10500	489							

[a] Minimum design factors: Collapse = 1.125; Burst = 1.2; Tension = 1.6 true

Table 5–10 Summary of 7 in. Production Casing Corrected for Combined Tension and Collapse[a]

						Section			Cumulative		Axial Load		Actual Design Factors		
Section	OD	ID	Weight	Grade	Conn.	Btm Depth	Length	Weight	Wt., Air	Buoyed Wt.	Bottom	Top	Collapse	Burst	Tension
4	7	6.184	29	P-110	LT&C	3500	3500	102	444	341	239.5	340.8	2.55	1.31	2.34
3	7	6.094	32	N-80	LT&C	9600	5870	188	342	263	49.7	237.1	1.125[b]*	1.2	2.84
2	7	6.094	32	P-110	LT&C	12000	2630	84	154	118	–34.3	49.7	1.14	2.13	High
1	7	6.004	35	P-110	LT&C	14000	2000	70	70	54	–112.3	–42.4	1.72	2.45	High
Total							14000	444							

[a] Minimum design factors: Collapse = 1.125 empty; Burst = 1.2 gas; Tension = 1.6 true

[b] Corrected for tension.

CHAPTER 6

Running Casing

6.1 Introduction

Many of the problems that occur with casing are not problems with design but problems with handling and running practices. Some companies have specific running practices, but they vary little from the basics. Several things must be kept in mind when transporting, handling, and running casing. Most of it falls in the category of common sense.

6.2 Transport and Handling

6.2.1 Transport to Location

Some casing gets damaged on the way to the location and on the location prior to running. There is no good reason for this to happen as often as it does. It is something that almost always could be avoided, but it still happens from time to time. Whether casing is loaded on trucks, boats, or barges, it must be adequately protected. This means not only care in handling while transferring from racks to trucks, to boat, to rig, all joints should have thread protectors in place and no cables or hooks should be used that can cause damage to the protectors or the pipe. On racks, trucks, or boats, the casing should be placed carefully with wood stripping between layers. The casing should be secured so that it cannot move during transport.

6.2.2 On Location

Once the casing is on location, it is off-loaded from trucks onto pipe racks or off workboats or barges onto the rig itself. In some cases, rack capacity is limited and the casing must remain on barges and transferred to the rig as it is being run. Whatever the procedure, it is imperative that the casing be subjected to as little transfer as possible to reduce the chances for damage. All transfer must be done using good practices to prevent damage to the casing. Another important consideration is that the final transfer, whether to the racks or onto a barge, must be done so that the casing is in the proper sequence in which it will be run into the hole. It is not acceptable to try to swap the order of pipe on the racks during the running process. Such an endeavor most likely will lead to errors. All transfers must be considered when the pipe is loaded at the pipe yard. For instance, if the casing must be loaded onto trucks for transport to a dock, loaded directly from the trucks onto workboats, and then off-loaded onto the rig, all these transfers must be taken into account, so that the order of the pipe does not have to be shuffled at the rig. There usually is too little rack space on most offshore rigs to do this, and it is much easier to do it in a pipe yard, as the pipe is loaded the first time.

Whether or not a company requires some type of electromagnetic inspection on location is a matter of policy that depends on the type of well being drilled. However, several things are essential:

- Casing should be drifted on location to ensure that no damage has caused a reduction of the internal diameter and that nothing is lodged inside the pipe.
- The thread protectors should be removed and the threads cleaned with a solvent to remove any unknown type of lubricant on the threads.
- The cleaned threads should be visually inspected.
- In offshore locations where bare metal rusts in a matter of minutes, the threads should be lubricated as soon as they are inspected with the same lubricant that will be used when the string is run in the hole.
- In most cases, the protectors on the pin end should be cleaned and reinstalled.
- Do not place any type of equipment, such as casing spiders or tongs, on the casing that is on the pipe rack. This is not always possible on many small offshore rigs, but it is a bad practice that should be avoided.

6.3 Pipe Measurements

One of the most critical aspects of running casing is the pipe tally or pipe measurement. There is no way to overemphasize the importance of this simple task. The success of the entire well depends on it being done correctly. So who is responsible for the measurements, the rig crew? Absolutely not! The final responsibility of the pipe measurements is the operator's representative on location, who is in charge of the well. If the operator's representative does not personally do the physical measurement, then he or she, at the very minimum, should witness and record a duplicate of the measurement. There is no excuse for botching this simple procedure, yet it continues to happen all too often.

As to the actual measurement procedure, there are many variations on how and when to do it. Most of the time it is done when the pipe is offloaded at the rig onto the rig pipe racks. The best methods involve removing the protectors (from both ends) and numbering each joint with a paint-type marker (not chalk) that will remain on the pipe until it goes in the hole. After a layer of pipe is measured and before it is covered with another layer, the recorded measurements should be inspected to ascertain that the joint numbers of the first and last casing joints on the rack correspond to the numbers of the recorded measurements. Then the measurements should be reviewed for any joints whose lengths vary significantly from the others. If a short or long joint is spotted in the tally, then that joint should be physically checked on the rack to be sure the recorded value is not a mistake. Most practical systems involve recording the joint measurements in a tally book or on some form that lists the joints in groups of 10. As the total length of each group is summed, the total length for each 10-joint group is 10 times the average length of joint in that group. If a mistake has been made in the measurements or the addition, it often is easy to spot using that method.

Accuracy is essential in pipe measurements, but it is unbelievable how many operators pay little attention to this phase of the process. Whatever system you use, it should be simple and consistent. The final responsibility for an accurate tally lies with the company representative on the location—not the roughnecks or roustabouts. One more time, there is no excuse for an incorrect casing tally!

6.4 Crossover Joints and Subs

When the pipe is measured, that is the time to check all the crossover joints (if required) as to correct threads and correct placement in the string on the racks or a separate location where they can be added to the string when required. Be sure the crossovers are included in the tally. There should be at least one spare for each type of crossover used. Crossovers for proprietary connections should be cut only by a machine shop or manufacturer licensed to cut that specific thread. The legal issue is one thing, but an improperly cut thread can cause failure of the string.

Crossover subs or couplings for API ST&C and LT&C threads need special comment. A short pin will make up into a long collar so no crossover normally is required when ST&C is run above LT&C. The reverse is not true, because a long pin will not make up into a short coupling. Some operators get around the crossover issue by purchasing an LT&C coupling and send it to the rig as a crossover. The idea is that, when it comes time to make up the LT&C pipe into the top of the ST&C, the short coupling is removed and the long coupling installed on the short pin. That sounds easy, but it is a bad practice. It often comes as a big surprise, but the short collar may not back off easily. We cannot predict the torque required to remove a coupling that was installed at the mill. It may come off easily if the pipe is relatively new, but if the pipe has been sitting on a rack in the hot sun for two or three years, it might require so much torque that the threads are galled and ruined in the process. This is not uncommon. It is also not uncommon to see a rig crew use a cutting torch to heat a coupling to get it off easier. So it is far better to have a dedicated crossover joint or sub (and a spare) for each place one is needed. The cost saved by purchasing a long coupling for a crossover is miniscule compared to the potential cost if something goes wrong.

6.5 Running the Casing

The running procedures are important, not only for the success of the well, but also for personnel safety. Many injuries occur during the running process because of the relatively large size and weight of the casing, the length of the joints, and the unfamiliarity of drilling crews with the procedure. For those fortunate enough to work on rigs that have automated pipe handling systems, it may seem hard to imagine the crude casing running procedures that are common to many drilling operations.

The running procedure itself must be looked on as a critical operation in the well. It should not be hurried but should be smooth and efficient. Typically, the worst thing that can happen during the running procedure is to have to stop for some reason. In many parts of the world, it might be possible to stop the operation for several hours or even a day without sticking the casing. In other parts of the world, if the operation is stopped for half an hour, the casing will never be moved again. For that reason, all the equipment must be in good working order and a certain amount of redundancy might be desirable.

6.5.1 Getting the Casing to the Rig Floor

Usually the pin protectors are removed before the pipe is picked up to the V-door of the rig, so as not to slow the makeup process on the rig floor. In this case, the pin should be protected with a quick-release, rubber protector until it is up on the rig floor ready to stab.

6.5.2 Stabbing

The stabbing process is critical to prevent damage to the casing. Not all rigs have or use adjustable stabbing boards. It is still quite common to see jury-rigged stabbing boards that are nothing more than some 2" × 8" boards tied to derrick cross members. Whatever means are used, it is important that they allow for accurate stabbing of the joints to prevent thread damage. In some cases, this means shelter from winds that can cause difficulty and misalignment. Some proprietary connections require clamp-on stabbing guides to protect sealing surfaces and threads during the stabbing process. If such guides are recommended, they always should be used.

6.5.3 Filling the Casing

In general, the casing should be filled with mud as it is being run into the well. An adequate fill line should be rigged up to assure that the filling operation will not slow the running process. In any event, you should visually assure the casing is full at least every few joints, even if it means slowing the running process until you see the mud at the surface inside the casing. It is especially important to be sure that the first joints of casing run are full because of the buoyancy effect if they are empty. If the first joints of casing are empty, they actually may begin to float or at least lag behind the elevators as they go in the hole. This is a recipe for disaster, because some casing tools with integral slips actually may open without a load and allow the casing to fall to the bottom. It has happened.

Some companies use differential or automatic-fill float equipment to aid or replace the surface fill procedure. Where it works, it is fine, but when it does fail (and it sometimes does), it can cause serious problems if you are not aware that it has failed and is not allowing fluid into the casing. Another objection that many operators have with this type of equipment is that it may allow hole debris to enter the casing at the bottom. Once on bottom and circulation is initiated, it may plug the float equipment, and there is no way to circulate it out, short of running pipe inside the casing to the float to clean it out. If debris should remain in the casing after circulation and is pushed down to the float collar with the bottom cement wiper plug and plugs the float, then one is left in the precarious position of having all the cement inside the casing and no way to pump it in either direction. Self-fill or differential-fill float equipment has been successful in hard rock areas, and it has failed mostly in areas of unconsolidated formations. If you use such equipment, just be aware of the possibilities for failure—it is much safer in most cases to fill from the surface.

6.5.4 Makeup Torque

All connections should be made up to the proper specified torque while running. Most casing crews have all the necessary values, but it is good practice to check that everyone is in agreement. The correct type of thread lubricant and clean threads are essential for getting the correct amount of torque. For critical applications, there are special services that measure both torque and the number of revolutions of the pipe to be sure that the maximum torque did not occur before the coupling was fully made up.

Another point about proper torque is its measurement. The torque of a typical casing tong is measured with a hydraulic transducer in the tong line. In other words, it actually measures tension in the tong line and not torque. The torque gauge is calibrated such that it multiplies the length of the tong arm times the tension in the tong line to give the torque reading. That only works if the tong line is perpendicular to the tong arm when the torque measurement is made. If the angle is more or less than 90°, then the actual torque is less than that shown on the gauge. A few degrees is not going to make an appreciable difference, but it is not uncommon to see casing tongs rigged up with a considerable deviation from the proper 90°.

One last but most important point about makeup: The best casing design with best quality pipe can fail if not properly made up on the rig.

6.5.5 Thread Locking

A true disaster associated with casing is the disengagement of the bottom joint (or several joints) after the casing has been cemented and operations have begun to drill out the cement inside the casing. The torque from the rotating bit drilling out the cement and float equipment in the bottom two joints starts to turn the casing, and the bottom joint backs out at a connection. Once this happens there usually is no remedy; the hole has been junked and must be abandoned. The reason that this sort of thing happens is that the cement around the bottom joints is incompetent, usually because it has not yet reached a satisfactory strength. It happens most often on surface casing, where the temperatures at the shoe are relatively low and the cement does not set as fast as expected, or the operator is in a hurry to start drilling and does not allow sufficient time for the cement to set. While those are cementing issues, which we do not cover here, something often is done in the running process of the casing to prevent such an event.

Most operators secure the connections on the bottom joints up to one joint above the float collar to prevent accidental backoff of the casing while drilling out cement. Chemical kits, consisting primarily of a thermoset polymer, are used to "glue" the connections. The resin and hardener are mixed and applied in place of thread lubricant to the cleaned connections on the float equipment and bottom joints. A couple of problems are associated with such a practice. One is that most use the compound only on the field makeup part of the connection. They assume that the mill end will not back out. This is a poor practice. If you are going to use the locking compound, remove the couplings and "lock" all the threads not just the field threads. The second problem is that, if something goes wrong and the casing string has to be pulled back out of the hole before reaching bottom, those connections cannot be broken out easily. That presents something of a dilemma, in that if you do it you are safe from backing off the pipe, but if you have to pull the pipe, you cannot easily undo it. You actually can heat the pipe with a welding torch to a temperature where the polymer will break down and the pipe can be backed out. Those joints should be replaced and not run back in the hole. An alternate procedure to the thread locking compound is that of tack welding the couplings on the lower joints. This was common practice for many years before the polymer compounds were available and is still common in some areas. However, welding on casing couplings can lead to serious problems and should be avoided at least for couplings with higher yield strengths than K-55.

When is thread locking necessary? As already mentioned, casing back-off while drilling float equipment is almost always caused by drilling out before the cement has had sufficient time to harden. Hence, the risk of a joint backing off is most acute on the surface casing or a cemented conductor casing, because the temperatures are relatively low and the cement sets relatively slowly. In the case of deeper intermediate strings, the issue is less critical because the temperatures are higher, and it often takes much longer to change out BOPs, drill strings, and other equipment to resume drilling. In the case of stage cementing equipment, there is little need to lock the threads of the stage tools, because for a joint to back-off, all the joints below it must also rotate and this is almost impossible for a stage tool that is a few thousand feet above the casing shoe.

6.5.6 Casing Handling Tools

A wide variety of the elevator and spider assemblies are available to run casing. Some elevators are what is called *square shouldered* (see Figure 6–1), they have no slip elements. Instead, they have an internal diameter that will fit around the casing body but is too small for a coupling to pass through; they have hinge openings. The spider may be similar to the elevator and hinged or large enough for the coupling to pass through with some type of slip assembly built in, or there may be just a simple set of manual slips.

Elevators and spiders increase in sophistication from there. We assume that anyone who runs casing knows to select an elevator and spider combination of sufficient strength to suspend the casing safely. There is one important point to make in this regard though. The elevator and spiders (see Figure 6–2 and Figure 6–3) normally used to run heavy casing strings are rated at 500 tons (1 million lbf) or even 1000 tons (2 million lbf) and have an internal slip assembly that is either manually activated with an external lever or is air or hydraulically actuated.

These are very good tools for running heavy strings of casing. The problem is that even a heavy string of casing is not "heavy" when it starts in the hole. The efficiency and ease with which the manual lever operates the slips is such that it is possible for someone on the rig floor to easily open the slips even with several hundred feet or so of casing suspended in the spider. A similar problem can occur when the pipe is in the elevator and an obstruction is hit, causing the load on the elevator to be momentarily released so that the slips jump open. The result in either case is a portion of a casing string dropping into the hole and going to the bottom. For this reason, it often is preferred to start a long string of casing in the hole with lower-rated tools, then switch over to the 500 ton tools when the casing is at the bottom of the surface casing or some other point where the running process can be paused to switch the elevator and spider. The possibility of such an event may sound remote, but a number of these instances inhabit many companies' annals of bad events. In one case, a casing crew member slipped and fell against the release lever on a spider and dropped 400 ft of 13⅜ in. casing to the bottom of a 5000 ft well. In another case, the crew was not filling the 7⅝ in. casing properly, and as the driller lowered the casing, it was buoyed enough that it did not descend at the same rate as the elevator; the elevator slips opened. No one

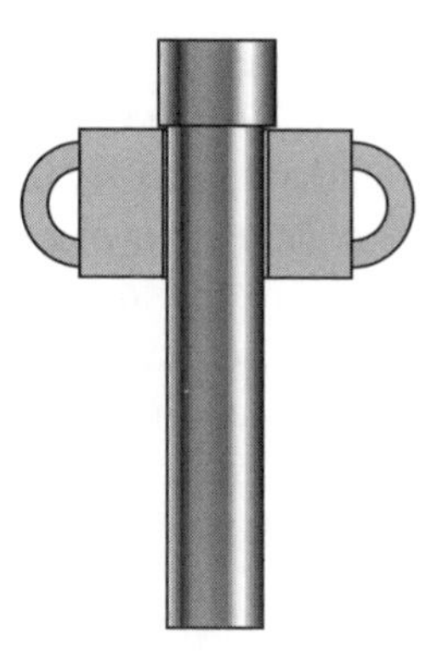

Figure 6–1 *A square-shouldered type elevator.*

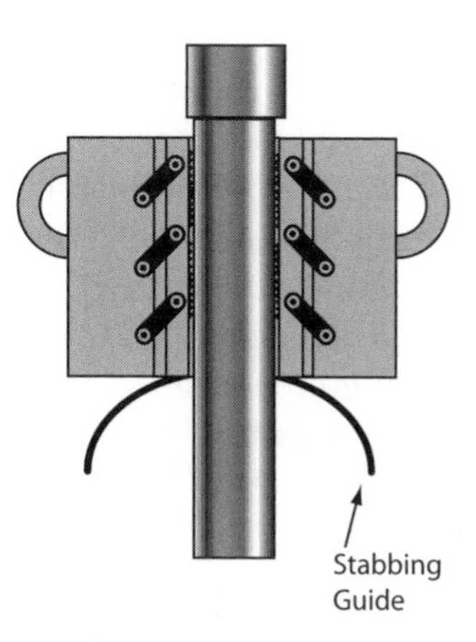

Figure 6–2 *A slip-type casing elevator.*

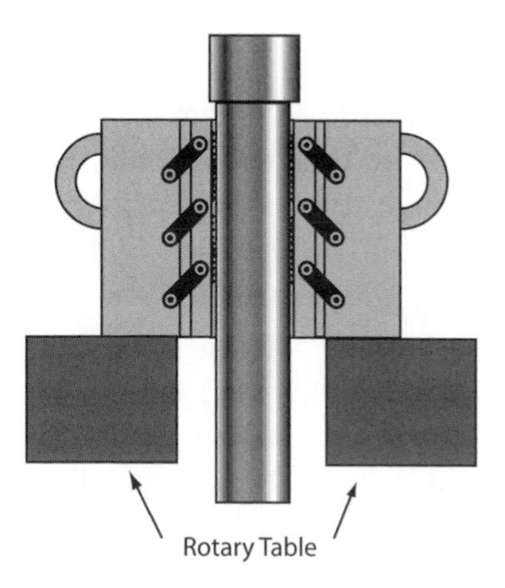

Figure 6–3 *A slip-type casing spider.*

realized the elevator slips were open until the driller stopped the elevator above the spider and the casing kept on going right through the spider before anyone had time to react. Approximately 1100 ft of 7⅝ in. casing fell 12,000 ft before it stopped. One other point about casing tools is that a spare elevator-spider combination should be on the rig, in case there is a problem with the primary tools. There will not be time to order a replacement if one fails in the process of running casing.

6.5.7 Running Casing in the Hole

Running Speed

Running casing is an intense operation; in cases where differential sticking is likely, it is even more so. There often is the temptation to run it too fast. But because casing is of a larger diameter than the drill string, the annular clearance is smaller and the displacement and surge pressures in the annulus usually are higher than when running a drill string in the hole. If a formation is fractured during the running process, then the tendency to differentially stick the casing off bottom is increased, and the chances of getting a good cement job usually are decreased considerably, depending on where the fracture occurs. There are formulas for calculating surge pressures. Almost no one ever uses them. A sure way to get into trouble running casing is to rely on some dubious value of the fracture pressure from some unknown source and a formula that may or may not model the actual mud conditions in the hole. The rule of thumb is that casing should be run in the hole at a slower rate than the drill string. Another is to observe the delay and rate at which the mud spills over the bell nipple. If there is a noticeable delay between the rate of displacement and the mud being displaced from the bore hole, then the casing is being run too fast for that particular mud. It is mostly a matter of experience and the known conditions of the specific bore hole. The point here is this: Do not attempt to run the casing too fast, it is not a race.

Getting to the Bottom

What should we do if the casing will not go to the bottom? When this happens (and it sometimes does), there is a tough decision to make. Should we rig up to circulate and try to wash past an obstruction or should we start out of the hole immediately? There are no firm rules on this, because there are so many variables. In many places, it is possible to install a circulating head (or top drive) and wash through an obstruction. In other situations, where differential sticking is prevalent, turning on the pump to circulate is the equivalent of saying, "This is where I want to stick my casing string."

One should always decide before starting in the hole what the risks are and what the decision will be, should something stop the casing from going to the bottom. It is much easier to make the decision before starting in the hole than when a problem arises during the running process.

Tagging Bottom

We might also mention the issue of tagging the bottom of the hole with the casing. Some say that it should not be done, because it can possibly plug the float shoe or even stick the casing in cuttings or fill on bottom. There is legitimacy to this line of thinking. Others routinely tag bottom to verify their pipe measurements. I was schooled in the "do not tag bottom" discipline and never had a problem, but I have seen a couple of cases where followers of that mode of thinking set the production casing shoe above the bottom of the pay zone, and that also can be problematic. In recent years, it has become common practice when setting liners in level 4 multilaterals (where the upper portion of the cemented liner is washed over and milled out flush with the junction) to tag bottom as a depth reference to avoid cementing the liner with a coupling in the window. If a liner coupling (or any liner connection) is in the window when the liner is cemented, the coupling or connection will be partially milled in the wash-over operation. The result is a loose section of milled pipe above the connection as well as a loose, partial coupling in the well bore. Do not make the foolish assumption that the cement will hold it in place. Laterals have been lost to reentry in such cases.

Another important point along these lines: We hate to admit that it happens, but sometimes we find that the casing stops 42 ft from bottom, or maybe even 83 ft or some such multiple of the pipe range being used. This sort of thing happens too often, and the embarrassment of having made a mistake in the tally or joint count is only secondary to the reality that that it could also end your current employment. Check your records quickly and make your decision, because the result of trying to wash pipe to bottom that is already on bottom may only compound your problems.

6.5.8 Highly Deviated Wells

Directional and highly deviated wells pose a special set of conditions. Bore-hole friction may be quite high, and bore-hole stability problems may complicate the situation even further. Unlike most nearly vertical wells, the hook load does not always increase as the casing nears bottom.

It often decreases as more casing enters the highly deviated portion and must be pushed in the hole (see Figure 6–4).

Obviously, if the hook load goes to zero, we have nothing more to push the casing with and the casing goes no further. (A top drive rig allows us to add additional force, but it may not be enough either.) If it is not on bottom, then our only hope is to be able to pull it out of the hole. Will our design allow the casing to be pulled out of the hole with all the bore-hole friction in this well? It is essential in highly deviated wells to incorporate bore-hole friction into our design and running procedures. We discuss bore-hole friction in Chapter 9.

6.6 Landing Practices

There is no standard practice for landing casing after it has been cemented. It is assumed that the casing is now fixed at the top of the cement. (The fixed point often is referred to as the *freeze point.*) The casing above the freeze point actually can buckle into a spiral or helix due to its weight, the weight of the fluids inside, or a change in temperature. Only in rare cases would this buckling actually result in damage to the casing, but it could cause wear problems in intermediate casing strings and difficulty in running production equipment in production strings. The severity of the buckled deformation is limited by the clearance between the casing and well-bore wall, which normally is relatively small but could be considerable in a washed out area. We discuss buckling in Chapter 8.

6.6.1 Common Landing Practices

Four landing procedures are common and were once mentioned as recommended practices in a number of publications (but no longer). Roughly they are as follows:

1. Land the casing with the same load on the wellhead as the hook load after cementing.
2. Land the casing with tension at the top of the cement, which is assumed to be the freeze point.
3. Land the casing with the neutral point (axial tension/compression) at the freeze point.
4. Land the casing with compression at the freeze point.

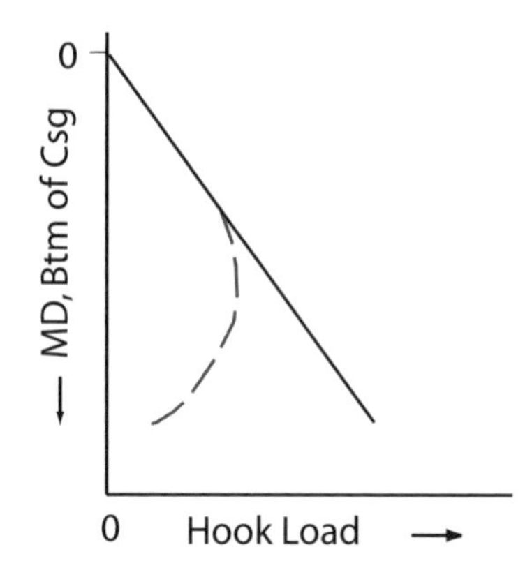

Figure 6–4 *Decreasing hook load in highly deviated well.*

You can see that some of those are the opposite of each other (the second and last), and none are in agreement. Some operating companies that have selected one of these (with possible variations) are adamant that theirs is the best method to use. The dilemma with all this is that, once the casing is on bottom and cemented, we are not really certain what happens down the hole when we land (or hang) the casing at the surface. In other cases, a limitation is placed on what the wellhead equipment can support. There is also the question of the type of hanger used—a slip-type hanger gives us considerable flexibility (if we can get it down into the casing head properly) whereas a mandrel-type hanger cannot be adjusted once the pipe is on bottom or cemented.

It is generally agreed on by most operators, though, that the casing should not buckle above the freeze point. This means that the effective axial load should be in tension if at all possible everywhere above the freeze point. Do not be misled into using the true axial load, the neutral point for buckling is the point where the effective axial load is equal to zero. Anything above that point should have an effective tensile load. (You learned to calculate the effective load in Chapter 2, so it should be no problem for you.) If the casing will be heated by circulating or produced fluids, the heat will expand the casing and reduce the tension, and possibly put part of the noncemented casing in compression. (Temperature effects are covered in Chapter 8.) If you use a mandrel-type hanger instead of a slip type, and in many situations you have no choice, then there is no way to adjust the axial load above the top of the cement after it has set. In that case, if possible, design the cementing job such that the cement top is well above the neutral point. You may also be able to rotate the pipe while cementing or before cementing. This will allow the pipe to overcome any residual frictional force from going in the hole and work the neutral point to the shoe (except in horizontal wells). However, before you elect to rotate the casing, be sure that the torque required for rotation does not exceed that maximum recommended makeup torque of the casing connections. It often exceeds the maximum for 8-rd couplings and non-shouldering-type connections.

One significant problem as far as casing buckling is concerned is hole washout and bad cement or no cement in the washed out interval. In the presence of heated circulating fluid or produced fluids, buckling can occur in this interval. This is not a landing problem but rather a cementing problem that must be addressed. Along these same lines is the presence of a stage tool in a casing string. Often a stage tool is used some distance above the top of the lower stage of cement. This means that an unsupported section of casing is fixed at both ends, and a significant rise in temperature can cause buckling of the casing in that unsupported interval.

6.6.2 Maximum Hanging Weight

There are limits on the amount of weight that may be hung on a casing hanger:

- Tensile strength of the casing string.
- Maximum support strength of wellhead and support casing.
- Support rating of the casing hanger.
- Collapse rating of casing when using a slip-type hanger.

The first limitation is a matter of proper casing design, so that if tension above the string weight is to be applied (e.g., preventing thermal buckling), the additional tension must be included in the design load. The second item is a matter of structural integrity of the supporting casings or platform and is beyond the scope of our discussion. The third item applies primarily to slip-type hangers, and one needs to refer to the hanger manufacturer's rating for the particular hanger to be used. Also, the weight and wellhead pressures could exceed the rating of the casing head so that the hanger actually causes the head to expand, but that is rare and most wellhead manufacturers have eliminated those problems from their equipment. The last item, concerning the collapse load of a slip-type hanger on casing, can be a serious problem for heavy strings of casing (see Figure 6–5). The weight

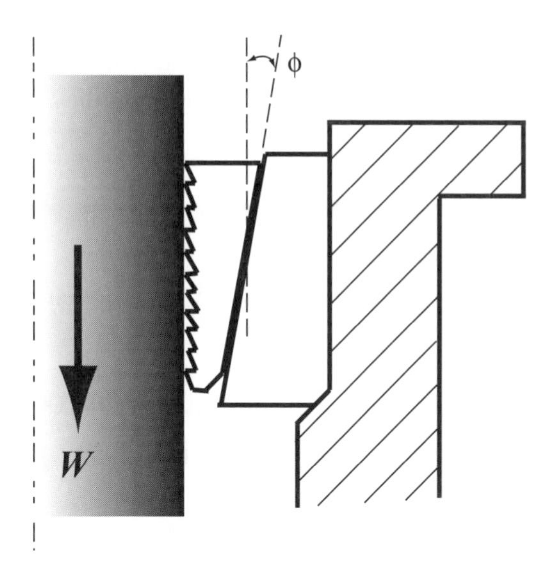

Figure 6–5 *Slip-type casing hanger.*

of the casing forces the slip segments downward, which in turn imposes a radial, compressive force on the casing. Such a force can exceed the collapse resistance of the casing.

A simple formula can be used to estimate the collapse load imposed by a slip-type casing hanger.

$$p_{\text{hngr}} = f_s \frac{W}{A_{\text{slips}} \tan \phi} \tag{6.1}$$

where

p_{hngr} = external casing pressure due to hanger

f_s = safety factor

W = hanging weight of casing

A_{slips} = gross contact area of slips

ϕ = taper of slips and hanger (measured from vertical)

To be as consistent as possible, the pressure load from this equation should be compared to the biaxial collapse pressure rating of the casing. As to the safety factor, it is a matter of company policy, but a commonly used safety factor is 2.0.

Example 6–1 Casing Hanger Collapse Load

From our continuing example, the 7 in. production casing has the following data:

Buoyed casing string weight at surface = 338,000 lbf

Type of 7 in. casing at surface = 29 lb/ft, P-110

Nominal collapse rating = 8530 psi

Hanger taper = 25°

Hanger slip length = 10 in.

Using a safety factor of 2.0, determine if the entire buoyed weight of the string can be hung on the hanger.

The biaxial collapse rating of the casing with 338,000 lbf tension is calculated from the formulas of the previous chapter and is 7280 psi:

$$p_{\text{hngr}} = 2.0\frac{338000}{\pi(7)(10)\tan 25} = 6592 \text{ lbf/in}^2$$

In this case, the casing may be hung safely with the full buoyed weight on the hanger without danger of collapse, since the equivalent collapse pressure load is less than the biaxial collapse rating of the casing. Whenever doing this type of calculation, it is important to know whether the angle of the slip segments is measured from the horizontal or vertical. If the angle is measured from the horizontal, it must be subtracted from 90° before using in this formula. Also important is to compare to the biaxial collapse rating of the casing rather than the published nominal collapse rating, although many assume the safety factor of 2.0 is sufficient to ignore the combined collapse-tension effect.

6.7 Closure

Running casing is as important as the casing design itself. If the casing is damaged or does not reach the bottom, the success of the entire well is jeopardized. We looked at some practical aspects in this chapter, and it is hoped that these may be of use to the reader. Many more aspects could have been discussed, but this chapter covers the essence.

This chapter concludes what might be called the basics of casing design and practices. Essentially, it is presented as a recipe for basic casing design. Although some of the issues are discussed in detail, little is said about where the strength values come from and what their limitations are. The remaining chapters of this text examine the mechanics of casing in more detail.

CHAPTER 7

Beyond Basic Casing Design

7.1 Introduction

The first six chapters of this text were written to provide a basic foundation in casing design. They more or less constitute a recipe, if you will, for basic casing design, which should serve for designing casing strings for the vast majority of wells drilled in the world. One need not be an engineer to do it successfully, and it is hoped that the preceding chapters were enlightening to those who are not. Beginning with this chapter, we abandon the recipe. From here on, we address a number of topics considered more advanced. The intent is not to try to teach a method for designing casing for critical wells but to help one understand the principles involved. More and more, we rely on software to do casing design. On the one hand, that is good, because it allows us to make sophisticated calculations and adjustments that take an excessive amount of time if done manually or that few actually understand how to do in practice. But, on the other hand, we have people using software to design casing for real wells who are clueless as to what the software is doing and what the results mean. This is not an exaggeration. In this chapter, we examine some of the topics that may fill a few gaps in the education of many petroleum engineers in regard to casing design. The purpose here is to impart a degree of understanding of some of the concepts and terminology for more advanced topics concerning casing and its use. This chapter begins with a brief discussion of design and then looks at some of the concepts of solid mechanics as applied to casing and oil-field tubulars in general. In later chapters, we discuss specific topics, formulas, and so forth.

7.2 Structural Design

The design of structures is almost as old as humanity itself. Whether the first "structures" were for shelter, tools, or weapons, the design process from its primitive beginnings has been around a long time. Almost everything we see each day is a structure of some sort. Casing also is a structure, though we may not often think of it in that context. It is a containment structure for the most part, and the design procedures we use are similar in many respects to those of more complex structures.

7.2.1 Deterministic and Probabilistic Design

There are two general approaches to designing casing or any type of structure. One is a deterministic design method, the process with which we are most familiar. We use published values for the minimum strengths and performance properties of the materials, hypothetical load scenarios based on observed and hypothetical criteria, and a set of formulas to calculate structural performance with those loads, then specify the types and sizes of structural materials required to safely sustain the loads. This is the method used for the design of most common static structures, such as bridges, skyscrapers, television transmission towers, drilling rig masts, and even oil-field casing strings. The other general approach is a probability-based design method, in which we use statistical test data for the strengths and properties of actual materials and probabilistic loading scenarios. This approach often is used in the design of structures subject to dynamic and cyclic loading such as airframes, turbine blades, and so forth, where fatigue failure is a significant or dominant factor. The probabilistic design criteria in these types of structures may also be weighted on the consequences of structural failure. In other words, the critical strengths and loads often are based on things like risk to human life, property value, the environment, and so forth. An example would be the blade of a gas turbine operating in some remote oil-field location as opposed to a jet engine turbine blade on an aircraft flying human passengers across a continent: in other words, a 0.1% probability of failure in 10,000 hours of service may be acceptable for the remote oil-field gas turbine, but that same failure probability in an aircraft engine design would likely have aircraft falling out of the sky almost daily, and that is not acceptable. This method also can be applied to static type structures: we see an example of this in the oil field in the design of pipelines, where the published standards for strength often are based on the human population density in the vicinity of the pipeline. A probability-based design must account for the consequences of failure in addition to the probability of a failure. And, to do that, one must have reliable limit data on actual materials or components to work with rather than some limit set by manufacturing standards that allow considerable tolerance. Obviously, the better data we have, the better both methods work, but it is especially important in probability-based design if any significant savings is to be realized.

7.2.2 Design Limits

One thing we must get very clear in our heads is that, when we design a casing string (or any other type of structure for that matter), we are not attempting to predict failure. Predicting actual failure is near impossible, even when we have the most complete data we can imagine: in the case of oil-field tubes and bore-hole conditions, predicting failure is impossible. So our goal is to select some design limits and select our casing such that the anticipated loads do not exceed those limits. Calculating design limits and predicting failure are separate and distinct processes.

A design limit is naturally linked to some strength property of the structural member, which is a tube in our case. Since we already stated that we cannot predict actual failure of the tube, there must be some other property of the tube that we can reliably predict or calculate. The historic design limit for casing, as well as most structures in the world, is the elastic yield point: the stress at which a material goes from elastic behavior to plastic behavior. The elastic yield point or yield stress (sometimes referred to as the *yield strength*) of a metal such as steel is well defined and relatively easy to determine experimentally. We can (and do) go beyond the yield strength in some cases of structural design, but working within the plastic regime is quite complex and generally avoided in all but the simplest cases. These usually are cases in which part of the material body remains elastic and the design limit might be selected as the point at which the entire body reaches the yield point (or possibly some point before that state is reached). These design limits typically are one-time limits and do not consider the effects of cyclical loading (which changes the yield stress value). An example is the new formula proposed for ductile rupture of casing, which we discuss in the next chapter. The only cyclical loading in the plastic regime for oil-field tubulars is in coiled tubing, and we all know (or should know) that coiled tubing has a very short service lifespan because of the cyclic loading in the plastic regime.

7.2.3 Design Comments

It is not likely in the near future that more than a small number of companies will be doing probability-based casing designs for more than a relatively few critical wells, as compared to the number of wells drilled in the world each year. Those companies doing this work have their own expertise and criteria for risk, and those criteria cannot necessarily carry over to other companies. For instance, years ago with the advent of tubingless completions, many looked at it as a way to save money. If you drilled a lot of these wells, you could save a sizeable amount of money. To be sure, there were failures, and the inexpensive wells completed with 2⅜ in. tubing/casing could not be effectively worked over with the equipment of the day. The larger companies accepted this risk, and the frequency of failed wells (which were usually plugged as expendable) was within reason. But, for a small operator that had only one well or two, a failure sometimes was cause for bankruptcy. That is not acceptable from the small operator's perspective. However, that said, the one benefit of the trend toward probabilistic casing design that every operator, big or small, has realized is a decided improvement in the design formulas and the ways in which certain variations in casing tolerances can be measured and accounted for, even in deterministic design. And even what we call a deterministic design is based on some probabilities of certain critical loading occurrences and possible failures of casing to meet minimum standards; the difference is that, in deterministic design, we do not attempt to quantify those probabilities (see Klever and Tallin, 2005).

One last comment should be made. A common mistake is to think that a deterministic design gives us a 100% safe structure at a higher cost, and a probabilistic design gives us a more cost-effective structure but at a slightly greater risk. While that may be true in some cases, it is not true in general. In fact, many of the probability-based designs are safer than some deterministic designs. Both methods have their place and applications.

7.3 Mechanics of Solids

The purpose of this section should be obvious by its title. In the petroleum engineering discipline, practitioners come from varied technical backgrounds; consequently, we see in the literature a variety of approaches to the mechanics of tubulars, some good, some not. One of the things that pervades all engineering disciplines, though, is the tendency to simplify as much as possible at the early levels. This is understandable. Consequently, as we advance to more complex considerations, we find that much of what we learned was not only an oversimplification but often quite misleading. And a big part of that is terminology. What we set out to do in this chapter is to impart a basic understanding of the mechanics of solids couched in the terminology of modern continuum mechanics. Some of what is covered here normally is at a graduate level for engineers, but the intent is to make it understandable to the undergraduate engineer, which is not unreasonable to attempt.

7.3.1 Index Notation

In a previous textbook I coauthored, I was told that it is verboten to explain index notation to undergraduates. I think this is nonsense. It takes only a few minutes to understand the basics of index notation for small deformations in Cartesian coordinate systems, and after a little practice and familiarity, one would wonder why this was never taught sooner. It greatly simplifies the appearance of the equations of mechanics and, I believe, enhances understanding. Index notation is simply this: Coordinates are numbered and noted with subscripts. That is it! That is the whole idea! It was invented by Einstein to make life easier in dealing with the geometric analogy of general relativity.[5] We use a much more modest version though.

Coordinates

Before we explain index notation, we need a coordinate system, because index notation always refers to a specific coordinate system. Figure 7–1 illustrates a Cartesian coordinate system similar to the one we employed in Chapter 2. Note, however, that the coordinate axes are no longer labeled x, y, z, but x_1, x_2, x_3. We just as easily could have labeled them y_1, y_2, y_3; z_1, z_2, z_3; x_1', x_2', x_3' ; or even X_1, X_2, X_3. In this notation, the letters, x, x', and X refer to specific coordinate systems, and we may have as many or as few as we need for a specific use (typically, we seldom need more than two). The numbered subscripts, 1, 2, and 3, refer to the three axes of those specific coordinate systems.

Once we have the coordinate system established, we can refer to a point in that coordinate system. So, instead of referring to a point by its three coordinates, x_1, x_2, x_3, we can simply refer to the point as

$$x_i \text{ where } i = 1, 2, 3$$

The letter index is assumed to include all three axes, so it is not necessary to write out the range of the index, i, except in cases where other subscripts might cause confusion.

We used a Cartesian coordinate system as an example, but we are not limited to Cartesian coordinates. For instance in a circular cylindrical

5. In some coordinate systems, the axes are not orthogonal. There, we must use both superscript and subscript indices. But we are not going to concern ourselves with that degree of complexity here. See Simmonds (1982) for a good foundation level book on vectors and tensors.

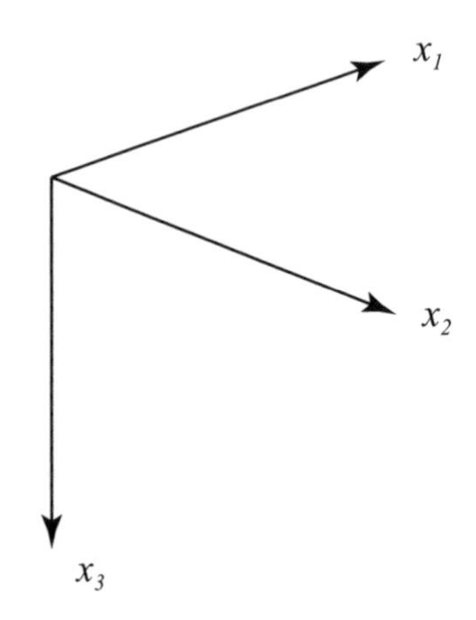

Figure 7–1 *A Cartesian coordinate system.*

coordinate system we may still number the axes as 1, 2, 3, instead of r, θ, z. The only thing we have to be careful about in other coordinate systems is the physical meaning of some of the quantities, but we do not concern ourselves with that for now.

Conventions of Index Notation

The power of the index notation is not merely the handy way to note coordinates, but in additional conventions that greatly reduce the amount of writing we have to do. We cover some of those now.

Summation Convention

Any index repeated in a product or quotient is automatically summed over its entire range. For example,

$$a_{ij}x_j = c_i \quad \rightarrow \quad \sum_{j=1}^{3} a_{ij}x_j = c_i \qquad i = 1,2,3$$

In the summation convention, a nonrepeated index is called a *free index*, and the repeated index is called a *dummy index*. In the example, i is a free index, and j is a dummy index.

Range Convention

Any free (not repeated) index is implied to take on all possible values of its range. For example,

$$a_{ij}\,x_j = c_i \quad \rightarrow \quad \begin{array}{l} a_{11}\,x_1 + a_{12}\,x_2 + a_{13}\,x_3 = c_1 \\ a_{21}\,x_1 + a_{22}\,x_2 + a_{23}\,x_3 = c_2 \\ a_{31}\,x_1 + a_{32}\,x_2 + a_{33}\,x_3 = c_3 \end{array}$$

Nuances

The dummy index may be changed without affecting the meaning. The following is acceptable because changing the dummy index does not change the meaning, since it only implies a sum over the range:

$$a_{ij}\,x_j = a_{ik}\,x_k$$

The free index may not be changed within an equation, unless we change all occurrences of that particular index. For example,

$$a_{ij}\,x_j = c_i \;\leftrightarrow\; a_{kj}\,x_j = c_k$$

The two equations are equivalent but the following is not:

$$a_{ij}\,x_j \neq a_{kj}\,x_j$$

because the free index is not consistent on both sides. A repeated index on a single variable is called a *contraction*:

$$a_{ii} = a_{11} + a_{22} + a_{33}$$

A dummy index can be repeated only once in a product or quotient. The following are meaningless:

$$C_{iiii}\,,\quad \in_{iii}\,,\quad a_i b_i c_i$$

We might think of indexed quantities as matrices, although that is not necessarily the purpose. Two examples follow:

$$a_i = \begin{Bmatrix} a_1 \\ a_2 \\ a_3 \end{Bmatrix} \qquad b_{ij} = \begin{bmatrix} b_{11} & b_{12} & b_{13} \\ b_{21} & b_{22} & b_{23} \\ b_{31} & b_{32} & b_{33} \end{bmatrix}$$

When thinking in terms of matrix algebra, the first index is the row number and the second is the column number. (Caution: While this is the most common practice, some use the opposite convention:)

$$c_i = a_j b_{ij} \;\rightarrow\; \begin{Bmatrix} c_1 \\ c_2 \\ c_3 \end{Bmatrix} = \begin{bmatrix} b_{11} & b_{12} & b_{13} \\ b_{21} & b_{22} & b_{23} \\ b_{31} & b_{32} & b_{33} \end{bmatrix} \begin{Bmatrix} a_1 \\ a_2 \\ a_3 \end{Bmatrix} = \begin{Bmatrix} a_1 b_{11} + a_2 b_{12} + a_3 b_{13} \\ a_1 b_{21} + a_2 b_{22} + a_3 b_{23} \\ a_1 b_{31} + a_2 b_{32} + a_3 b_{33} \end{Bmatrix}$$

In index notation, it makes no difference whether we write

$$c_i = a_j b_{ij} \qquad \text{or} \qquad c_i = b_{ij} a_j$$

but it does make a difference as to the order of the indices; for instance, in general,

$$a_j b_{ij} \neq a_j b_{ji}$$

Partial Derivatives

Partial derivatives with respect to a spatial coordinate occur frequently in solid mechanics, and index notation allows for a shortcut in notation by using a comma to denote partial differentiation.

$$f_{i,j} \equiv \frac{\partial f_i}{\partial x_j} \qquad f_{i,jk} \equiv \frac{\partial^2 f_i}{\partial x_j \partial x_k}$$

We also can use summations in derivatives; for instance,

$$u_{i,ij} = \begin{bmatrix} \frac{\partial^2 u_1}{\partial x_1 \partial x_1} + \frac{\partial^2 u_2}{\partial x_2 \partial x_1} + \frac{\partial^2 u_3}{\partial x_3 \partial x_1} \\ \frac{\partial^2 u_1}{\partial x_1 \partial x_2} + \frac{\partial^2 u_2}{\partial x_2 \partial x_2} + \frac{\partial^2 u_3}{\partial x_3 \partial x_2} \\ \frac{\partial^2 u_1}{\partial x_1 \partial x_3} + \frac{\partial^2 u_2}{\partial x_2 \partial x_3} + \frac{\partial^2 u_3}{\partial x_3 \partial x_3} \end{bmatrix}$$

And, even more,

$$u_{i,k}\, u_{k,j} = \begin{bmatrix} \frac{\partial u_1}{\partial x_1}\frac{\partial u_1}{\partial x_1} + \frac{\partial u_1}{\partial x_2}\frac{\partial u_2}{\partial x_1} + \frac{\partial u_1}{\partial x_3}\frac{\partial u_3}{\partial x_1} & \frac{\partial u_2}{\partial x_1}\frac{\partial u_1}{\partial x_1} + \frac{\partial u_2}{\partial x_2}\frac{\partial u_2}{\partial x_1} + \frac{\partial u_2}{\partial x_3}\frac{\partial u_3}{\partial x_1} & \frac{\partial u_3}{\partial x_1}\frac{\partial u_1}{\partial x_1} + \frac{\partial u_3}{\partial x_2}\frac{\partial u_2}{\partial x_1} + \frac{\partial u_3}{\partial x_3}\frac{\partial u_3}{\partial x_1} \\ \frac{\partial u_1}{\partial x_1}\frac{\partial u_1}{\partial x_2} + \frac{\partial u_1}{\partial x_2}\frac{\partial u_2}{\partial x_2} + \frac{\partial u_1}{\partial x_3}\frac{\partial u_3}{\partial x_2} & \frac{\partial u_2}{\partial x_1}\frac{\partial u_1}{\partial x_2} + \frac{\partial u_2}{\partial x_2}\frac{\partial u_2}{\partial x_2} + \frac{\partial u_2}{\partial x_3}\frac{\partial u_3}{\partial x_2} & \frac{\partial u_3}{\partial x_1}\frac{\partial u_1}{\partial x_2} + \frac{\partial u_3}{\partial x_2}\frac{\partial u_2}{\partial x_2} + \frac{\partial u_3}{\partial x_3}\frac{\partial u_3}{\partial x_2} \\ \frac{\partial u_1}{\partial x_1}\frac{\partial u_1}{\partial x_3} + \frac{\partial u_1}{\partial x_2}\frac{\partial u_2}{\partial x_3} + \frac{\partial u_1}{\partial x_3}\frac{\partial u_3}{\partial x_3} & \frac{\partial u_2}{\partial x_1}\frac{\partial u_1}{\partial x_3} + \frac{\partial u_2}{\partial x_2}\frac{\partial u_2}{\partial x_3} + \frac{\partial u_2}{\partial x_3}\frac{\partial u_3}{\partial x_3} & \frac{\partial u_3}{\partial x_1}\frac{\partial u_1}{\partial x_3} + \frac{\partial u_3}{\partial x_2}\frac{\partial u_2}{\partial x_3} + \frac{\partial u_3}{\partial x_3}\frac{\partial u_3}{\partial x_3} \end{bmatrix}$$

Special Symbols

Some symbols in index notation have special meaning. These are handy tools that we use later.

The Kronecker delta symbol is used time and again. It takes on values of either 1 or zero:

$$\delta_{ij} \equiv \begin{cases} 1 & i = j \\ 0 & i \neq j \end{cases}$$

The permutation symbol is used primarily for cross products and determinates:

$$\epsilon_{ijk} \equiv \frac{1}{2}(i-j)(j-k)(k-i)$$

It takes on values of 0, 1, or −1, depending on the values of the indices:

$$\epsilon_{ijk} = \begin{cases} 1 & \text{even permutation, (1,2,3) or (2,3,1) or (3,1,2)} \\ -1 & \text{odd permutation, (3,2,1) or (2,1,3) or (1,3,2)} \\ 0 & i = j \text{ or } j = k \text{ or } k = i \end{cases}$$

To better understand what is meant by even and odd permutations, think of a triangle with the vertices numbered sequentially 1, 2, 3 in a clockwise direction. If one starts at any of the three vertices and proceeds around the triangle in a clockwise direction the sequence of numbers is an even permutation. If one goes around the triangle in a counterclockwise direction then the sequence of numbers is an odd permutation. This is illustrated in Figure 7–2.

This is enough to get us started. We will add to this as needed in context of certain applications. The important things to remember for now are the ideas of the indices and the summation convention.

7.3.2 Coordinate Systems

In the previous section, we mentioned coordinate systems in relation to indices. Now, we need to address a little more about coordinate systems

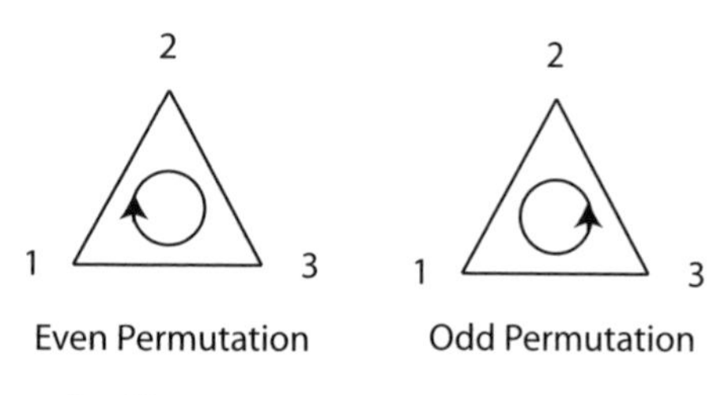

Figure 7–2 *Even and odd permutations.*

7.3.2 Coordinate Systems

In the previous section, we mentioned coordinate systems in relation to indices. Now, we need to address a little more about coordinate systems and the transforms necessary to transform quantities in one coordinate system to another. For instance, a vector, such as a displacement **u**, is independent of any coordinate system. The conventional way of writing a displacement vector is $\vec{u}$, another way is **u**. The first is used when we are writing by hand, since we do not have a boldface font, and the second convention is used when we are typing it in print, where we do have the capability of using boldface fonts. In either case, this is a vector or directed line segment with both a direction and magnitude, and as written, it is independent of any coordinate system. Once we select a coordinate system and write it in index notation as u_i, we are essentially now referring to the three components of a vector, and those components are not independent of a coordinate system. The numerical values of the components depend on our choice of coordinate system. Again, it is the same vector no matter what coordinate system we select, but the numerical values of its components depend on the particular coordinate system. It could be said that mathematicians generally prefer to work with direct notation, such as $\mathbf{u} = \hat{\mathbf{n}}\mathbf{T}$, which is coordinate independent, because they are seldom concerned about actual values, but engineers prefer to work with index notation, such as $u_i = \hat{n}_j T_{ij}$, which refers to a specific coordinate system, because they are concerned about the numerical values and directions.

Coordinate systems have a set of vectors called the *basis vectors*. In simple terms, basis vectors point in the directions of the coordinate axes. The basis vectors are linearly independent, meaning that they are not made up of linear combinations of each other. That is to say, we cannot add two basis vectors or scalar multiples of each other in any combination to get the third. The most common coordinate system used in solid mechanics is a Cartesian coordinate system of three orthogonal axes, meaning they all are perpendicular to each other. For a Cartesian coordinate system, it is common to label a standard set of unit basis vectors as

$$\{\mathbf{e}_1, \mathbf{e}_2, \mathbf{e}_3\}$$

or, another way of putting it is

$$\mathbf{e}_1 \sim (1,0,0), \quad \mathbf{e}_2 \sim (0,1,0), \quad \mathbf{e}_3 \sim (0,0,1)$$

The magnitude of a unit basis vector is, of course, unity:

$$|\mathbf{e}_i| = 1$$

The magnitude of a vector is calculated by taking the square root of the sum of the squares of its components; for example, for the basis vector $\mathbf{e}_1$, its magnitude is given by

$$|\mathbf{e}_1| \equiv \sqrt{1^2 + 0^2 + 0^2} = 1$$

Any vector, $\mathbf{u}$ for instance, can be written in terms of its components and Cartesian base vectors:

$$\mathbf{u} = u_i \mathbf{e}_i = u_1 \mathbf{e}_1 + u_2 \mathbf{e}_2 + u_3 \mathbf{e}_3$$

Note the summation convention over all three components. The magnitude (or length) of the vector $\mathbf{u}$ is defined like before as

$$|\mathbf{u}| \equiv \sqrt{u_1^2 + u_2^2 + u_3^2}$$

Note that the superscripts denote the component is squared, while the subscript is the index number. To avoid confusion, it is sometimes customary to enclose the component in parentheses and place the superscript on the outside as, $(u_1)^2$. If we wanted to convert **u** into a unit vector, we would do so by dividing each of its components by its magnitude:

$$\hat{\mathbf{u}} \equiv \frac{\mathbf{u}}{|\mathbf{u}|} = \left(\frac{u_1}{\sqrt{u_1^2 + u_2^2 + u_3^2}} \mathbf{e}_1, \frac{u_2}{\sqrt{u_1^2 + u_2^2 + u_3^2}} \mathbf{e}_2, \frac{u_3}{\sqrt{u_1^2 + u_2^2 + u_3^2}} \mathbf{e}_3 \right)$$

A unit vector can be thought of simply as a direction.

We can add two vectors by adding their respective components:

$$\begin{aligned} \mathbf{u} + \mathbf{v} &= u_i \mathbf{e}_i + v_i \mathbf{e}_i \\ &= \left(u_1 \mathbf{e}_1 + u_2 \mathbf{e}_2 + u_3 \mathbf{e}_3\right) + \left(v_1 \mathbf{e}_1 + v_2 \mathbf{e}_2 + v_3 \mathbf{e}_3\right) \\ &= \left(u_1 + v_1\right)\mathbf{e}_1 + \left(u_2 + v_2\right)\mathbf{e}_2 + \left(u_3 + v_3\right)\mathbf{e}_3 \\ &= \left(u_i + v_i\right)\mathbf{e}_i \end{aligned}$$

We also can multiply a vector by a scalar. A scalar may be thought of as simply a number. Examples of scalars are temperature, time, energy, work, density, to name just a few. If α is a scalar, then

$$\begin{aligned} \alpha \mathbf{u} &= \alpha u_i \mathbf{e}_i \\ &= \alpha\left(u_1 \mathbf{e}_1 + u_2 \mathbf{e}_2 + u_3 \mathbf{e}_3\right) \\ &= \alpha u_1 \mathbf{e}_1 + \alpha u_2 \mathbf{e}_2 + \alpha u_3 \mathbf{e}_3 \end{aligned}$$

It is important to note that scalars are not constants and their values may depend on their location within a coordinate system. For example, temperature may be a function of its position in a coordinate system, in other words its magnitude may depend on its location, but it does not have a direction. Partial derivatives of scalars with respect to the coordinates are no longer scalars but are vectors. An example is a temperature gradient $T_{,i}$, which is a vector.

Before we go further, there are some things we need to say about vectors. When we said the basis set was linearly independent, that implies a couple of important things. First,

$$\mathbf{u} = \mathbf{0} \;\Rightarrow\; u_i\mathbf{e}_i = u_1\mathbf{e}_1 + u_2\mathbf{e}_2 + u_3\mathbf{e}_3 = 0 \quad \Leftrightarrow \quad u_i = 0 \quad i = 1,2,3$$

That means that a vector, **u**, is equal to zero if and only if all its components are zero. The double ended arrow above means iff (mathematical shorthand for *if and only if*). Another consequence of the linear independence is the dot product of the basis vectors:

$$\mathbf{e}_i \cdot \mathbf{e}_j = \left|\mathbf{e}_i\right| \cdot \left|\mathbf{e}_j\right| \cos\left(\mathbf{e}_i, \mathbf{e}_j\right) = \delta_{ij}$$

The dot product then is the product of the magnitudes (in this case 1) times the cosine of the angle between the two vectors. It is easy to see that, for a Cartesian coordinate system, the angle between the three base vectors is $90°$, hence, the cosine is always 0 unless the two vectors are the same vector, in which case the angle between them is $0°$ and the cosine is 1. An example of the dot product between two vectors in a Cartesian coordinate system would work like this:

$$\begin{aligned}
\mathbf{u} \cdot \mathbf{v} &= u_i\mathbf{e}_i \cdot v_j\mathbf{e}_j \\
&= u_i v_j \left(\mathbf{e}_i \cdot \mathbf{e}_j\right) \\
&= u_i v_j \delta_{ij} \\
&= u_i v_i \\
&= u_1 v_1 + u_2 v_2 + u_3 v_3
\end{aligned}$$

The properties of the Kronecker delta are seen in this dot product. The physical significance of a dot product in continuum mechanics is that it transforms two vectors into a scalar quantity. For example, the dot product of a force vector with a distance vector results in a scalar quantity called *work*. A scalar has magnitude but no direction. We might think of a dot product of a vector being a linear transformation that transforms a vector into a scalar, as in a force transforms a distance into work.

One other product that we might have occasion to use is the cross product. The cross product between two basis vectors is given by

$$\mathbf{e}_i \times \mathbf{e}_j = e_{ijk}\mathbf{e}_k$$

In the cross product, we use he permutation operator. The result is a vector perpendicular to the two base vectors in the cross product. Let us see how this works with two general vectors:

$$\begin{aligned}
\mathbf{u} \times \mathbf{v} &= u_i\mathbf{e}_i \times v_j\mathbf{e}_j \\
&= u_i v_j \left(\mathbf{e}_i \times \mathbf{e}_j\right) \\
&= u_i v_j e_{ijk}\mathbf{e}_k \\
&= u_1v_2(1)\mathbf{e}_3 + u_2v_3(1)\mathbf{e}_1 + u_3v_1(1)\mathbf{e}_2 + u_3v_2(-1)\mathbf{e}_1 + u_2v_1(-1)\mathbf{e}_3 + u_1v_3(-1)\mathbf{e}_2 \\
&= \left(u_2v_3 - u_3v_2\right)\mathbf{e}_1 + \left(u_3v_1 - u_1v_3\right)\mathbf{e}_2 + \left(u_1v_2 - u_2v_1\right)\mathbf{e}_3
\end{aligned}$$

Note that this contains all three even permutations and all three odd permutations. The use of the permutation operator made this a lot shorter than had we included all the terms whose cross products are zero. The resulting vector is perpendicular to the two vectors in the cross product. If we reverse the order of the cross product, the resulting vector is still perpendicular to the two but in the opposite direction:

$$\mathbf{u} \times \mathbf{v} = -\mathbf{v} \times \mathbf{u}$$

The cross product transforms a vector into another vector. An example would be that the cross product with a force vector transforms a distance vector into a torque vector. Note that work (the result of the dot product of force and distance) is a scalar, but torque (the result of the cross product of force and distance) is a vector even though they both have the same physical units, lbf · ft or N · m.

7.3.3 The Continuum

All matter and perhaps even space itself is composed of particles (or strings) that are discontinuous. This is fine for physicists studying things small, but it is not very useful for modeling the behavior of things such as derricks, drilling fluids, casing, and so forth. In other words, we need a different approach for modeling the things of everyday life to make them mathematically tenable. We do this by assuming the existence of a continuum. What we mean is that the materials and bodies we work with are continuous to what ever scale we need apply. While it may not seem important for now, we can say that all quantities in which we are interested are continuously differentiable to any extent we require at any point within the body. We also assume that Newtonian mechanics apply and inertial reference frames are available so that the laws of nature are not violated. That is all a bit of formality that we need not worry about for oilfield casing, but you should know that there are limitations to our assumptions on some scales and at some speeds.

In Newtonian mechanics, we work in a vector space that consists of three spatial dimensions called a *Euclidian space*, E_3, and a real line, R, called the time line. The letter R stands for all real numbers, in other words scalars like time, and we assume that time is always increasing along this line of real numbers. This vector space may be formally represented as $E_3 \times R$. This is a special case or a limiting case of general relativity.

That is enough background in continuum mechanics for now.

Deformation and Strain

All real materials are deformable. It is perfectly appropriate in many cases to assume that certain things behave as rigid bodies and do not deform, since in many cases we are not interested in deformation or the magnitude of deformation does not affect our observations or calculations. An example would be picking a casing string up off bottom in a well to reciprocate the pipe while cementing. When we first begin to pull on the pipe, it is stretching. The top is moving, but the casing shoe is not. Once we pull a certain amount, the entire string is moving. If we are trying to determine the maximum load to reciprocate the pipe, then we are not interested in the load as the pipe is stretching before the entire string is moving. We are interested only in the load due to gravity and friction when the entire string is moving. This is a rigid body motion and the deformation or stretch in the pipe has no significance in this context. A rigid body motion may be a *translation* and/or a *rotation in space*. A body moves from one position to another without deformation. A rotary table rotates while drilling, and that is a rigid body motion for most applications. Certainly, there is some small amount of deformation in the individual parts, but that is of no interest if we are interested in penetration rate as a function of rotary speed. If we are interested in bit speed as a function of rotary speed, however, we might consider the deformation of the drill pipe in torsion as a possible fluctuating variable. If we are interested in the collapse pressure of a casing string, the collapse value is based on a deformation of the casing.

Deformation by itself is meaningless. For instance, if our casing string is stuck, we pick up on it to try to free it, and we pull maybe 6 ft or 2 m through the rotary table. Is that significant? We cannot answer such an important question with only the amount of information given. If it is stuck at a depth of 10,000 ft, then a 6 ft or 2 m stretch is not very much. But, if the pipe rams in the BOP are closed on top of a coupling unbeknownst to us, then 6 ft of deformation is quite significant. So, what we need is a measure that gives us some idea of how significant the deformation is. We could divide the stretch by the original length to give us a measure. In the first case, 6 divided by 10,000 gives us a measure of 0.0006 or 0.06%. For the second case, assuming the pipe rams are 15 ft below the rotary table, we would have a value of 0.40 or 40%. Now, we know how serious the stretch is. In the second case, we know that the pipe would have long since parted, and of course, we would have already known that had happened. The 6 ft "stretch" in that case would have been an observation made after everyone stopped running and returned to the rig floor.

One way to measure deformation, then, is simply the stretch divided by the original length. That is a simple measure of strain and quite useful at low values for uniaxial deformations. However, if we were to measure the wall thickness of the pipe very accurately, we also would find that, as the pipe stretched and got longer, the wall thickness decreased slightly. A simple definition of strain is

$$\varepsilon_{ij} = \frac{1}{2}\left(u_{i,j} + u_{j,i}\right) \tag{7.1}$$

where u_i is the deformation in one of the three axis directions. This definition of strain, called the *Cauchy infinitesimal strain*, is the measure most commonly used. Suppose, for our pipe example, the x_3 coordinate is vertical and downward and the pipe is weightless, so that the strain is uniform along the entire length of the tube, that is, $\mathrm{d}u/\mathrm{d}x_3$ is a constant all along the length of the pipe, then

$$\varepsilon_{33} = \frac{1}{2}\left(u_{3,3} + u_{3,3}\right) = \frac{1}{2}\left(\frac{\partial u_3}{\partial x_3} + \frac{\partial u_3}{\partial x_3}\right) = \frac{1}{2}\left(0.0006 + 0.0006\right) = 0.0006$$

This is fairly straightforward, but note that there are a total of nine strain components instead of one, and the others may not be zero. In fact, not all will be zero. Because, as we said, when we stretch the pipe in one direction, there is a change in the other dimensions as well. Those other changes may be insignificant or they may not. In addition to the strain definition given by equation (7.1), the Cauchy infinitesimal strain, there are other definitions of strain. We save those for later. The strain may be written as a matrix:

$$\varepsilon_{ij} = \begin{bmatrix} \varepsilon_{11} & \varepsilon_{12} & \varepsilon_{13} \\ \varepsilon_{21} & \varepsilon_{22} & \varepsilon_{23} \\ \varepsilon_{31} & \varepsilon_{32} & \varepsilon_{33} \end{bmatrix}$$

Larger Deformations

We are not going to cover large deformations in this text, but we should mention a few things that are important. Most all of the engineering problems we solve are based on small deformations and infinitesimal strains. The world gets a lot more complicated when we consider finite or large deformations and finite or large strains. For instance, look at Figure 7–3. Here, we see a simple cantilevered beam with a single load at the end. The common solution to this problem is based on very small deformations of the beam, such that the end load is always perpendicular to the beam and the length does not change significantly enough to affect our calculations. This is a one-dimensional problem, and it is easy to solve for the displacements of the beam.

What if we were to make the load on the beam such that the beam is deformed further, as in Figure 7–4? We now see that the direction of the end load is still in the vertical direction, but part of the beam is no longer horizontal. In this case, we see that the loading now has components of force along the axis of the beam as well as transverse to the beam. So, in addition to the vertical deformation, there is also a longitudinal deformation. Furthermore, the axis of the beam no longer coincides with the x-axis of our coordinate system. In this case, there is a finite deformation and our strain measure begins to take on a different meaning.

Look at another illustration. Suppose we take a bar as in the next figure, Figure 7–5(a) and stretch it in the X_1-direction by a fraction of its original length, say 0.05. We might say that the strain in the bar is $\varepsilon_{11} = 0.05$. But what if, in addition to stretching the bar by the same amount, we also rotate it

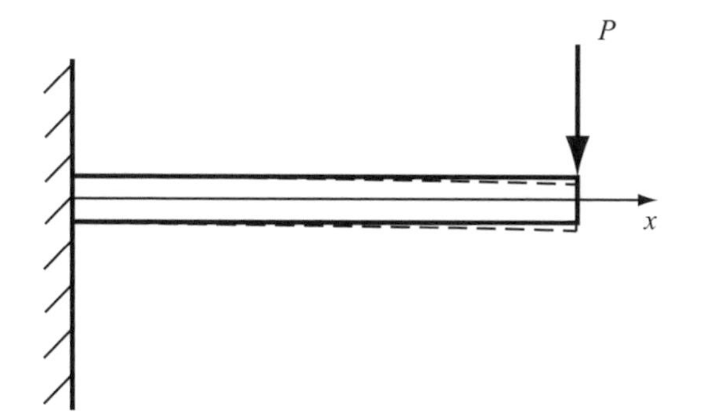

Figure 7–3 *Small defection of beam.*

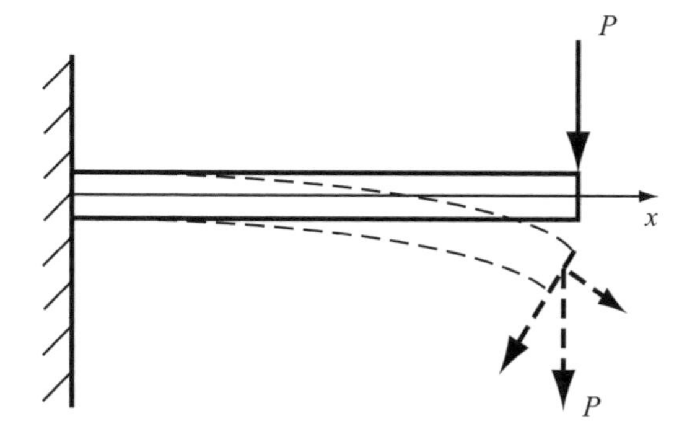

Figure 7–4 *Large defection of beam.*

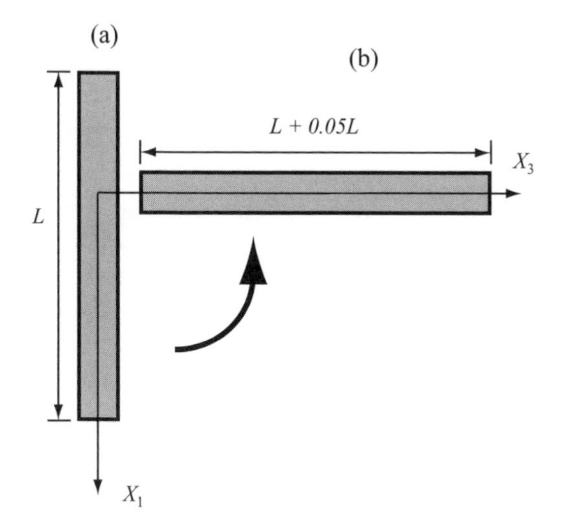

Figure 7–5 *Large uniaxial strain: (a) original position, (b) final position.*

90° as in Figure 7–5(b)? Now, what is the axial strain in the bar? Is it $\varepsilon_{11} = 0.05$? Or is it $\varepsilon_{33} = 0.05$? It is obvious that the strain is now in the X_3-direction, so can we say that the latter is the correct measure of strain? What if we cause the strain before we rotate it? Or what if we are stretching the bar as we rotate it? So the question is this: If the bar has moved in respect to the coordinates as well as having been deformed, then how do we account for this and still make sense of it? There are several ways to do this. One way is to take the strain with respect to coordinates in the undeformed configuration: another way is to take the strain with respect to a set coordinates in the deformed configuration. So here are two more ways to define strain:

$$E_{ij} = \frac{1}{2}\left(u_{i,j} + u_{j,i} + u_{k,i}u_{k,j}\right) = \frac{1}{2}\left(\frac{\partial u_i}{\partial X_j} + \frac{\partial u_j}{\partial X_i} + \frac{\partial u_k}{\partial X_i}\frac{\partial u_k}{\partial X_j}\right) \tag{7.2}$$

$$e_{ij} = \frac{1}{2}\left(u_{i,j} + u_{j,i} - u_{k,i}u_{k,j}\right) = \frac{1}{2}\left(\frac{\partial u_i}{\partial x_j} + \frac{\partial u_j}{\partial x_i} - \frac{\partial u_k}{\partial x_i}\frac{\partial u_k}{\partial x_j}\right) \tag{7.3}$$

The first of these strain measures is called the *Lagrangian strain*. The partial derivatives are taken with respect to the undeformed coordinates, X_i; and the second strain measure is called the *Eulerian strain*, where the partial derivatives are taken with respect to the deformed coordinates, x_i. We can even have a coordinate system that deforms with the body itself and define the strain in that deformed coordinate system, whose axes are no longer straight but curved as the body is deformed. There are many ways to measure strain.

The subject of finite and large deformations can get as complicated as you can possibly imagine. For instance, if we were to try to model the inflation of a large balloon, it is one thing to start out with the initial configuration of the balloon in a partially inflated state and continue the inflation process until its diameter has grown to twice its initial dimension. It is quite another to start out with the balloon folded in an open box as the initial configuration when we begin the inflation process. The point of mentioning large deformations is not to confuse you, but to help you understand the assumptions we make in using small strains. And the reason this is necessary is that it is not uncommon to see small strain assumptions "stretched" beyond their limitations in a number of applications. It is important that this is thoroughly implanted into your thinking, so that whenever you encounter applications involving casing deformation, for instance, you are aware of the limitations of small strain assumptions. The reason we tend to adopt simplifying assumptions so readily is because those are the only problems we know how to solve easily. But there are many times when these assumptions are adopted inappropriately simply because one does not know any better, and that is dangerous. Equation (7.1) is the equation for infinitesimal strain only.

Stress

Our early concept of stress was most likely that it is a "distributed load" as opposed to a "point load." That is all right for many simple engineering calculations, but it is quite misleading when we advance to more complicated problems. First of all, we need to recognize that, in the real world, there is no such thing as a point load. A point load is a mathematical convenience that exists only in theory and calculations. All real loads are distributed loads. Think about it this way; if we could apply a true point load of 100 lbf to the surface of a steel block, what would happen? If it is truly a point load, the contact area shrinks to zero, the pressure exerted by the load goes to infinity, and the steel block fails at the point of contact. Of course, this does not happen in the real world, because the contact area is not zero, so the load actually is distributed over some area, even though it may be very small. So, even though true point loads do not exist, we use them in our calculations for convenience. Now, back to the distributed load, is a distributed load we typically measure in lbf/in.^{2} or Pa a stress? No, it is not. Just because it has the same units of stress does not make it a stress.

What is a distributed load then? Unfortunately, some call it a *stress vector*. This is a bit of sloppy terminology that has been used for so long it has taken root. If you want to use that term, you have plenty of company, but it can be confusing to call two totally different things stress. A distributed load is a vector, it has magnitude and direction, but stress is never a vector. The proper name for a distributed load is a traction or traction vector, and it is a directional load (force) distributed over some area of contact. Something else worth noting is that you cannot apply stress. You can apply only traction; stress is a result of the traction. But, most important, remember that stress itself is never a vector. Here is an example to illustrate why it is not. Suppose we have a bar in a Cartesian coordinate system (see Figure 7–6) and we apply a traction (distributed load) to the end of the bar of, say, 100 lbf/in.^{2} (or 100 Pa, take your pick) in the x_1-direction; that is, $F_1 = 100$, ($F_2 = F_3 = 0$). We assume that the bar is weightless to keep things simple, and we can intuitively say something like, "the stress in the bar is obviously also 100 (lbf/in.^{2} or Pa)." And that is true; it is a uniaxial stress in the "x_1-direction" in our coordinate system; that is, $\sigma_{11} = 100$.

A block of material at some location in the bar obviously has stress components of $\sigma_{11} = 100$, $\sigma_{22} = \sigma_{33} = 0$. These are principal stress components, and there are no snear components. Even though we have not shown an x_3-coordinate, we include it to illustrate the transformation of vectors and tensors from one coordinate system to another. Now, suppose we change the coordinate system to look at things in a different frame of reference. Figure 7–7 shows the new coordinate system, which is rotated some angle, $30°$, counterclockwise from the original coordinate system.

In this new coordinate system, which we call the x_1', x_2', x_3' coordinate system, we may express the traction vector as

$$F_{1'} = F_1 \cos\theta + F_2 \sin\theta = 100\cos 30 + 0 = 87$$

$$F_{2'} = -F_1 \sin\theta + F_2 \cos\theta = -100\sin 30 + 0 = -50$$

This is a simple vector transformation from one Cartesian coordinate system to another. We could have written it more formally as

$$\begin{Bmatrix} F_{1'} \\ F_{2'} \\ F_{3'} \end{Bmatrix} = \begin{bmatrix} \cos\theta & \sin\theta & 0 \\ -\sin\theta & \cos\theta & 0 \\ 0 & 0 & 1 \end{bmatrix} \begin{Bmatrix} F_1 \\ F_2 \\ F_3 \end{Bmatrix} = \begin{bmatrix} \cos 30 & \sin 30 & 0 \\ -\sin 30 & \cos 30 & 0 \\ 0 & 0 & 1 \end{bmatrix} \begin{Bmatrix} 100 \\ 0 \\ 0 \end{Bmatrix} = \begin{Bmatrix} 87 \\ -50 \\ 0 \end{Bmatrix}$$

Now what about the stress in the bar? Can we transform the uniaxial stress component just as we did the traction vector to give the new stress components on a block of material in our new coordinate system? After all, it is measured in lbf/in.^{2} or Pa and it has only one direction; isn't it just like a vector? In other words, can we now say

$$\sigma_{11'} = 87 \quad ?$$

$$\sigma_{22'} = -50 \quad ?$$

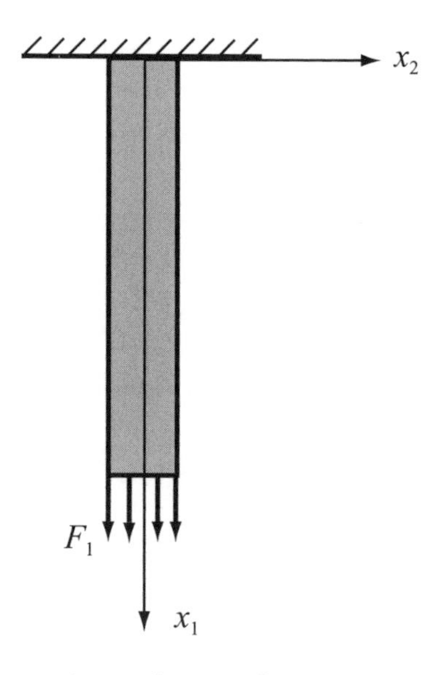

Figure 7–6 *A simple traction and uniaxial stress.*

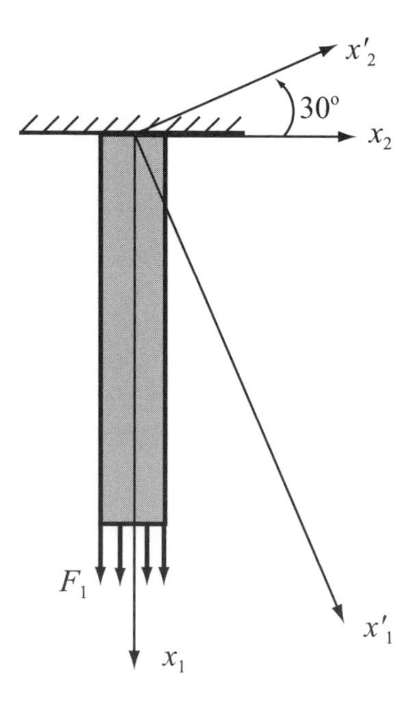

Figure 7–7 *A 30° coordinate rotation.*

Absolutely not! Even though the stress field has only a single component in that original coordinate system, it does not transform as a vector because it is not a vector, it is a second-order tensor. Here is the transform of the stress field done properly:

$$\begin{bmatrix} \sigma_{1'1'} & \sigma_{1'2'} & \sigma_{1'3'} \\ \sigma_{2'1'} & \sigma_{2'2'} & \sigma_{2'3'} \\ \sigma_{3'1'} & \sigma_{3'2'} & \sigma_{3'3'} \end{bmatrix} = \begin{bmatrix} \cos\theta & \sin\theta & 0 \\ -\sin\theta & \cos\theta & 0 \\ 0 & 0 & 1 \end{bmatrix} \begin{bmatrix} \sigma_{11} & \sigma_{12} & \sigma_{13} \\ \sigma_{21} & \sigma_{22} & \sigma_{23} \\ \sigma_{31} & \sigma_{32} & \sigma_{33} \end{bmatrix} \begin{bmatrix} \cos\theta & -\sin\theta & 0 \\ \sin\theta & \cos\theta & 0 \\ 0 & 0 & 1 \end{bmatrix}$$

$$\begin{bmatrix} \sigma_{1'1'} & \sigma_{1'2'} & \sigma_{1'3'} \\ \sigma_{2'1'} & \sigma_{2'2'} & \sigma_{2'3'} \\ \sigma_{3'1'} & \sigma_{3'2'} & \sigma_{3'3'} \end{bmatrix} = \begin{bmatrix} \cos 30 & \sin 30 & 0 \\ -\sin 30 & \cos 30 & 0 \\ 0 & 0 & 1 \end{bmatrix} \begin{bmatrix} 100 & 0 & 0 \\ 0 & 0 & 0 \\ 0 & 0 & 0 \end{bmatrix} \begin{bmatrix} \cos 30 & -\sin 30 & 0 \\ \sin 30 & \cos 30 & 0 \\ 0 & 0 & 1 \end{bmatrix}$$

$$\begin{bmatrix} \sigma_{1'1'} & \sigma_{1'2'} & \sigma_{1'3'} \\ \sigma_{2'1'} & \sigma_{2'2'} & \sigma_{2'3'} \\ \sigma_{3'1'} & \sigma_{3'2'} & \sigma_{3'3'} \end{bmatrix} = \begin{bmatrix} 75 & -43 & 0 \\ -43 & 25 & 0 \\ 0 & 0 & 0 \end{bmatrix}$$

You can see now that the stress components in the new coordinate system are:

$$\sigma_{1'1'} = 75$$

$$\sigma_{2'2'} = 25$$

$$\sigma_{1'2'} = \sigma_{2'1'} = -43$$

This is nothing at all like a vector. For this reason, we should never use terminology that confuses stress with vector quantities. If you are not familiar with these coordinate transforms, it is not necessary that you learn it to understand this section. Essentially, the transformation of a vector goes like this:

$$u_{i'} = a_{i'j}\, u_j$$

which is exactly what we did when we transformed the traction vector. The components of the transformation $a_{i'j}$ are equal to the cosine of the angles between the coordinates of one coordinate system with the other. For example, $a_{1'1}$ is the cosine of the angle between x_1 and $x_{1'}$, which is $\theta = 30$ in this case, and $a_{2'1}$ is the cosine of the angle between x_1 and $x_{2'}$, which is $\theta + 90 = 120$, but rather than add angles, it is easier to just use θ so that $\cos(\theta + 90) = -\sin\theta$ is used instead. Now, a second-order tensor transforms as

$$\sigma_{ij} = a_{ik} a_{jl} \sigma_{kl}$$

where the primes have been omitted as being understood to be on the *i* and *j* subscripts. If you are interested, the references at the end of the chapter provide detailed explanation.

We might also mention that the actual stress at a point in a material does not depend on our selection of any particular coordinate system, even though its component values depend on which coordinate system we select. In the previous example, we saw that the individual stress components have differing values in different coordinate systems, but the actual stress at that point in the material is not changed by our selection of coordinate systems. When we refer to the stress at a number of points, we typically refer to a stress field (like we refer to a magnetic field).

It is not completely obvious from our example, but the stress tensor is symmetric, in that $\sigma_{ij} = \sigma_{ji}$. This is true except in the case of body moments (such as a magnetic field), which do not appear in our considerations. One way to check your work when you transform a symmetric tensor, such as stress, is that it always is symmetric in any orthogonal coordinate system.

We said that stress is a tensor. What is a tensor? That is a good question for which there is not a good answer. Or at least, there is no good answer that would ever satisfy an engineer. A tensor is a mathematical quantity that transforms according to certain rules (which were illustrated previously, but we lack space to explain here). Also, when we speak of a tensor, we typically are talking about a second-order tensor, such as stress or strain. There are other tensors of different orders as well. Table 7–1 shows some.

By these definitions, even a scalar is a tensor (of zero order), a vector is a tensor (of first order), so you can see that we have even become a bit sloppy when we commonly use the term *tensor* to mean a second-order tensor.

Table 7–1 Example Tensors of Different Orders

Tensor Order	Common Name	Some Examples
0	Scalar	Distance, s
		Temperature, T
		Speed, $\dot{s}$
		Bank balance, $
1	Vector	Force, f_i
		Velocity, v_i
		Acceleration, $\dot{v}_i$
		Traction, t_i
2	Tensor	Stress, σ_{ij}
		Strain, ε_{ij}
3	—	Permutation operator, ϵ_{ijk}
4	—	Elastic constant, C_{ijkl}

One final point before we move on. It is misleading to assume that we can always determine a uniaxial stress by dividing a uniaxial load by the cross-sectional area of the material body. This is true only for a prismatic bar, that is, one with a constant cross section, but it is not true near the ends of the bar, where the loads are applied. We cannot determine the axial stress in a tube under a thread cut into the tube by dividing the axial load by the cross-sectional area under the thread. Likewise, we cannot calculate the uniaxial stress in a coupling, connection, or upset in a tube in a similar fashion. Whenever the cross-sectional area of a tube changes, the stress field at the change and in the near vicinity of the change is more complicated than a single uniaxial component. In the case of a connection, additional complication comes from the addition of tangential, radial, and shear stress components due to the connection itself. Saint-Venant's principle, however, effectively states that, at some distance away from the ends of a long tube (or point of change in diameter), the stress field becomes uniaxial, and there we are safe in dividing the cross-sectional area by the axial load to get the uniaxial stress component. For all practical purposes, that distance is relatively short in oil-field casing, but in a strict interpretation of the theory, it is valid only for tubes of infinite length.

Stress Invariants

Some things about a stress tensor are invariant no matter how we may rotate our coordinate system. These are called *stress invariants*, and three are associated with a symmetric stress tensor:

$$I_1 = \sigma_{ii}$$

$$I_2 = \frac{1}{2}\left[(I_1)^2 - \sigma_{ij}\sigma_{ji}\right]$$

$$I_3 = \frac{1}{3}\left[3I_1I_2 - (I_1)^3 + \sigma_{ij}\sigma_{jk}\sigma_{ki}\right] \tag{7.4}$$

You might well ask, of what use are those three invariants other than to check my arithmetic? There are a number of times when those are useful, but one use that is of importance is to find the three principal stresses. We can expand the determinant in the following equation to get a cubic equation, from which we can solve for the three principal stress components:

$$\det\left(\sigma_{ji} - \sigma\delta_{ji}\right) = 0 \tag{7.5}$$

where σ is a principal stress component and δ_{ji} is the Kronecker delta. If we go to the trouble to expand that determinant, we get a characteristic equation whose coefficients are the three stress invariants:

$$-\sigma^3 + I_1\sigma^2 - I_2\sigma + I_3 = 0 \tag{7.6}$$

When the stress tensor is symmetric, that cubic equation has three real roots, which are the three principal stress components. Also, any cubic equation always has a closed form solution, meaning there is a formula we can use to find the three principal stress components. To throw in a couple of other terms here, we can say that, from equation (7.5), σ (without indices) represents an eigenvalue whose three solutions (eigenvalues) in this equation are the numerical values of the three principal stress components. Their directions (which must be determined by other methods) are eigenvectors. If that makes sense to you, fine; if not, do not worry about it because we are not going to use those terms. (Refer to any text on linear algebra or one of the mechanics references at the end of this chapter if you want to learn more.)

Deviatoric Stress

To understand a yield stress and plastic material behavior, it is necessary to learn about one other type of stress, the deviatoric stress. The stress tensor may be decomposed into a spherical (or hydrostatic) stress and a deviatoric stress. The spherical stress is that part of the stress tensor that is basically equal in all directions, that is, just like hydrostatic pressure. The deviatoric stress is what is left after the spherical stress is taken out. One way of thinking about it is that the spherical stress might be said to be the part of the stress tensor trying to compress a material body (or pull it apart) uniformly in all directions and the deviatoric part of the stress is what attempts to distort its shape.

In terms of the three principal stresses the spherical stress is

$$\sigma_{\text{spherical}} = \frac{\sigma_1 + \sigma_2 + \sigma_3}{3} \tag{7.7}$$

We could then calculate the three principal deviatoric stress components by subtracting the spherical stress from each principal stress component:

$$\sigma_1' = \sigma_1 - \frac{\sigma_1 + \sigma_2 + \sigma_3}{3}$$

$$\sigma_2' = \sigma_2 - \frac{\sigma_1 + \sigma_2 + \sigma_3}{3}$$

$$\sigma_3' = \sigma_3 - \frac{\sigma_1 + \sigma_2 + \sigma_3}{3} \tag{7.8}$$

If we do not have the principal stress components, we can calculate the deviatoric stress from the stress tensor components as

$$\sigma'_{ij} = \sigma_{ij} - \delta_{ij} \frac{\sigma_{kk}}{3} \tag{7.9}$$

There are several things to note here. The off-diagonal components of the deviatoric stress tensor are the same as the regular stress tensor. The only components that are changed are the ones on the diagonal. Each of those has subtracted from it one third of the sum of the diagonal components, which is $I_1/3$. Now, the deviatoric stress also has invariants, and one of these is extremely important:

$$I'_1 = \sigma_{ii} = 0$$

$$I'_2 = \frac{1}{2}\left[(I'_1)^2 - \sigma'_{ij}\sigma'_{ji}\right] = -\frac{1}{2}\sigma'_{ij}\sigma'_{ji}$$

$$I'_3 = \frac{1}{3}\left[3(I'_1)(I'_2) - (I'_1)^2 + \sigma'_{ij}\sigma'_{jk}\sigma'_{ki}\right] = \frac{1}{3}\sigma'_{ij}\sigma'_{jk}\sigma'_{ki} \tag{7.10}$$

Rather than use the same notation as the regular stress invariants, it is customary to define these deviatoric invariants as follows:

$$J_1 \equiv I'_1 = 0$$

$$J_2 \equiv -I'_2 = \frac{1}{2}\sigma'_{ij}\sigma'_{ji}$$

$$J_3 \equiv I'_3 = \frac{1}{3}\sigma'_{ij}\sigma'_{jk}\sigma'_{ki} \tag{7.11}$$

The important invariant here is J_2, which we use later when we discuss yield stress. And, like before, we can use these invariants to find the three principal deviatoric stress components by solving the cubic equation:

$$(\sigma')^3 + J_2\sigma' + J_3 = 0 \tag{7.12}$$

This is similar to equation (7.6) except for that J_1 is zero and does not appear. We also show an example of the deviatoric stress when we talk about yield stress.

7.3.4 Sign Convention

Perhaps it is a bit remiss to leave this until the end of the section on continuum mechanics, but it is quite important. We showed no derivations of much of what was discussed in this section; otherwise, it would be apparent that tensile stress is positive in sign and compressive stress is negative. That is hardly worth mentioning, except that hydrostatic pressure actually is a negative stress, and we are not accustomed to thinking of it that way. Pressure is so universally accepted as a positive quantity that, most of the time, there is a builtin "workaround" just to accommodate that convention, such as $-p\delta_{ij}$ as in $\sigma_{ij} = -p\delta_{ij}$ In rock mechanics, an opposite sign convention is convenient, since most of the rocks in the earth's crust are in compression, and putting a negative sign in front of every stress value could be a significant nuisance. That is well and good, and you may be wondering what difference it makes as long as we understand what convention we are using. Well, it does make a difference when you get deeper into the computational aspects of mechanics. If you take that all the way back to the basic kinematics of continuum mechanics, you will find that an opposite sign convention leads to motion in the opposite direction from the arrow on a velocity vector, for instance. It is one thing to assume an opposite sign convention in simple calculations but quite another to assume that motion is in the opposite direction of velocity vectors. When it comes to serious computation, almost no one uses odd sign conventions. Tension is positive, compression is negative.

7.4 Material Behavior

Suppose we have a solid cube of some material that measures 1 m on each edge, and it is lying on a flat surface. We know the weight of the cube. We apply a downward force of a known magnitude on top of that object. If we are asked, "What is the force on the flat surface?" we have an easy problem. It is Newton's third law, and the answer is the force on the surface is the weight of the cube plus the force we applied to the top of the cube. If we approach it slightly differently, and instead of measuring the force we apply on top, we measure how much we actually compress that cube vertically, by say 0.2% of its original height (see Figure 7–8). We now have a different problem in determining the resultant force on the flat surface.

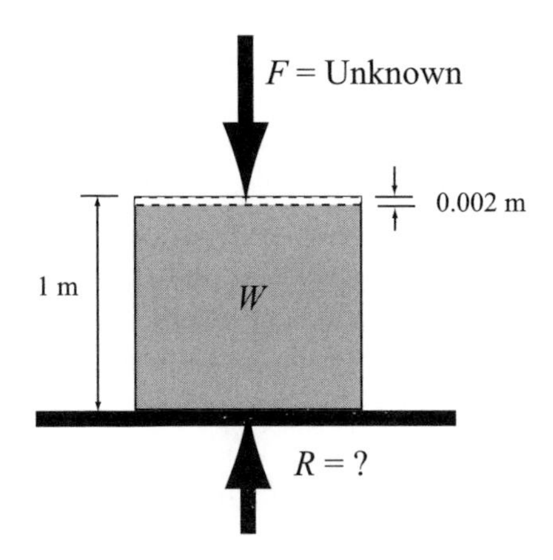

Figure 7–8 *Force and reaction.*

We know that it is equal to the weight of the cube plus whatever force is necessary to compress the cube by 0.2%. To solve this problem, we need to know something about the material of which the cube is made, and more specifically, how the material responds to a compressive load. Such a relationship is called a *constitutive equation*, and in this case, it is a simple one-dimensional version of Hooke's law:

$$\sigma_z = E\frac{\mathrm{d}u}{\mathrm{d}z}$$

This simple relationship relates a load (stress), to a deformation gradient (strain), by means of a constant, Young's modulus, which is a property of the material. Many constitutive relationships are in every day use, although we do not often think of them as such. Here are two more one-dimensional examples.

$$q = -k\frac{\mathrm{d}T}{\mathrm{d}x}$$

$$v = \frac{k}{\mu}\frac{\mathrm{d}p}{\mathrm{d}x}$$

The first is Fourier's law of heat conduction, which relates heat flux to a temperature gradient by means of a material property, the conductivity.

The second is the Darcy flow, which relates fluid flux to a pressure gradient with two material properties, permeability and viscosity.

There are many types of material behavior, for example, elastic, plastic, viscoelastic, viscoplastic. Elastic behavior could be subdivided into linear elastic, nonlinear elastic, hyperelastic, and viscoelastic. If one were to apply a load to a material and plot the load curve (load versus deformation) such as a stress-strain curve, one could not really understand much about a material's behavior. Only when the load is removed can we begin to understand its behavior. For example, look at the material load curve in Figure 7–9. What type of behavior is this? We might be tempted to say it is elastic-plastic, since it looks like what we often see to illustrate how a metal behaves elastically up to a yield point and then becomes plastic beyond the yield point. But, in truth, all we can say for this example without more information is that its behavior is nonlinear.

Now let us reveal the unloading behavior on that same material, Figure 7–10. We see that it returns to the same point where we started. So, when the load is removed, it has no permanent deformation. It is elastic.

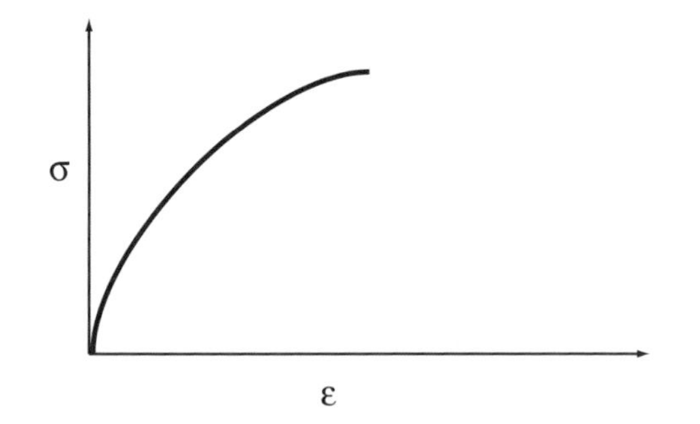

Figure 7–9 *Nonlinear material behavior.*

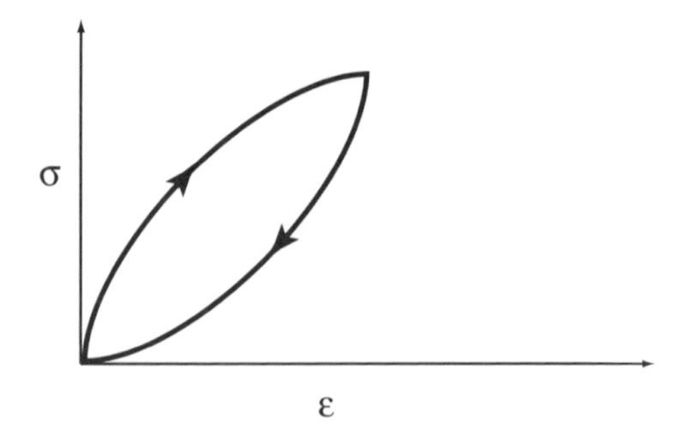

Figure 7–10 *Nonlinear loading and unloading curve.*

The defining criteria for elastic behavior is that, if a body is subjected to a load and the load is later removed, it will return to its original state. But we notice something else about this material load curve. It did not return by the same path with which it was loaded. What could explain that? Time. It is a rate-dependent behavior; in other words, it is viscoelastic. The shape of this load curve depends on the loading and unloading rate. The loading and unloading curve does not tell us the rate at which the load was applied or removed. For example, we could load the same material very slowly and remove the load, also very slowly. We might get a loading and unloading curve that would look like Figure 7–11. If we were not aware of the previous curves, we might think it is a linear elastic material. And, it is, albeit at very slow rates of loading. Many materials that we might consider linear elastic actually are rate-dependent materials. Formation rock in most wells is an example.

Appearances can be deceiving. We are not going to concern ourselves with rate-dependent materials here, but the point is important, do not jump to conclusions that are not justified.

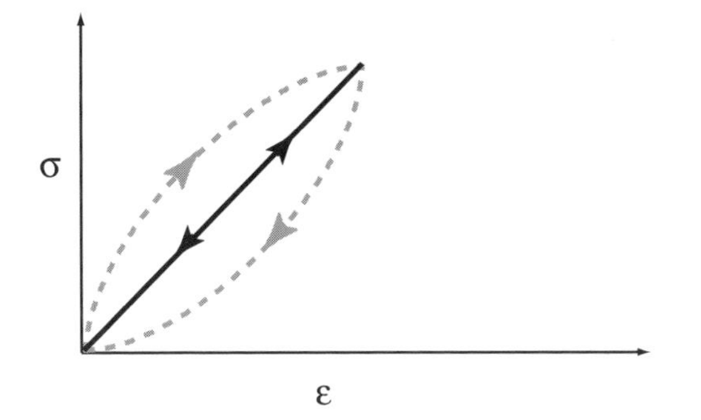

Figure 7–11 *Slow loading and unloading of a rate-dependent material. Linear elastic?*

7.4.1 Elasticity

We said that an elastic material is one in which the material returns to its original state after an applied load is removed. A linear elastic material[6] is one whose loading path and unloading path are the same straight line. The constitutive behavior of a linear elastic material is modeled by Hooke's law,

6. We should note that, when we refer to an elastic material, we are speaking to only a particular range of a particular material's behavior. Materials may exhibit different types of behavior in certain load ranges.

as mentioned before. It is called a *law*, but it is not a law of physics or nature. It is just a convenient relationship that models the behavior of certain materials in a very limited range of deformation. We might even say that we are very fortunate that it does, even though its range is quite small, because almost every structure or machine we come into contact with daily is designed using this constitutive relationship. Written in three-dimensional tensor form it is

$$\sigma_{ij} = C_{ijkl}\,\varepsilon_{kl} \tag{7.13}$$

If we were to write that out in matrix form, the stress and strain tensors would each contain 9 terms and the elastic modulus would contain 81 terms. Without considering body moments, all of them are symmetric. For an isotropic material, one whose material properties are the same in all directions, equation (7.13) can be simplified considerably. We can use a contracted notation (called *Voigt notation*) and write the stress and strain in vector form and the elastic modulus as a 6 × 6 matrix, and even include the thermoelastic terms:

$$\begin{Bmatrix} \sigma_{11} \\ \sigma_{22} \\ \sigma_{33} \\ \sigma_{23} \\ \sigma_{13} \\ \sigma_{12} \end{Bmatrix} = \frac{E}{(1+\nu)(1-2\nu)} \begin{bmatrix} 1-\nu & \nu & \nu & 0 & 0 & 0 \\ \nu & 1-\nu & \nu & 0 & 0 & 0 \\ \nu & \nu & 1-\nu & 0 & 0 & 0 \\ 0 & 0 & 0 & \dfrac{1-2\nu}{2} & 0 & 0 \\ 0 & 0 & 0 & 0 & \dfrac{1-2\nu}{2} & 0 \\ 0 & 0 & 0 & 0 & 0 & \dfrac{1-2\nu}{2} \end{bmatrix} \begin{Bmatrix} \varepsilon_{11} \\ \varepsilon_{22} \\ \varepsilon_{33} \\ \varepsilon_{23} \\ \varepsilon_{13} \\ \varepsilon_{12} \end{Bmatrix} - \frac{E\alpha\Delta T}{1-2\nu} \begin{Bmatrix} 1 \\ 1 \\ 1 \\ 0 \\ 0 \\ 0 \end{Bmatrix} \tag{7.14}$$

where we have now included Poisson's ratio, v, and the coefficient of thermal expansion, α . This is a bit more complicated than the simple one-dimensional form, but often a number of simplifying assumptions, such as plane strain or plane stress, can be adopted to reduce it to two dimensions for some problems. We are not going to use this constitutive relationship, but merely show it so that you can see what it looks like.

7.4.2 Plasticity

We defined an elastic material as one in which, when an applied load is removed, the material goes back to its original state. A material is said to behave plastically when an applied load is removed and it does not go back to its original state. In other words, it has undergone some permanent deformation in the loading process. Plasticity is a complex topic and there are exceptions to just about everything, but we are concerned primarily with steel in casing here, so we are going to confine our discussion to that limited scope. Steel behaves more or less as a linear elastic material up to a point, called the *yield point*. When loaded beyond that point, its behavior is said to be plastic, but the elasticity has not disappeared, it is still exhibited when the loads are removed. Figure 7–12 shows a load curve for elastic-plastic behavior.

The loading in this curve is such that the sample is linear elastic up to its yield point, Y, then deforms plastically until the load is removed, at which time, it unloads elastically. The difference between the initial strain (zero here) and the final strain is the permanent deformation of the material in a plastic mode. It is of significance at this time to make mental note of this elastic unloading. When a metal, such as steel, is deformed plastically to a certain size or shape, it will always display some amount of elastic "rebound" or deformation when the load is removed. When coiled tubing is bent plastically onto a reel, a lot of elastic energy is stored on that reel, and if you were to release the end of the tube, you would witness an amount of elastic unloading you might not otherwise imagine.

In many cases, the behavior of the metal at the yield point is somewhat more complicated than that in Figure 7–12. Sometimes the elastic behavior becomes nonlinear before the yield stress is reached, and that point is called the *proportional limit*. Ductile steels often exhibit this behavior and materials like cast iron seldom exhibit a distinctive yield point. There are other cases where the stress actually decreases slightly after the yield stress is reached; hence, there is an upper yield stress and a lower yield stress. This is typical of some steel alloys. In many cases, the yield point is indistinct on a stress-strain chart, so the yield point is

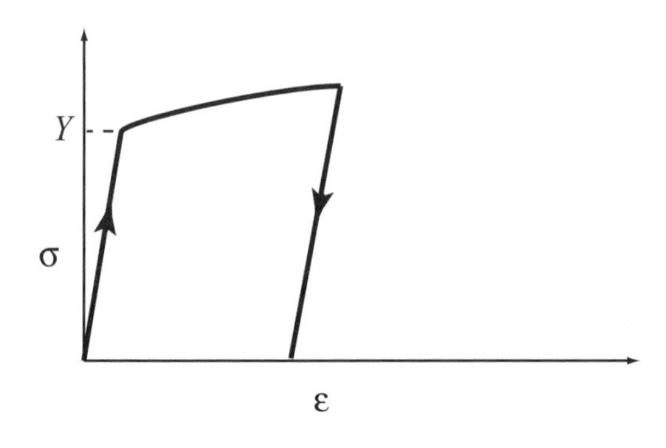

Figure 7–12 *Elastic-plastic behavior of steel with a uniaxial load.*

defined as some arbitrary point offset from the proportional limit by a specified amount of strain (API does this). We do not concern ourselves with those details, but rather assume that, as long as we do not exceed the published yield stress for the casing material, its behavior is linear and elastic to that point.

What happens if we continue to load the sample? Figure 7–13 shows the uniaxial loading of a sample all the way to failure. At this point, we need to explain a few things. First of all, in this figure, we are looking at a uniaxial load curve. It is plotted as stress versus strain, which we already discussed but we need to understand a bit more.

The samples of material used in these tests are relatively small compared to the massive machine in which they are tested. The samples usually look like the one shown in Figure 7–14, sometimes colloquially referred to as a *dog bone* sample. The sample is placed in a large machine that pulls it in tension, as the load and stretch are recorded.

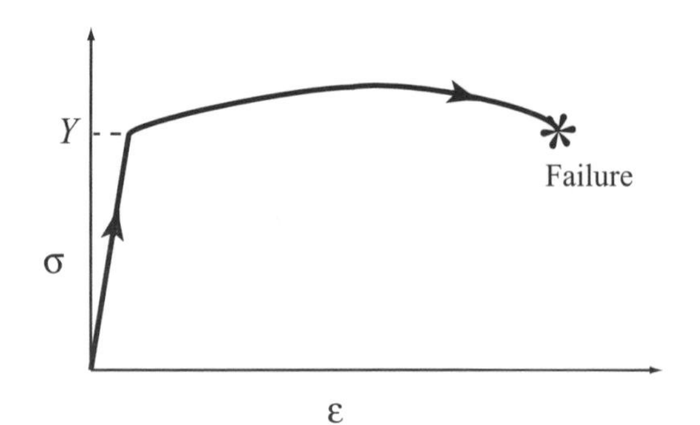

Figure 7–13 *A uniaxial load curve to failure.*

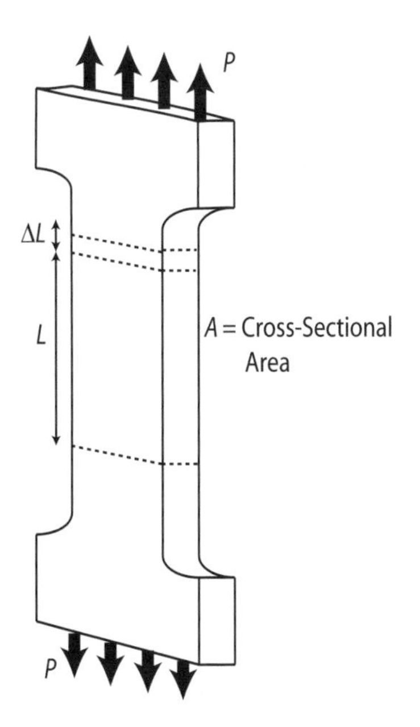

Figure 7–14 *A test sample of steel.*

The ends are larger than the test portion, so the gripping effects of the machine do not affect the stress in the thinner test portion. The cross-sectional area of the test portion is measured accurately before the test begins. The stretch in the sample is measured by the machine as a displacement, and the load is measured with a pressure transducer, These are recorded on a chart similar to that shown in Figure 7–13. The machine itself is massive in relation to the size of the sample, but nevertheless it has some amount of elastic stretch while pulling tension, as does the wider portion of the sample, so usually an electronic strain gauge is attached to the sample in the thinner area to measure the stretch. The results may be plotted as load versus stretch, which is the raw data from the machine, but that is not of much use unless one is testing a particular structural element or part as opposed to a sample of material like we are considering here. We use the following relationships to get the values for a stress-strain plot:

$$\sigma = \frac{F}{A} \qquad e = \frac{\ell - \ell_o}{\ell_o} = \frac{\Delta\ell}{\ell_o}$$

where the stress is the machine-measured load divided by the cross-sectional area of the sample, and the strain is the stretch in the sample section divided by the original length of the sample section. These are plotted as in Figure 7–13. Similar tests may be run in compression or cyclical loading.

In looking at Figure 7–13, we see that the material deforms elastically up to the yield point, then begins its plastic deformation, in which a lot of strain takes place with little increase in stress, yet the stress continues to increase up to a point, where it begins to decline until the sample fails. The part of the curve where the stress continues to increase in the plastic range is called *strain hardening*. But what about the part of the curve where it starts to decline. Is this "strain softening" then? This part of the curve can be misleading because of the way we defined the stress and strain. Our definition holds only for very small deformations. If we actually measured the cross-sectional area of the sample as it stretches, we would find that the area is getting smaller. So, what we have plotted on the vertical axis is not the true stress, but rather the load divided by the original cross-sectional area of the sample, which is commonly referred to as the *engineering stress*. And, if we looked at the failed sample, we would see that the most significant decrease in cross-sectional area occurred in only a small part of the length of the sample and the cross-sectional area at the point of failure is possibly less than half the original area. If we were to plot the true stress, we would find that, in many cases, it continues to increase right up to the point of failure; and for some materials, the increase is rather drastic just before failure. We might also note that our measure of strain is no longer valid either, since the reduction in cross section (called *necking-down*) is quite localized, so that the uniaxial strain in the region of failure is apparently a lot greater than elsewhere along the sample length. The true strain is not $\Delta\ell/\ell_o$ but rather $d\ell/\ell$, which leads to a true uniaxial strain measure:

$$\varepsilon = \ln\left(\frac{\ell}{\ell_o}\right) = \ln\left(1 + \frac{\ell - \ell_o}{\ell_o}\right) \tag{7.15}$$

We might also note that the appearance of the sample itself began to change just before failure, in that visible bands of discoloration or surface texture began to appear on the surface of the sample where the area reduction was most pronounced, giving the appearance of some change in the metal itself. These are called *Lüder's bands*, and that is exactly what they indicate. Two things are of importance here. If we tested several samples of the same material, all would fail at different values. That is the nature of materials; we cannot predict the actual load value at which the material fails. We can determine a range of values, but we cannot predict the exact value for any particular sample. If all of the samples were cut in precisely the same way under the same conditions and we could perform rigorous inspections on them, we could get pretty close, but the point is we could not predict the exact failure strength. The second point is this. We said that, if we used a measure of true stress and strain, then we could say that the material never got "weaker" before it failed, and that is often true. However, the sample itself got weaker before it failed, and that is important. In terms of total load, it failed at a load less than the maximum load it was subjected to in the test. If it were a structural member or casing, that would be important. So, while the true stress is important, the load on a structure does not know to reduce itself when the cross-sectional area of a structural member is reduced due to plastic deformation. We never want to go beyond that maximum load value, no matter how we measure stress or strain.

Now, let us go back to the issue of strain hardening mentioned earlier. Strain hardening in a metal, for the most part, is caused by defects in its crystalline structure, called *dislocations*. All steels have some amount of this type of defect. But *defect* might be a severe term to use in an oil-field context, so perhaps a dislocation should be thought of as an imperfection in the crystalline lattice of a metal. A dislocation is a missing bond at a lattice junction. When a metal is stressed beyond its initial yield point, these dislocations begin to move or migrate. No material physically moves (at least, on an observable scale), but where a bond is missing at a lattice junction, a bond forms, and the missing bond is transferred to the next junction, sort of like the game of musical chairs, where there is one fewer chair than the number of people present. When one person gets up, another takes his or her place. What causes the strain-hardening effect is that as these dislocations begin to accumulate at grain boundaries, there is no other place for them to go, and they begin to resist the deformation of the material. There are other contributing factors, but that is the main one. While most structural steels are strain hardening materials to some degree, some are not. Some brittle steels exhibit very little strain hardening, and failure strength is very close to the yield strength. Some soft steels behave more like what is known as an *elastic–perfectly-plastic material*, in that once yield stress is reached, the material continues to elongate to failure with no additional stress required. The steel used for coiled tubing closely approximates this latter behavior. Also some strain-softening materials are such that, once yield is reached, continued elongation requires less and less stress.

For a strain-hardening type of steel, if we stop loading a sample before we reach the maximum and remove the load, we know that it unloads elastically as shown in Figure 7–12. What happens if apply a load again? The answer is that it goes right back up the same path as the unloading path. Furthermore, it does not yield until it reaches the value at which the load was removed before. It may vary with a bit of hysteresis and become slightly nonlinear before yielding again, but after yield, it continues on the same path, as if the unloading never took place. The result, though, is that, on re-loading, it actually yields at a higher value than the original yield value at the start of the test. We could repeat this process any number of times, and each time, we increase the yield value until we reach the top of the curve. So, for a strain-hardening material, we can increase the yield strength by cold working it. Or can we? It all depends on what our application is. Let us look at compression.

If we were to test an identical sample in compression, it would be a mirror image (and upside down) of the one we just showed. And we could compress it to the same point and increase its yield strength in compression just as we did in tension. The question is, then, if we work our sample in tension first then test it in compression, will the yield point in compression be increased just like the one in tension? The answer is no, it would not. In fact, if we increased the yield in tension, then the yield in compression actually would be less than the initial yield. This is called the *Bauschinger effect*. In other words, we cannot have it both ways. Is the amount of reduction in compression equal to the amount of increase in tension? Again the answer is no, the reduction in the compressive yield stress generally is less than the increase in tensile yield stress. We can show this with an ellipse, which we call a *yield surface* for now. We are not plotting strain now but principal stress components in two directions. Up is tension, down is compression. We do not concern ourselves with the transverse component for now. We know that, when we increase the tensile load, we increase the yield stress in that direction. One way to think about it would be that the yield surface gets larger. In other words, it expands by the amount we go beyond the initial yield stress, as shown in Figure 7–15. This is known as *isotropic hardening*; the yield surface grows uniformly in all directions.

This means that we also increased the yield strength in compression by the same amount; and we already said that does not happen. Isotropic hardening does not happen, but it is useful in a few applications when repeated

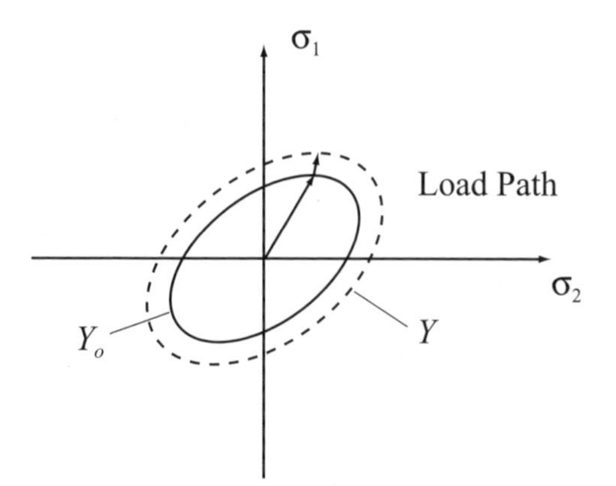

Figure 7–15 *Isotropic hardening, the yield surface grows uniformly.*

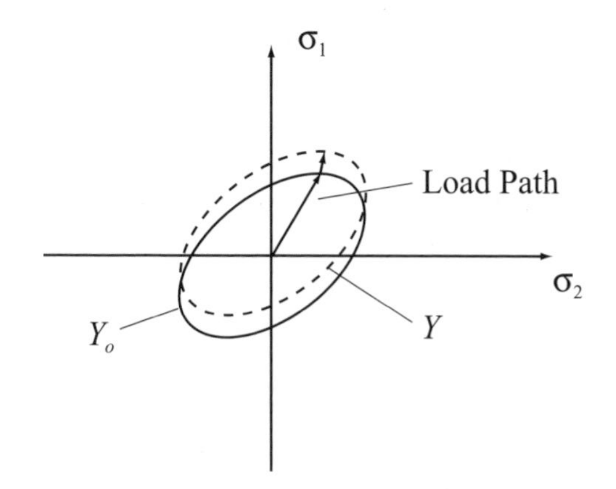

Figure 7–16 *Kinematic hardening.*

loading does not cause yielding in different directions. Another possibility might be that the yield surface actually moves as shown in Figure 7–16. Here, the yield surface stays the same size, but it moves as the stress exceeds the yield stress. This model is called *kinematic hardening*.

For kinematic hardening in the simple uniaxial case we are discussing, the reduction in yield stress in compression is of exactly the same magnitude as the increase in the tensile yield stress. We already said that does not happen either. What actually takes place is something in between. The yield surface grows, but it also moves. This is called *combined hardening*. And what all this amounts to is this: When working in the plastic regime, we have to keep up with the growth of the yield surface and keep track of where it is. This usually is done with something called *internal state variables*, which are defined by a flow rule to account for the translation of the yield surface and a growth law that accounts for its expansion or hardening. In the simple incremental theory of plasticity, one internal state variable is a second-order tensor that tracks the translation of the yield surface, and the other is a scalar that keeps track of the size of the yield surface. Take into account also that we are not talking about the three-dimensional space we live in, but rather a nine-dimensional stress space. Another point of complication is that, when the stress is on the yield surface (it cannot go past it), the most common plasticity theory requires that plastic strain can take place only in a direction normal to this nine-dimensional hypersurface. In some plasticity theories, the yield surface is not regular and smooth, like the ellipse we illustrated, and loading paths also may change shape in addition to size and location. Additionally, in larger deformations such as the necking discussed in the uniaxial test previously mentioned, localized rotations within the material begin to occur. These rotations must be tracked, too. At some point, we begin to kid ourselves as to what we know how to do in this strange space. So we are going to drop back a notch or two for now and think about how to stay out of it. Obviously, if we stay inside that yield surface with our stress and do not bump into it, we should stay out of serious trouble. That is the topic of the next section, but let us look at one more scenario first.

One of the most difficult aspects of materials that have been loaded beyond the initial yield point is predicting their properties at a later time, because the loading beyond initial yield makes it a history-dependent material. You cannot predict its future behavior unless you know the history of its loading. For example, suppose we take two identical samples, both with an initial yield strength of 80,000 lbf/in.^{2}. We subject them both to a uniaxial load of 90,000 lbf/in.^{2}. In other words, we work hardened them in tension to increase the yield strength (in tension). We give the two samples to Alice and Bob, who are engineers for a company looking for higher-strength casing for some special high-pressure applications. We tell them we developed a process for increasing the yield of N-80 material to 90,000 lbf/in.^{2} yield strength, and we can sell them this new material at a cost only slightly above the cost of N-80. Bob takes his sample to a lab to verify our claim. He subjects it to a uniaxial tensile test and finds that it indeed has a yield strength of 90,000 lbf/in.^{2}, as we claimed. He is impressed. But Alice, also skeptical, has an application where the casing will be in compression due to thermal loading while producing hot gas from a deep well. Consequently, she asks the lab to test her sample in compression. She finds that the yield is not 90,000 lbf/in.^{2}, as we claimed, nor is it even 80,000 lbf/in.^{2}. In fact, her lab tests show the yield in compression is only 78,000 lbf/in.^{2}, which is lower than N-80. She thinks we are frauds.

That example was for a simple uniaxial test. If we had subjected a large sheet of steel to the same uniaxial stress then cut samples from it in various directions, the results would have been even more alarming. The point is that, when a metal is loaded beyond the initial yield point, we cannot possibly predict its behavior, unless we know the history of the loading. And, that not only means the exact loads but also the exact sequence. One oil-field example would be a well like the one with which Alice was concerned. If, in compression due to thermal expansion, as in Alice's well, suppose a portion of the production casing string yields. Take into account that, during the heated stage, the packer fluid also is heated and expands, causing a high internal pressure in the casing. Assume that the casing actually yields at the inner wall with a combined axial compressive stress, a radial compressive stress, and a tangential tensile stress, but because of strain hardening, it does not actually fail in this particular case. Then, the well is shut in for a few days and cools to where that portion of the casing is now in tension and even has a net differential pressure from the outside. Now, the state of stress at the inner wall is axial tension, radial compression, and tangential compression. What is the yield strength of the casing under that load? All we can honestly say is that it is different than it was before the yield occurred and probably less, but without knowing the exact history of the loading, we cannot predict the yield in the current state. We may be able to get close enough for "oil-field use" with some assumptions—or we may not.

The history dependence and the changes in yield are the primary reasons we try to stay out of the plastic regime in most engineering design. We can do fairly well with one-time loading, but when the loading is cyclical and even varies with the cycles, it gets extremely difficult to get any meaningful results. John Bell, who did significant experimental work with large deformation plasticity at Johns Hopkins University, once made the comment that, if you subject a material to varying loads beyond the initial yield point, it may become impossible to even find the yield point experimentally. What that translates to is that incremental plasticity theory works much better as a mathematical concept than it does with real materials.

7.5 Yield Criteria

A yield criterion is a collection of assumptions and formulas that define the limit of elastic behavior under any possible state of stress. We call that limit a *yield surface*, although it is a surface in a mathematical sense only. There have been many yield criteria over the years, but two have proven quite successful, especially for metals such as steel. The oldest of these is that of Tresca, dating back to 1864. It is a piecewise linear model that in cross section is a hexagon. It found a lot of use before computers became available for computations and still enjoys some use because it lends itself to a number of closed-form solutions that otherwise require a computer and numerical solutions. The other yield criterion that is most used and best models materials like steel for small deformations is the one widely attributed to von Mises in 1913, who developed it from theoretical concepts. The idea behind this yield criterion was first mentioned by James Clerc Maxwell in 1858 but never published except in his private letters and again by M.T. Huber in 1905. Huber's work was published, but it went unnoticed because it was published in a journal that was not widely read outside his home country. About 11 years after von Mises's publication, Hencky did some work with plasticity, especially in the range beyond the yield point, and his name became associated with it too. So, variously, you will hear it referred to as the *von Mises yield criterion, the Maxwell-Huber-Mises yield criterion, the Huber-Mises yield criterion*, or the *Hencky-Mises yield criterion*. That should give fair due to all those who contributed their skills to the idea, but we call it the *von Mises yield criterion* for the sake of brevity. The von Mises yield criterion is circular in cross section, which makes more sense than the Tresca criterion, but it does not lend itself well to many closed-form solutions of elastic-plastic problems.

The von Mises yield criterion may be stated mathematically as

$$Y \geq \sqrt{3J_2} \tag{7.16}$$

where Y is the uniaxial yield strength of the material and J_2 is the second deviatoric stress invariant, defined previously. The yield condition is

$$\begin{aligned} Y &> \sqrt{3J_2} \quad \rightarrow \quad \text{no yield} \\ Y &= \sqrt{3J_2} \quad \rightarrow \quad \text{yield} \\ Y &< \sqrt{3J_2} \quad \rightarrow \quad \text{not possible} \end{aligned} \tag{7.17}$$

The first two conditions are easily understood, but the third begs comment. This means that the combined stress combination, $\sqrt{3J_2}$, can never be greater than the yield, because the yield always changes so that that value never goes beyond the yield surface. So Y in this context refers to the actual yield and not the initial yield point.

Now, comes a bit of semantics. Suppose we calculate J_2 from our load data on our casing at some point, plug it into the formula, and find that it is greater than the published yield strength of our casing. In that case, we can say our load is such that the initial yield would be exceeded. But this is important to understand, it does not tell us by how much. If we used elasticity formulas to calculate the stress in the pipe, and the resulting value of $\sqrt{3J_2}$ is greater than the yield strength of the pipe, then the values we calculated for the radial and tangential stress and axial stress are purely fictitious.

It is time to discuss the quantity $\sqrt{3J_2}$. What is it? It has the units of stress, traction, pressure—is it any of those things? No, it is not. It is a scalar quantity with the same units as stress, or traction, or pressure. It is definitely not a stress. You might well ask, then, how can we compare a scalar to the uniaxial yield stress and get anything meaningful? The truth is we are not comparing it to the uniaxial stress, we are comparing it to the value of $\sqrt{3J_2}$ in which $\sigma_{11} = Y$ and all other stress components are zero. We, in fact, are comparing it to another scalar. In continuum mechanics, $\sqrt{3J_2}$ may be written in several different forms and it is never referred to by any particular name. In the petroleum industry, it has come to be known as the *von Mises stress*. But, as we have stated, it is not a stress. And, in particular, if it is greater than the initial yield strength of the pipe, then it is purely fictitious as a physical quantity if we calculated it from elastic assumptions. If you want to call it the *von Misses stress*, the *equivalent von Mises stress*, or the *fictitious stress*, go right ahead, as long as you understand what it is and what it is not. Here, we adopt a way to use it that should not violate anyone's sensitivity. We define a yield indicator:

$$\Psi \equiv \sqrt{3J_2} = \left\{ \frac{1}{2} \left[(\sigma_1 - \sigma_2)^2 + (\sigma_2 - \sigma_3)^2 + (\sigma_3 - \sigma_1)^2 \right] \right\}^{\frac{1}{2}} \tag{7.18}$$

where Ψ is our yield indicator and $\sigma_1, \sigma_2, \sigma_3$ are the three principal stress components, assuming the material is linear elastic. We also state that $Y \equiv Y_o$, in other words Y is the initial yield strength, and we do not concern ourselves with changes in the yield strength due to work hardening. The yield condition is

$$Y > \Psi \quad \rightarrow \quad \text{no yield}$$

$$Y \leq \Psi \quad \rightarrow \quad \text{yield} \tag{7.19}$$

This may seem a lot like nitpicking. It is.

The best way to comprehend all this is to see a picture of it. Figure 7–17 shows a plot of the von Mises yield surface plotted in principal stress space, which fortunately has only three coordinate axes.

The von Mises yield surface is a cylinder in this space. The central axis of the surface is along the line, $\sigma_1 = \sigma_2 = \sigma_3$. In theory, the ends do not terminate, but we can assume that it extends as far as any load we could practically imagine. (Be careful to not confuse this cylinder with a section of pipe.) The meaning is that any combination of principal stress components that plots on the inside of this cylindrical yield surface does not cause yield and anything on the surface or outside does cause yield. The radius of the cylinder is the yield stress. As discussed in the last section, this yield surface grows and its axis moves about as various combinations of stress exceed its initial boundaries. As it grows and moves, it may no longer retain its shape as a cylinder and may even develop corners. Most plasticity theory demands that the outside of the surface remain convex though. But, as we said, our primary interest is not what happens outside this yield surface but that we keep our casing stress inside it. Another thing that should be apparent with this surface is that the central axis is a hydrostatic stress. You can see that, no matter the magnitude of the hydrostatic stress, it always plots within the cylindrical yield surface, it makes no difference whether it is 1000 lbf/in.^{2} or 10 × 10^6 lbf/in.^{2}. Hydrostatic pressure or tension cannot cause yield in materials that can be modeled by the von Mises yield criterion. Figure 7–18 shows a case where we calculate a load on casing using elastic assumptions, and we plot that point in principal stress space. You can see that it is outside the yield surface (meaning that the material will yield), and our yield indicator is the distance from the central axis to that point. This gives some physical meaning to our yield indicator. So the yield indicator or $\sqrt{3J_2}$ is actually a distance measure in the principal stress space. A distance measure (sometimes called a *metric*) is not a vector but merely a scalar. It always is positive. You can see that Y also is

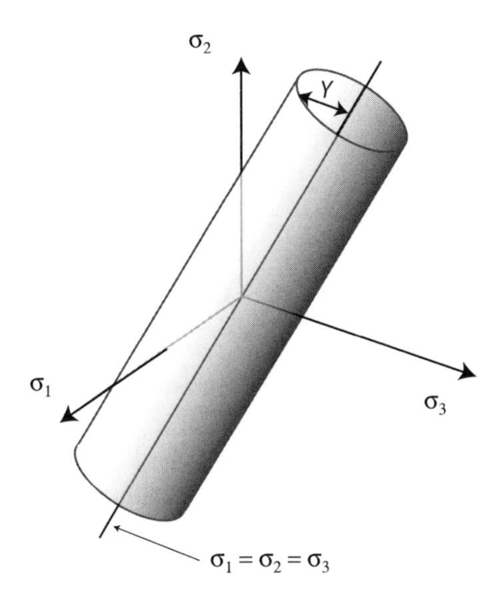

Figure 7–17 *The von Mises yield criterion in principal stress space.*

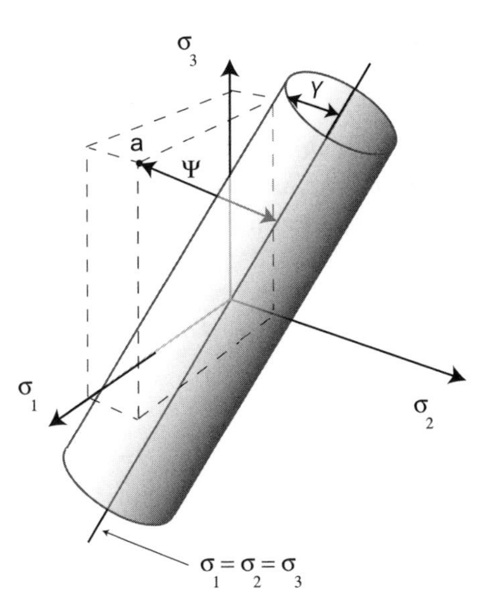

Figure 7–18 *Combined load outside the yield surface.*

a distance measure in this space, and although it has the same value of, say, σ_{11} at the yield point in a uniaxial stress-strain test, it is not the same thing. Maybe that does not justify the nitpicking, but at least you can see the point.

As mentioned before, we can write the von Mises yield criterion in several forms, such as equation (7.18), which is in terms of principal stress components. If the principal stress components correspond to the coordinate axes we can write it in terms of the coordinate axes components:

$$\Psi = \left\{ \frac{1}{2}\left[(\sigma_\theta - \sigma_r)^2 + (\sigma_r - \sigma_z)^2 + (\sigma_z - \sigma_\theta)^2 \right] \right\}^{\frac{1}{2}} \tag{7.20}$$

If the principal stress components do not correspond with the coordinate axes, as when torsion is present, then we have a choice. We can resolve the stress into principal stress components and use equation (7.18) or we use the expanded version of the von Mises criterion when the axial components of stress are not the principal stress components:

$$\Psi = \left\{ \frac{1}{2}\left[(\sigma_\theta - \sigma_r)^2 + (\sigma_r - \sigma_z)^2 + (\sigma_z - \sigma_\theta)^2 \right] + 3\left(\sigma_{r\theta}^2 + \sigma_{rz}^2 + \sigma_{\theta z}^2\right) \right\}^{\frac{1}{2}} \tag{7.21}$$

In this case, we normally consider only the shear stress due to torsion, $\sigma_{r\theta}$ and the other two components of shear are assumed to be zero. If we want to get the principal stress components and use the principal stress formula, then

$$\sigma_1 = \frac{\sigma_\theta + \sigma_r}{2} + \sqrt{\left(\frac{\sigma_\theta - \sigma_r}{2}\right) + \sigma_{r\theta}^2}$$

$$\sigma_2 = \frac{\sigma_\theta + \sigma_r}{2} - \sqrt{\left(\frac{\sigma_\theta - \sigma_r}{2}\right) + \sigma_{r\theta}^2}$$

$$\sigma_3 = \sigma_r \tag{7.22}$$

When we work with principal stress components, it is immaterial how they are numbered, but the custom is that the largest value is number 1 and the smallest is number 3. There are a number of other ways to show the von Mises criterion in equation form. Usually, it is shown in a form that makes sense as to how it relates to deviatoric stress, octahedral stress, principal stress, and the like. One note of caution: Whenever doing manual calculations with the von Mises yield criterion, no matter which formula you use, be careful with the signs, so that you do not run amok. Tension is positive; compression is negative.

Now that we have seen the von Mises yield criteria in equation form and in a graph, let us examine how it applies to casing design. First of all, we hear terms like *biaxial casing design* and *triaxial casing design.* Those terms refer to the coordinate axes and the state in which the principal stress components are aligned with those coordinate axes. And, in terms of casing, the coordinate axes usually are circular cylindrical coordinates (Figure 7–20 in next section). So biaxial casing design refers to two principal stresses aligned with two coordinate axes. The two principal stresses referred to are the axial stress component and the tangential stress component. In a biaxial sense, the radial stress component is ignored. Triaxial design refers to all three. If there is a nonzero shear stress component, as in rotational torsion, then neither of those terms apply as the principal stress components are not aligned with the coordinate axes. Although it is not technically correct, some refer to any three-dimensional stress state as *triaxial*, whether the principal stresses are aligned with the coordinate (or pipe) axes or not, and that is okay as long as we understand the meaning.

In Chapter 5, we employed an ellipse in a two-dimensional chart to correct for the combination of tension and collapse pressure. Essentially, that was a biaxial approach, and that chart is another way to visualize the von Mises yield criterion. If the principal stress components are the tangential, radial, and axial stress components, that is, no shear components, then we can use equation (7.20) to derive the equation for the ellipse. On the yield surface, $\Psi = Y$, the yield indicator is equal to the yield stress:

$$Y = \left\{ \frac{1}{2}\left[\left(\sigma_\theta - \sigma_r\right)^2 + \left(\sigma_r - \sigma_z\right)^2 + \left(\sigma_z - \sigma_\theta\right)^2 \right] \right\}^{\frac{1}{2}}$$

$$Y^2 = \frac{1}{2}\left[\left(\sigma_\theta - \sigma_r\right)^2 + \left(\sigma_r - \sigma_z\right)^2 + \left(\sigma_z - \sigma_\theta\right)^2 \right]$$

$$Y^2 = \left(\sigma_\theta - \sigma_r\right)^2 - \left(\sigma_\theta - \sigma_r\right)\left(\sigma_z - \sigma_r\right) + \left(\sigma_z - \sigma_r\right)^2$$

With a bit of algebra, we have gotten the von Mises yield surface into an elliptic equation of the form

$$r^2 = x^2 - xy + y^2$$

This is a quadratic equation we can solve as

$$x = \frac{y}{2} \pm \sqrt{r^2 - \frac{3y^4}{4}}$$

which we can put in a more useful dimensionless form:

$$\frac{x}{r} = \frac{y}{2r} \pm \sqrt{1 - \frac{3y^2}{4r^2}}$$

We can write this in terms of the principal stress components:

$$\frac{\sigma_z - \sigma_r}{Y} = \frac{\sigma_\theta - \sigma_r}{2Y} \pm \sqrt{1 - \frac{3(\sigma_\theta - \sigma_r)^2}{4Y^2}} \qquad (7.23)$$

or, equivalently,

$$\frac{\sigma_\theta - \sigma_r}{Y} = \frac{\sigma_z - \sigma_r}{2Y} \pm \sqrt{1 - \frac{3(\sigma_z - \sigma_r)^2}{4Y^2}} \qquad (7.24)$$

We can plot either of those as in Figure 7–19, which is exactly what we used in Chapter 5.

So far we have not accomplished much. We have taken a three-dimensional surface and plotted it in two dimensions. Since the radial stress is the negative value of the pressure on the wall of the tube, we can substitute that into the equations. Then, if we know the pressure and one other component, we can find the value of the third at the yield surface. However, this is no easier than calculating it, but it does provide a useful

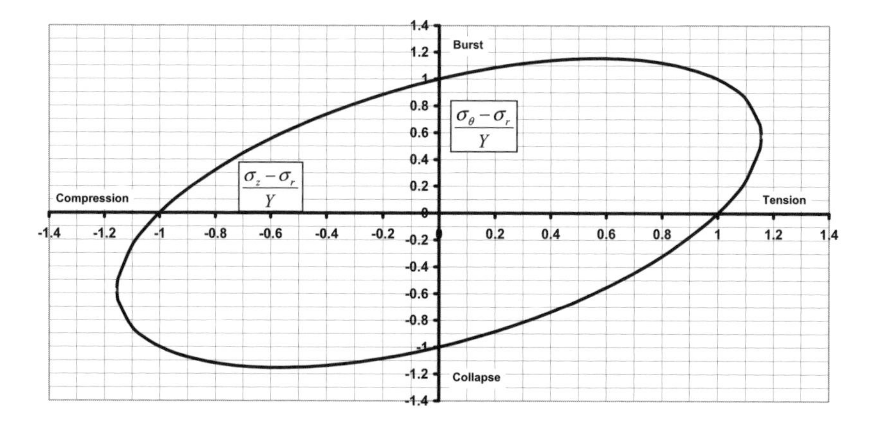

Figure 7–19 *The von Mises yield criterion in two dimensions.*

way of illustrating an elastic stress state in relation to the yield surface. The basis of biaxial casing design is that we ignore the radial stress, which usually is small compared to the other two components, and we have the following equivalent equations:

$$\frac{\sigma_z}{Y} = \frac{\sigma_\theta}{2Y} \pm \sqrt{1 - \frac{3(\sigma_\theta)^2}{4Y^2}} \tag{7.25}$$

$$\frac{\sigma_\theta}{Y} = \frac{\sigma_z}{2Y} \pm \sqrt{1 - \frac{3(\sigma_z)^2}{4Y^2}} \tag{7.26}$$

Those are the biaxial design formulas (we need only one), then we could plot them and the plot would be exactly as before, except we assume the radial stress or pressure at the wall is zero. One always should remember that the underlying assumption of biaxial casing design is that the radial stress component is relatively small compared to the tangential and axial stress components, and that assumption is not always true.

Before leaving this section, it might be worth illustrating the von Misses yield criterion with an example to show how the hydrostatic pressure has no effect on yield.

Example 7–1

Suppose a block of steel in the shape of a cube has a yield strength of 40,000 lbf/in². It is loaded three-dimensionally in tension such that the three principal stress components are $\sigma_1 = 80000\,\text{lbf/in}^2$, $\sigma_2 = 50000\,\text{lbf/in}^2$, and $\sigma_3 = 50000\,\text{lbf/in}^2$. Each of those components is greater than the yield strength of the steel. Will it yield? We calculate the deviatoric stresses to see the hydrostatic pressure effects:

$$\sigma_1' = 80000 - \frac{80000 + 50000 + 50000}{3} = 20000$$

$$\sigma_2' = 50000 - \frac{80000 + 50000 + 50000}{3} = -10000$$

$$\sigma_3' = 50000 - \frac{80000 + 50000 + 50000}{3} = -10000$$

Then we calculate J_2 :

$$J_2 = \frac{1}{2}\left[(\sigma_1')^2 + (\sigma_2')^2 + (\sigma_3')^2\right]$$

$$= \frac{1}{2}\left[(20000)^2 + (-10000)^2 + (-10000)^2\right]$$

$$= 300 \times 10^6$$

Then, we substitute into the yield criterion to calculate the yield indicator:

$$\Psi = \sqrt{3J_2}$$

$$\Psi = \sqrt{3(300 \times 10^6)}$$

$$\Psi = 30000$$

And, we check the yield condition:

$$Y > \Psi \ ?$$

$$40000 > 30000 \quad \rightarrow \quad \text{no yield}$$

For this particular example we did not have to even make the calculations, because we could see that two of the components were 50,000 and the third was greater than that, so the hydrostatic stress is 50,000. We subtract that value from 80,000 and it leaves us with the equivalent of 30,000 in one direction.

Before leaving this section, we offer a word of caution. The von Mises yield criterion is not based on physics, it is more of what might be called *phenomenological*, in that it describes what is observed rather than any underlying principle of physics. Before you go away overly enthralled with it, we should offer a quotation from Bernard Budianski of Harvard University, one of the world's foremost solid mechanics experts, "no one really believes Mises' theory is really right. It might be good enough, but it is not really right" (Budianski, 1983).

7.6 Mechanics of Tubes

Casing is a tube. A tube may serve as a beam, a column, a pressure vessel, or any combination of the three from the standpoint of mechanics. A casing string, to some extent, is all of those. The stress in a tube is of particular interest to us in light of the yield criterion we just discussed, since that is what we need to know to determine if the tube is in an elastic range or not. We are fortunate in many respects that a tube is a fairly easy structure to analyze in an elastic state. In general, we need be concerned with only four stress components for our applications:

- Axial stress.
- Radial stress.
- Tangential stress.
- Torsional stress.

To understand these stresses, we first need a convenient coordinate system. One that fits our needs quite well is a circular cylindrical coordinate system, as shown in Figure 7–20.

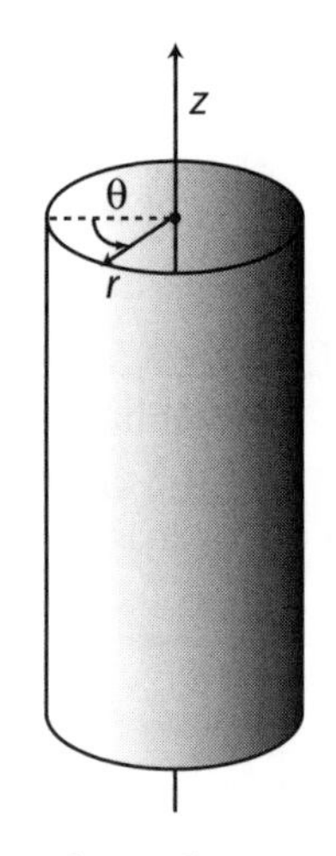

Figure 7–20 *Circular cylindrical coordinate system for use with casing.*

There are three orthogonal coordinate axes, as with a Cartesian coordinate system. The axial coordinate, z, runs along the central axis of the tube; the radial coordinate, r, that runs from the central axis out in any direction; and the tangential coordinate, θ, which is an angular measure about the central axis as measured from some arbitrary reference point. The main difference between this and a Cartesian coordinate system is that two of the coordinate measures are in standard length units and one is an angular measure. Hence, the physical meaning of θ is not the same as z or r. The measure of physical distance in the θ coordinate is $r\theta$.

Axial Stress

The axial stress in a tube that is not bent is merely the axial load divided by the cross-sectional area of the tube:

$$\sigma_z = \frac{F_z}{A_t} = \frac{F_z}{\pi\left(r_o^2 - r_i^2\right)} = \frac{F_z}{\frac{\pi}{4}\left(d_o^2 - d_i^2\right)} \tag{7.27}$$

where

σ_z = axial load stress component
F_z = axial load
r_o = pipe radius, outside wall
r_i = pipe radius, inside wall
d_o = pipe diameter, outside wall
d_i = pipe diameter, inside wall

We always assume that the pipe is straight and stress free before it is run in a well. If the pipe is in a curved bore hole, there will be additional stresses in the axial direction due to bending, which we will discuss later.

Radial and Tangential Stress

We should all be thankful to Lamé, who in 1852, worked out the elastic stress solutions in tubes due to internal and external pressure. His solutions for the axial stress due to the pressure depend on whether the tubes are open on the ends, capped on the ends, or the ends are fixed (plane strain). The Lamé solutions for the stress components due to pressure are

$$\sigma_r = \frac{r_i^2 r_o^2\left(p_o - p_i\right)}{r_o^2 - r_i^2}\frac{1}{r^2} + \frac{p_o r_o^2 - p_i r_i^2}{r_o^2 - r_i^2} \tag{7.28}$$

$$\sigma_\theta = -\frac{r_i^2 r_o^2\left(p_o - p_i\right)}{r_o^2 - r_i^2}\frac{1}{r^2} + \frac{p_o r_o^2 - p_i r_i^2}{r_o^2 - r_i^2} \tag{7.29}$$

$$\sigma_z = \begin{cases} \dfrac{p_i r_i^2}{r_o^2 - r_i^2} - \dfrac{p_o r_o^2}{r_o^2 - r_i^2} & \text{capped ends} \\ 0 & \text{open ends} \\ \nu(\sigma_\theta + \sigma_r) & \text{fixed ends} \end{cases} \tag{7.30}$$

where the nonsubscripted r is the radius at some point in the wall of the pipe, where we want to calculate the stresses. The axial stress due to the pressure is added to the axial stress in the pipe due to gravity and borehole friction if any. Be careful and think about equation (7.30) before you start adding pressure effects to the axial stress due to gravity and friction. Capped-end conditions affect casing only when one end of the casing is free to move. When casing is run in a well, the only time the capped-end condition normally applies is when the top wiper plug is bumped during cementing. After that, it is fixed at the top by the wellhead and below by cement. If you want to consider the wellhead a cap, then you must also consider that, for it to move enough to cause an axial stress change in the casing string, it must move every tubular string in the well, some of which are cemented to the surface. (Some thermal wells allow for wellhead movement.) The second term in the capped-end formula is for external pressure that acts on the free end, pressure that may or may not be present. If the ends are fixed, you must know if they were fixed before the pressure was applied or afterward because the fixed-end axial stress equation is only valid for changes in the axial stress due to pressure applied after the ends are fixed.

Those formulas are useful as they stand, but as it turns out we are not really interested in the stress at various points within the wall of the cylinder, because the maximum stress always is at one of the walls if there is pressure. Which one? It is not intuitive, but whether the greater pressure is internal or external, yield always occurs at the inner wall first. (Work it out, if you do not believe it.) If we substitute r_i in place of r to get the Lamé solutions at the inner wall, we find they are greatly simplified:

$$\sigma_r = -p_i \tag{7.31}$$

$$\sigma_\theta = \frac{p_i\left(r_o^2 + r_i^2\right) - 2p_o r_o^2}{r_o^2 - r_i^2} \tag{7.32}$$

The axial stress does not change from equation (7.30). The sum of the radial and tangential stress in that equation is a stress invariant through the wall of the tube, so it does not matter if they are calculated at the inner wall or outer wall, so long as both are calculated at the same place. One more caveat. Often you may see the tangential stress equation without the second term in the numerator. That is typical of pressure vessels where there is no external pressure. Many use that form and use the difference between the internal and external pressure as the internal pressure. Don't do that! Yes, it gives close results, but it is a sloppy practice.

There are times when we might want to calculate the radial and tangential stress at the outer wall. This is because, in bending and torsion, the tube yields first at the outer wall, so if we are checking for yield, we may want to check both the inner and outer walls. The radial and tangential stress components at the outer wall are

$$\sigma_r = -p_o \tag{7.33}$$

$$\sigma_\theta = \frac{-p_o\left(r_o^2 + r_i^2\right) + 2p_i r_i^2}{r_o^2 - r_i^2} \tag{7.34}$$

Remember that these are elasticity solutions based on a linear elastic material. They are not valid beyond the yield point.

Torsion

We do not often rotate casing in a well bore. It does help in attaining a good primary cement job, but often the friction in the well bore is such that the torque required to rotate the casing exceeds the maximum recommended makeup torque of the connections such as ST&C, LT&C, buttress, and so on. However, many times, liners are rotated while cementing and casing can be rotated with some proprietary connections or special stop rings inserted to prevent overtorqueing of nonshouldered connections, like buttress, ST&C, or LT&C. The equation for the shear stress due to torsion in a pipe body is given by

$$\sigma_{r\theta} = \frac{2r\tau}{\pi\left(r_o^4 - r_i^4\right)} \tag{7.35}$$

where $r = r_o$ to calculate the shear stress at the outer wall where it is a maximum or $r = r_i$ at the inner wall. The torque in the casing is τ and it must be in consistent units. In oil-field units it usually is in lbf · ft and it must be multiplied by 12 to change it to lbf · in. to be consistent with the radii units. In SI units, the torque will be in Joules (N · m) so the radii must be in meters so the stress will be in Pa.

Bending Stress

The bending of casing in curved well bores is discussed in detail in Chapter 9, so all we present here is a formula for bending stress to complete this section:

$$\sigma_{\mathrm{b}} = \pm E\frac{r}{R} \tag{7.36}$$

The radius of the pipe is $r = r_o$ at the outside wall where the bending stress is a maximum, and $r = r_i$ if it is desired to calculate the bending stress at the inner wall. The radius of the well bore curvature is R. It should be in the same units as the radius of the pipe. The bending stress has a plus sign on the convex side of its curvature as that portion is in tension. On the concave side, it is negative because it is in compression. It is added to the axial stress for determining yield, but it should be remembered that the bending stress is a maximum only along a line running parallel to the central axis in the plane of curvature on the convex side and the concave side. The values calculated in the bending equation are not the bending stress at other points around the circumference of the tube.

7.7 Closure

We have barely scratched the surface of solid mechanics, but we have gone far beyond what most petroleum engineers and other engineers coming into petroleum engineering from other disciplines, such as electrical or chemical engineering, normally have been exposed to at an undergraduate level. This should help you understand much of what is written in the literature on the behavior of casing. Some of the terminology appearing in the petroleum literature is a bit convoluted at times, but most of it represents the honest efforts of those dedicated to trying to solve the problems of casing loading and design.

A number of references are available for those who want to learn more about the mechanics of solids. Some are a bit advanced, but most are readable and understandable. The book by Fung (1965) is excellent for solids, although some of the terminology has been supplanted since it was published. The book on continuum mechanics by Malvern (1977) remains the all-time basic standard, and he covers fluids in addition to solids. As to plasticity, the book by Chakrabarty (1987) is easy to read and understand, although it may be hard to find now, but his newer book, Chakrabarty (2004), has a good overview of plasticity in the first chapter. The book by Hill (1950) on plasticity is still available, since it has been reprinted many times (1998 being the latest). The basics of plasticity also are covered by Fung and by Malvern in their books. As far as elasticity is concerned the book by Timoshenko (1934) is an all-time classic on almost everyone's bookshelf, although it is not a good book for learning about elasticity, since it is mostly a collection of elastic solutions (and in that respect it has no peer). As to learning about elasticity, many books are on the market, and the one by Boresi and Chong (1987) is a good one, although almost everyone swears by some favorite from which he or she learned elasticity.

7.8 References

Boresi, A. P., and K. P. Chong. (1987). *Elasticity in Engineering Mechanics.* New York: Elsevier.

Budiansky, B. (1983). "Comments by Session Chairmen." In *Theoretical Foundations for Large-Scale Computations for Nonlinear Material Behavior*, ed. S. Nemat-Nasser, R. J. Asaro, and G. Hagemier, pp. 383–400. Dordrecht, the Netherlands: Martinus Nijhoff Publishers.

Chakrabarty, J. (1987). *Theory of Plasticity.* New York: McGraw-Hill.

Chakrabarty, J. (2004). *Applied Plasticity.* Berlin: Springer.

Fung, Y. C. (1965). *Foundations of Solid Mechanics.* Englewood Cliffs, NJ: Prentice-Hall.

Hill, R. (1950). *The Mathematical Theory of Plasticity.* Oxford: Oxford University Press.

Klever, F. J., and A. G. Tallin. (2005). T*he Role of Idealization Uncertainty in Understanding Design Margins.* SPE paper 97574. Richardson, TX: Society of Petroleum Engineers.

Malvern, Lawrence E. (1977). *Introduction to the Mechanics of a Continuous Medium.* Englewood Cliffs, NJ: Prentice-Hall.

Simmonds, James G. (1982). *A Brief on Tensor Analysis.* New York, Heidelberg, Berlin: Springer-Verlag.

Timoshenko, S. P., and J. H. Goodier. (1970). *Theory of Elasticity*, 2d ed. New York: McGraw-Hill.

CHAPTER 8

Casing Design Performance

8.1 Introduction

In earlier chapters, we took the published values of casing strengths a face value. In this chapter, we look at some of the formulas from which those strengths are calculated. We are going to see where they came from, their limitations, and some of the formulas being considered to replace them. We are going to look at combined loading and its affects, lateral instability (buckling), and the effects of temperature on casing design.

Reiterating what was mentioned in the preceding chapter, we cannot actually predict failure of a joint of casing. We are not interested in the actual value at which a joint of casing fails. What we are interested in is a value we can use as a design limit. A particular joint of casing may fail at that limit or it may not; the important consideration is that it not fail before that point. Historically, yield strength has been used as a design limit, and that continues. However, some of the formulas used in the past were based on some simplifying assumptions and tests that did not realistically model actual loading. For example, the API collapse formulas are based on collapse tests on very short samples of casing, and it has been found that joint length plays a part in collapse, to the extent that end conditions affect the collapse of short tubes. So, much of what is covered in this chapter might be characterized as design strength formulas and calculations. Most are still based on yield strength or test data. While it may be perfectly acceptable to use conservative formulas in the vast majority of wells drilled in the world, there are deeper, high-pressure, high-temperature wells where such conservative formulas might greatly increase well costs by requiring much higher strengths than is necessary or even available. We examine the current formulas and discuss some of the proposed changes.

8.2 Tensile Design Strength

Casing failure in tension is not common. When it does occur it usually occurs at a connection, and the connection is usually ST&C or LT&C. The failure in those types of connections usually is a result of pullout rather than fracture of the casing. In some cases, the pullout is the result of a split coupling due to hydrogen embrittlement or over torqueing during makeup. As far as tensile strength of casing is concerned, it is specified in two ways by the API and ISO, pipe body yield and joint strength. The first is the value of the pipe body yield strength exclusive of threads, expressed as axial tension (or compression) rather than axial stress, and the second is the value of the connection strength, always refers to tension and not compression. The pipe body yield values use the minimum yield strength as the design strength limit. The joint strength is based the minimum value from formulas that use the minimum ultimate strength and the yield strength separately or in combination.

Pipe Body Yield

Pipe body strength at yield is the yield stress of the metal multiplied by the specified cross-sectional area of the tube.

$$F_{max} = Y A_t \tag{8.1}$$

where

F_{max} = pipe body strength at yield, lbf or N

Y = yield strength of pipe, lbf/in^2 or Pa

$A_t = \frac{\pi}{4}\left(d_o^2 - d_i^2\right)$ = cross sectional area of tube, in^2 or m^2

d_o = outside diameter of tube, in. or m

d_i = inside diameter of tube, in. or m

For example, the pipe body yield for 7 in. 23 lb/ft N-80 casing is

$$F_{max} = 80000\frac{\pi}{4}\left(7^2 - 6.366^2\right) \approx 532000 \text{ lbf}$$

which is the value shown in the published tables. This value is valid for either tension or compression. For ST&C and LT&C couplings, the strength of the connection usually is less than that of the pipe body. So one must be aware of the joint strength as well.

Joint Strength

Joint strength is usually the yield strength of a casing connection in tension; however, formulas also are available for the fracture strength of connections, although they are not used often in casing design. The calculation of joint strength depends on the specific type of coupling and includes things such as the makeup length of the threads, the cross-sectional area of the tube under the last full thread, and so forth. Furthermore, the resulting formulas have been adjusted to fit actual test results with various samples of casing connections. The resulting formulas for ST&C, LT&C, buttress, and extreme-line are listed in API 5C3 and ISO 10400. Additionally, one must refer to API 5B to get some of the necessary thread dimensions for use in the formulas. Proprietary thread manufacturers, in general, do not publish their formulas but only the connection strength values. For those reasons, we do not include any joint strength formulas here. Almost no one actually uses them, since the results are published in many tables. The thing we must point out as a precautionary note is that one should always check both the body strength and connection strength for any casing design. And remember, the strength of connections in compression is not addressed by API standards.

8.3 Burst Design Strength

The historical API formula for what we commonly refer to as burst is not actually a formula for burst or pipe rupture but a yield formula for internal pressure. It is based on a thin-wall tube that assumes yield takes place across the entire wall thickness at a single pressure. It also includes a design factor to account for the 12.5% variation in wall thickness allowed by API casing specifications (API Spec 5C2 or ISO 11960). The result is a formula for the internal yield pressure:

$$p = 0.875Y\frac{d_o - d_i}{d_o} \tag{8.2}$$

or, in terms of wall thickness, t,

$$p = 0.875Y\frac{2t}{d_o} \tag{8.3}$$

This formula is adequate for basic casing design, but it leaves a lot to be desired. For one thing, it assumes that yield occurs throughout the pipe wall at the same time, which it does not. What is lost by the assumption of a thin-wall cylinder with a uniform stress throughout the wall of the cylinder? We can use the Lamé elastic formulas along with a yield criterion to determine the yield at the inner wall for a thick-wall tube and compare the results. In the absence of any axial load or axial constraints, that is, the pipe is free to move axially and the ends are not capped, the Lamé formulas (see the previous chapter) give the stress components at the inner wall as

$$\sigma_r = -p_i$$

$$\sigma_\theta = \frac{p_i\left(r_o^2 + r_i^2\right) - 2p_o r_o^2}{\left(r_o^2 - r_i^2\right)} \tag{8.4}$$

At yield, the following holds for the von Mises yield criterion:

$$Y = \left\{\frac{1}{2}\left[\left(\sigma_\theta - \sigma_r\right)^2 + \left(\sigma_r\right)^2 + \left(-\sigma_r\right)^2\right]\right\}^{\frac{1}{2}} \tag{8.5}$$

If the external pressure is zero, that is, $p_o = 0$, then we can solve this equation for the internal pressure at yield in terms of the yield strength of the pipe and the internal and external diameter of the pipe. The result is

$$p_i = Y\frac{d_o^2 - d_i^2}{\sqrt{3d_o^4 + d_i^4}} \tag{8.6}$$

This equation is the thick-wall equivalent of equation (8.2), except it does not account for a tolerance in the wall thickness. The wall thickness is accounted for in ISO 10400 by calculating an internal diameter based on the tolerance in the wall thickness such that, for this equation,

$$\tilde{d}_i \equiv d_o - 0.875(d_o - d_i) \tag{8.7}$$

This definition uses the standard API tolerance; however, other values could be used for specific situations where the actual tolerance is known as opposed to the specified maximum tolerance of 12.5%.

Example 8–1

Comparing equations (8.2) and (8.6) for 7 in. 26 lbf/ft P-110 casing, the first gives an internal yield of

$$p = 0.875Y\frac{d_o - d_i}{d_o} = 0.875(110000)\frac{7 - 6.276}{7} \approx 9960 \text{ lbf/in}^2$$

which is the current value shown in most published tables. Equation (8.6), along with the 12.5% tolerance, gives

$$\tilde{d}_i = d_o - 0.875(d_o - d_i) = 7 - 0.875(7 - 6.276) = 6.367 \text{ in.}$$

$$p_i = Y\frac{d_o^2 - \tilde{d}_i^2}{\sqrt{3d_o^4 - \tilde{d}_i^4}} = 110000\frac{7^2 - 6.367^2}{\sqrt{3(7)^4 + 6.367^4}} \approx 9900 \text{ lbf/in}^2$$

That is only a slight difference for this example, and the difference will vary with wall thickness. The formula based on thick wall tubes will usually be slightly conservative. The problem with the API formula, though, is that it is valid only if the pipe has no axial stress. It also assumes that any external pressure can be accounted for by subtracting it from the internal pressure and using the difference, $\Delta p = p_i - p_o$, as the internal pressure (as long as $\Delta p > 0$). Some do the same thing with the Lamé formula for tangential stress, which is not good practice, because it does not give the same result as when both internal and external pressures are used.

Ductile Rupture

The new ISO 10400, currently up for adoption, includes new formulas for ductile rupture. Few, if any, oil-field tubulars actually fail in rupture when the internal wall surface reaches the yield point, which is the basis for the conventional API/ISO formulas for "burst" or, more correctly, internal yield. If the material were perfectly plastic, it would quickly yield all the way through the wall thickness as the pressure is increased and rupture, but that is not the way most oil-field tubulars behave (coiled tubing excepted). Unless they are very brittle, they are made of a strain-hardening material, so that once the yield stress has been exceeded, the stress still increases before ultimate failure occurs, as discussed in Chapter 7. New ISO formulas take that into account by actually modeling the material behavior in the plastic regime. In addition, certain defect sizes are taken into account, so that the pipe may more realistically model actual casing, and even fracture growth is considered. To use these formulas, one must have more data than one normally has when looking at a pipe inventory list of available casing for a specific application. In particular, one needs inspection results for the casing to be used or at least have a specific inspection in mind and know that casing not meeting those standards will be culled from this particular application. We are going to present only some basics of the formulas here, but before using these formulas, one should definitely read the discussion in ISO 10400 (Appendix B) and some of the associated references. The primary ductile rupture formula in ISO 10400 includes a capped-end effect such that the internal pressure generates an axial strain due to the pressure effect on the end caps. For a capped-end effect to be realized, one end of the casing must be free to move in relation to the other end, so that the axial stress is a function of the internal pressure. This almost never happens in a well, since one end of the casing is cemented (fixed) and the other attached to a wellhead. Unless the wellhead is free to move, the capped-end effect does not occur. One could argue that the wellhead may move, but its movement is considerably restricted by the other strings of pipe also attached to the wellhead, in addition to the fact that the conductor and surface casing are usually cemented at the surface. For casing, the pressure effects are almost always those of fixed ends not capped ends. That being said then, the design formula for the internal pressure at ductile rupture with capped ends is

$$p = 2k_1 U \frac{\tilde{t} - k_2 \delta}{d_o - \tilde{t} + k_2 \delta} \tag{8.8}$$

where

$p =$ internal pressure at ductile rupture

$U =$ minimum tensile strength of casing

$d_o =$ outside diameter of casing

$\tilde{t} =$ reduced wall thickness due to tolerance, e.g. (0.875t)

$k_1 = \left[\left(\frac{1}{2}\right)^{n+1} + \left(\frac{1}{\sqrt{3}}\right)^{n+1}\right] =$ correction factor

$k_2 =$ burst strength factor, 1.0 for QT or 13Cr steels, default 2.0 for others not specifically measured

$\delta =$ depth of imperfection in wall thickness

$n =$ dimensionless hardening index for true stress/strain from uniaxial tensile test

This formula is called a design formula, which calculates rupture assuming certain minimum values for the quantities in the formula, such as the minimum tensile strength. This is derived from a limit state formula, which predicts rupture for a specific sample where the quantities just listed are known exactly for that sample. Some of these quantities can be calculated from measurements or tests. Table 8–1 lists suggested values for the hardening index, n. A hardening index, as used here, is a means of approximating a uniaxial stress-strain curve for a particular material, with a curve fit to an idealized material, such as a Ramberg-Osgood material (Ramberg and Osgood, 1943) or a Ludwik power-law material (Ludwik, 1909). The units of equation (8.8) are either in inches for length measurements and lbf/in^2 for ultimate yield and pressure or mm for length measurements and Pa (or kPa or Mpa) for ultimate yield and pressure. Consequently, no conversion factors are needed.

We use this formula to determine the ductile rupture pressure for the same casing in the previous example.

Table 8–1 Suggested Hardening Index Values (from ISO 10400)

API Grade	Hardening Parameter, n
H-40	0.14
J-55	0.12
K-55	0.12
M-65	0.12
N-80	0.10
L-80	0.10
C-90	0.10
C-95	0.09
T-95	0.09
P-110	0.08
Q-125	0.07

Example 8–2

Using the same casing as in the previous example and assuming our casing inspection finds all defects in excess of 5% of the wall thickness, we determine the variable quantities for use in equation (P-110 casing has a minimum ultimate tensile strength of 125,000 lbf/in.^{2}):

$$\tilde{t} = 0.875t = 0.875(0.362) = 0.317 \text{ in.}$$

For the hardening index, we use $n = 0.08$ for P-110 and calculate the correction factor:

$$k_1 = \left[\left(\frac{1}{2}\right)^{n+1} + \left(\frac{1}{\sqrt{3}}\right)^{n+1}\right] = \left[\left(\frac{1}{2}\right)^{0.08+1} + \left(\frac{1}{\sqrt{3}}\right)^{0.08+1}\right] = 1.026$$

The depth of the wall thickness defect is

$$\delta = 0.05(0.362) = 0.0181 \text{ in.}$$

For the burst strength factor we use 2.0, the default value. We substitute these values into equation (8.8):

$$p = 2k_1U \frac{\tilde{t} - k_2\delta}{d_o - \tilde{t} + k_2\delta} = 2(1.026)(125000)\frac{0.317 - 2.0(0.0181)}{7.0 - 0.317 + 2.0(0.0181)}$$

$$p \approx 10{,}720 \text{ lbf/in}^2$$

We can see that this value is about 8% higher than the current API formula. Part of this is because of accounting for strain hardening of the material. But part of it is because of the axial stress due to the capped-end effect. As we saw earlier with the von Mises yield criterion, internal yield is higher in the presence of axial tension, which is what we have with a capped-end effect.

An additional ISO ductile rupture formula accounts for external pressure and arbitrary axial loading. It is a limit state formula that is much more complicated, and it is not shown here. Finally though, one should note that the ductile rupture formulas take into account material behavior in a plastic regime. In our discussion on plastic behavior in the previous chapter, we pointed out that materials become history dependent in this regime, so these formulas are valid only if the loading exceeds the yield stress and proceeds to rupture. If the loading stops once yield has been exceeded but short of the rupture value, then the formulas are valid for subsequent loading only if the loading path is exactly the same as before, once the new yield value is exceeded. Therefore, the use of the ductile rupture formulas for cyclic loading that might exceed the yield of the casing is questionable.

8.4 Collapse Design Strength

Casing collapse probably is the most common type of failure after corrosion and wear. There have been many cases of collapse due to defective casing joints, but the cause often is one of not accounting for the actual tension and subjected to a collapse load. Perfectly round casing with a uniform wall thickness is quite resistant to collapse pressure. If there is a variation in the wall thickness because of eccentricity or defects, the cross section is ovalized, or the collapse loading is other than hydrostatic pressure (e.g., bore-hole stability problems), then failure from collapse usually occurs at lesser loads. Collapse also depends on outside support of the casing (see Figure 8–1) in that casing with little outside clearance or partially supported by cement may begin to collapse, but the partial support may prevent total collapse.

Sometimes partially collapsed casing may be restored with an internal casing roller, but fully collapsed casing, especially with tubing inside usually is a total loss. It may be possible to mill out, and sometimes even recover, a collapsed portion of the casing. When a tube begins to buckle in collapse (collapse is a form of buckling), the buckle propagates along the tube at a much lower pressure than what caused the initial collapse. The buckle propagation pressure may be on the order of 60% of the collapse pressure if the casing is unsupported. For an undersea, welded pipeline, the entire line may collapse because of a defect in one joint. Fortunately, in a casing string, a hydrostatically induced collapse typically is limited to one joint, because as the buckle propagates, it usually stops at a coupling. In the case of a threaded coupling, the propagation stops because the collapse of the pipe inside a coupling does not transfer the load to the other pin inside the coupling. It also opens the interior of the casing to external pressure, so that the pressure differential is relieved. In the latter case, it becomes what is known as a *wet buckle*, and it tends to stop at that point. In the case of integral joints, the propagating buckle usually is stopped by the increased thickness of the pipe upset. In that case, the upset serves as a "buckle arrestor." In the case of flush joint casing, the buckle might continue to propagate until the pressure differential is less than the buckle propagation pressure, or it may stop if it becomes a wet buckle due to an opening at one of the collapsed connections. An extreme example is coiled tubing that has no connections; if it collapses, the buckle propagates until the pressure differential is less than the propagation pressure. There are formulas for collapse strength but not for buckle propagation. Buckles have been observed to propagate at pressures of about 65% or so of the initial collapse pressure (see Yeh and Kyriakides, 1986; Kyriakides, Babcock, and Elyada, 1984; Chater and Hutchinson, 1984). For this reason, many undersea pipelines now include rings, called buckle arrestors, welded to the pipe at various intervals so that, if a collapse

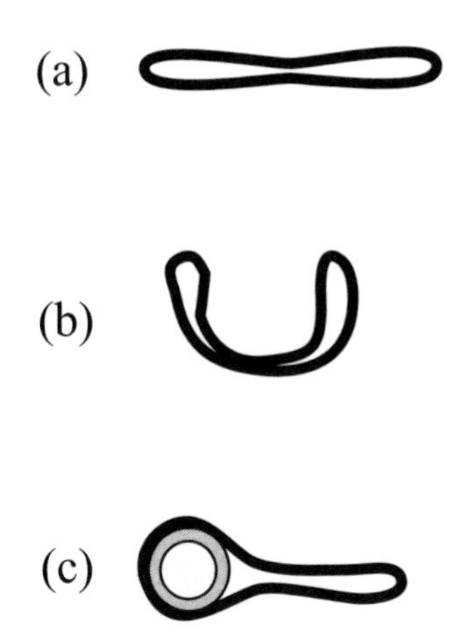

Figure 8–1 *Collapse modes for casing: (a) unsupported collapse, (b) partially supported collapse (e.g., cement or coupling), (c) unsupported collapse with tubing inside.*

should occur, it will not propagate the full length of the submerged pipeline.

Current API Collapse Formula

The API formula for collapse is not a single formula, but rather four:

- Yield strength collapse formula.
- Plastic collapse formula.
- Transition collapse formula.
- Elastic collapse formula.

Each formula has a range for which it is valid, depending on the yield strength of the material and the ratio of the outside diameter to the wall thickness.

Yield collapse formula:

$$p_{\mathrm{YC}} = 2Y\left[\frac{(d_o/t)-1}{(d_o/t)^2}\right] \tag{8.9}$$

Valid range:

$$(d_o/t) \le \frac{A-2+\sqrt{(A-2)^2+8(B+C/Y)}}{2(B+C/Y)} \tag{8.10}$$

Plastic collapse formula:

$$p_{\text{PC}} = Y\left[\frac{A}{(d_o/t)} - B\right] - C \tag{8.11}$$

Valid range:

$$\frac{A-2+\sqrt{(A-2)^2+8(B+C/Y)}}{2(B+C/Y)} < (d_o/t) \le \frac{Y(A-F)}{C+Y(B-G)} \tag{8.12}$$

Transition collapse formula:

$$p_{\text{TC}} = Y\left[\frac{F}{(d_o/t)} - G\right] \tag{8.13}$$

Valid range:

$$\frac{Y(A-F)}{C+Y(B-G)} < (d_o/t) \le \frac{2+B/A}{3B/A} \tag{8.14}$$

Elastic collapse formula:

$$p_{EC} = \frac{46.95 \times 10^6}{(d_o/t)\left[(d_o/t) - 1\right]^2} \tag{8.15}$$

Valid range:

$$(d_o/t) > \frac{2 + B/A}{3B/A} \tag{8.16}$$

where

d_o = outside diameter
t = nominal wall thickness
Y = yield stress of pipe
p_{YC} = collapse pressure, yield pressure formula
p_{PC} = collapse pressure, plastic formula
p_{TC} = collapse pressure, transition formula
p_{EC} = collapse pressure, elastic formula

$A = 2.8762 + 0.10679 \times 10^{-5} Y + 0.21301 \times 10^{-10} Y^2 - 0.53132 \times 10^{-16} Y^3$

$B = 0.026233 + 0.50609 \times 10^{-6} Y$

$C = -465.93 + 0.030867Y - 0.10483 \times 10^{-7} Y^2 + 0.36989 \times 10^{-13} Y^3$

$$F = \frac{46.95 \times 10^6 \left[\dfrac{3B/A}{2 + (B/A)}\right]^3}{Y\left[\dfrac{3B/A}{2 + (B/A)} - (B/A)\right]\left[1 - \dfrac{3B/A}{2 + (B/A)}\right]^2}$$

$G = F\,B/A$

The units in these formulas are inches and lbf/in.[2]. For SI units, where yield stress is in MPa, the collapse formulas and range formulas are the same, but the API constants may be calculated from the following formulas:

$$A = 2.8672 + 0.15489 \times 10^{-3} Y + 0.44809 \times 10^{-6} Y^2 - 0.16211 \times 10^{-9} Y^3$$

$$B = 0.026233 + 0.73402 \times 10^{-4} Y$$

$$C = -3.2125 + 0.030867 Y - 0.15204 \times 10^{-5} Y^2 + 0.77810 \times 10^{-9} Y^3$$

$$F = \frac{3.237 \times 10^5 \left(\dfrac{\dfrac{3B}{A}}{2 + \dfrac{B}{A}} \right)^3}{Y \left(\dfrac{\dfrac{3B}{A}}{2 + \dfrac{B}{A}} - \dfrac{B}{A} \right) \left(1 - \dfrac{\dfrac{3B}{A}}{2 + \dfrac{B}{A}} \right)^2}$$

$$G = \frac{F\,B}{A}$$

Numerical values of the constants are listed in tables in API 5C3 and elsewhere. They are of little use though, because the values calculated with those constants are already published in API 5C2 and many other sources. However, when one finds it necessary to calculate collapse for some casing not having a standard yield value (the only time one would need the values of the constants), the table values of the constants are of no use.

The yield collapse formula is based on the external pressure that causes yield to occur at the inner wall of the casing, so that the yield strength is the design limit. Usually, the pipe will not collapse at that pressure. The elastic collapse formula is based on elastic stability and does not depend on the yield strength of the casing. The plastic collapse and the transition collapse formulas are based on tests done on casing samples, and the formulas essentially are curves fitted to the test results. The end of each range for the various formulas is the intersection of the curves for each formula.

There are difficulties with these collapse formulas other than that they are not valid for collapse in combination with tension; for instance, only the elastic collapse formula is valid in tension. The collapse tests were made with very short sections of casing, and work done in recent years has shown that the values in those tests were affected by the end conditions. In the ISO 10400 standard currently up for adoption, a new approach is recommended, although currently it appears only in an informational appendix. The new formulas are based on work originally done by Tamano, Mimaki, and Yanagimoto (1983) and recently published (Klever and Tamano, 2004). The full formula has in it provisions for inclusion of defects, ovality, eccentricity, and so forth. Although currently not adopted, it is an improvement on the API formulas, plus it has the advantage of being able to include known data about the specific pipe for a casing string and can be used in probabilistic casing design methods. The formula essentially contains a yield collapse formula and an elastic collapse formula and accounts for the transition between those two:

$$p_{\mathrm{clps}} = \frac{p_{\mathrm{elas}} + p_{\mathrm{yld}} - \left[\left(p_{\mathrm{elas}} - p_{\mathrm{yld}}\right)^2 + 4 p_{\mathrm{elas}} p_{\mathrm{yld}} H_{\mathrm{t}}\right]^{\frac{1}{2}}}{2\left(1 - H_{\mathrm{t}}\right)} \tag{8.17}$$

where

$$p_{\mathrm{elas}} = \frac{0.825(2E)}{\left(1-\nu^2\right)\left(\frac{d_o}{t}\right)\left(\frac{d_o}{t} - 1\right)^2} \tag{8.18}$$

is the elastic collapse portion, and

$$p_{\mathrm{yld}} = 2k_{\mathrm{y}}Y\left(\frac{t}{d_o}\right)\left(1 - \frac{t}{2d_o}\right) \tag{8.19}$$

is the yield collapse portion.

In these formulas, k_y is a bias factor for yield collapse, and H_t is a decrement factor for the transition region between elastic and plastic collapse. These values can be determined from tests or one can use the values in the tables of ISO/DIS 10400, as is done in Table 8–2.

These formulas are used to calculate the collapse value of the casing, which is not the same value as the old API formulas from API 5C3. Once we have that value, we may use it and the tension in another formula to calculate a reduced yield strength. Then, we use the reduced yield strength in the preceding formulas to calculate a reduced collapse rating due to the tension. The formula for determining the reduced yield is

Table 8–2 Common yield bias and transition decrement factors (after ISO/DIS 10400, 2004)

	Cold Rotary, Straightened		Hot Rotary, Straightened	
API Grade	H_t	k_y	H_t	k_y
H-40	0.164	0.910	n/a	n/a
J-55	0.164	0.890	n/a	n/a
K-55	0.164	0.890	n/a	n/a
M-65	0.164	0.880	n/a	n/a
L-80	0.164	0.855	0.104	0.865
L-80 9Cr	0.164	0.830	n/a	n/a
L-80 13Cr	0.164	0.830	n/a	n/a
N-80 as rolled	0.164	0.870	n/a	n/a
N-80 Q&T	0.164	0.870	0.104	0.870
C-90	n/a	n/a	0.104	0.850
C-95	0.164	0.840	0.104	0.855
T-95	n/a	n/a	0.104	0.855
P-110	0.164	0.855	0.104	0.855
Q-125	n/a	n/a	0.104	0.850

$$\tilde{Y} = \frac{1}{2}\left(\sqrt{4Y^2 - 3\left(\frac{F_z}{A_t}\right)^2} - \frac{F_z}{A_t}\right) \tag{8.20}$$

The procedure for calculating reduced collapse with the new, proposed method is much easier than with the traditional API method. We show an example of both methods of collapse with tension in the next section, but for now, we look at an example of a collapse calculation with the proposed new formula in the absence of tension.

Example 8–3 Example of New Formulas

We apply these new formulas to determine the collapse rating of 7 in. 32 lb/ft N-80 casing. We calculate the wall thickness first:

$$t = \frac{1}{2}(d_o - d_i) = \frac{1}{2}(7.000 - 6.094) = 0.453 \text{ in}$$

Next, we calculate the elastic collapse using equation (8.18):

$$p_{\text{elas}} = \frac{0.825(2E)}{(1-\nu^2)\left(\frac{d_o}{t}\right)\left(\frac{d_o}{t} - 1\right)^2}$$

$$= \frac{0.825(2)(30\times10^6)}{(1-0.28^2)\left(\frac{7.000}{0.453}\right)\left(\frac{7.000}{0.453} - 1\right)^2}$$

$$= 16641 \text{ lbf/in}^2$$

then, the yield collapse, using equation (8.19):

$$p_{\text{yld}} = 2k_{\text{y}}Y\left(\frac{t}{d_o}\right)\left(1-\frac{t}{2d_o}\right)$$

$$= 2(0.870)(80000)\left(\frac{0.453}{7.000}\right)\left(1-\frac{0.453}{2(7.000)}\right)$$

$$= 8717 \text{ lbf/in}^2$$

Now, we use these values in equation (8.17) to calculate the collapse rating of the casing:

$$p_{\text{clps}} = \frac{p_{\text{elas}} + p_{\text{yld}} - \left[\left(p_{\text{elas}} - p_{\text{yld}}\right)^2 + 4p_{\text{elas}}p_{\text{yld}}H_{\text{t}}\right]^{\frac{1}{2}}}{2\left(1-H_{\text{t}}\right)}$$

$$= \frac{16641 + 8717 - \left[(16641-8717)^2 + 4(16641)(8717)(0.164)\right]^{\frac{1}{2}}}{2(1-0.164)}$$

$$\approx 7650 \text{ lbf/in}^2$$

This is the value of collapse we would read from a table that uses the proposed new collapse formula (recall that API rounds values to the nearest 10 lbf/in.^{2}). It is a bit lower than the current published value of 8600 lbf/in.^{2} calculated using the current API formulas. Partly, this reflects that many of the early API tests used short tube samples, which generally gave higher collapse values. Also, the new formula distinguishes between cold and hot rotary straightened pipe. We used the cold value here, but N-80 is one grade that can be straightened by either method.

8.5 Combined Loads

Almost always, casing is subjected to some type of combined loading. Here are the most common possibilities:

- Tensile and compressive loads due to gravitational forces, bore-hole friction, hydrostatic forces, and bending forces.
- Collapse and burst loads due to hydrostatic pressures.
- Torsion loads due to bore-hole friction.

There are various ways to calculate a design limit for combined loading. Most of them work, but some are quite misleading and can cause serious problems if one does not understand the limitations. We are going to look at a simple method that has been around for more than 150 years and in publication for almost 100 years. It has proven effective throughout all those years in all engineering design applications.

8.5.1 A Yield-Based Approach

What we would like to have is some method of quantifying the combined loads into a single value to compare with some simple strength or stress value for the material of the tube. For example, if Y is the yield stress determined from a uniaxial test and Ψ represents the combined load, we might compare them thus:

$$\begin{aligned} Y > \Psi &\quad \rightarrow \quad \text{no yield} \\ Y \le \Psi &\quad \rightarrow \quad \text{yield} \end{aligned} \tag{8.21}$$

This is exactly what we looked at with the von Mises yield criterion in the previous chapter. The only practical difficulty we have at this point is that casing loads generally are known in terms of axial force, pressure (internal and external), and possibly torque. We need stress values for the von Mises yield criterion. We showed formulas for those stress components in the section on tubular mechanics, also in the previous chapter. Let us look at an example.

Example 8–4 Example of Combined Loads

Suppose we have a point in a casing string where the internal pressure is 4,000 lbf/in.², the external pressure is 4,000 lbf/in.², and the true tension in the pipe at this point is 160,000 lbf. Our casing is 7 in. 23 lb/ft, K-55. It has an internal diameter of 6.366 in. We increase the pressure at the surface to 3000 lbf/in.² to test it for a planned stimulation (the internal pressure at our point of interest will be 7000 lbf/in.²). Will the pipe yield under this load?

First of all, we want to determine where the pipe will yield first, at the inner wall or the outer wall. Internal or external pressure always causes yield at the inner wall first, as mentioned in the last chapter. We have no bending or torque in the pipe, which always causes yield at the outer wall first. So, we require the yield condition at the inner wall. Second is that the test pressure of 3000 lbf/in.² is applied after the casing is cemented and hung off at the surface (i.e., the ends are fixed), so we have to account for that in the axial stress.

Determine the axial stress before test pressure:

$$\sigma_{z_o} = \frac{F_z}{A_t} = \frac{160000}{\frac{\pi}{4}\left(7^2 - 6.366^2\right)} = 24040 \ \text{lbf/in}^2$$

Determine the radial stress before test pressure using the Lamé equation for the inner wall:

$$\sigma_{r_o} = -p_i = -4000 \ \text{lbf/in}^2$$

Determine the radial stress after the test pressure is applied:

$$\sigma_r = -7000 \ \text{lbf/in}^2$$

Determine the tangential stress before test pressure using the Lamé for thinner wall:

$$\sigma_{\theta_o} = \frac{p_i\left(r_o^2 + r_i^2\right) - 2p_o r_o^2}{\left(r_o^2 - r_i^2\right)} = \frac{4000\left[\left(\frac{7}{2}\right)^2 + \left(\frac{6.366}{2}\right)^2\right] - 2(4000)\left(\frac{6.366}{2}\right)^2}{\left(\frac{7}{2}\right)^2 - \left(\frac{6.366}{2}\right)^2}$$

$$= 4000 \text{ lbf/in}^2$$

Determine the tangential stress after test pressure is applied:

$$\sigma_\theta = \frac{7000\left[\left(\frac{7}{2}\right)^2 + \left(\frac{6.366}{2}\right)^2\right] - 2(4000)\left(\frac{6.366}{2}\right)^2}{\left(\frac{7}{2}\right)^2 - \left(\frac{6.366}{2}\right)^2} = 35690 \text{ lbf/in}^2$$

Determine the incremental radial and tangential stress due to the test pressure:

$$\Delta\sigma_r = \sigma_r - \sigma_{r_o} = -7000 - (-4000) = -3000 \text{ lbf/in}^2$$

$$\Delta\sigma_\theta = \sigma_\theta - \sigma_{\theta_o} = 35690 - 4000 = 31690 \text{ lbf/in}^2$$

Then, using the Lamé equation for fixed end tubes, we calculate the change in axial stress due to the test pressure:

$$\Delta\sigma_z = \nu\left(\Delta\sigma_\theta + \Delta\sigma_r\right) = 0.28(31690 - 3000) = 8033 \text{ lbf/in}^2$$

The axial stress including the test pressure effects is

$$\sigma_z = \sigma_{z_o} + \Delta\sigma_z = 24040 + 8033 = 32073 \text{ lbf/in}^2$$

Now, using the three stress components calculated in the presence of the test pressure, we want to determine whether or not yield will occur. Since there is no torsion, these values are principal stress components and may be plugged directly in to the von Mises yield formula:

$$\Psi = \left\{ \frac{1}{2}\left[\left(\sigma_\theta - \sigma_r\right)^2 + \left(\sigma_r - \sigma_z\right)^2 + \left(\sigma_z - \sigma_\theta\right)^2 \right] \right\}^{\frac{1}{2}}$$

$$\Psi = \left\{ \frac{1}{2}\left[(35690 + 7000)^2 + (-7000 - 32073)^2 + (32073 - 35690)^2 \right] \right\}^{\frac{1}{2}}$$

$$\Psi = 41001 \text{ lbf/in}^2$$

Finally, check the yield condition:

$$Y = 55000 \text{ lbf/in}^2$$

$$\Psi = 41001 \text{ lbf/in}^2$$

$$Y > \Psi \quad \rightarrow \quad \text{no yield}$$

In this case, there is no yield. The combined load in this example is approximately 75% of the yield strength of the pipe. Questions might arise: How close to the yield strength would we allow if we were aware of these calculations? What would be a reasonable limit? What are the recommended design factors for combined loading? Those are good questions, and there are no good answers. Some operators would go up to 80% of the yield in a case like this, which would amount to a design factor of 1.25. That might be acceptable for a one-time occurrence, where the specifics are known in detail. If we were spot checking a conventional casing design for a well that had not been drilled, we might want to think again. In those cases, our confidence level might be somewhat less, so we might set a 1.6 design factor as an absolute minimum. You are on your own in this area, unless your company has some particular policy.

One other caveat about using yield stress as a limiting point in combined loading is that it does not account for collapse at loads lower than the yield strength of the casing. You should always check the collapse using the combined load collapse formulas of API or ISO, which are covered in the next two sections.

8.5.2 Current API-Based Approach to Combined Loading

The current API method does not account for combined loads in combination with burst. Combined collapse and tension is considered, based on a simplification of the von Mises yield criterion. A strict yield-based approach like we just covered does not account for the possibility that collapse for some casing may occur before the yield stress actually is reached. In those cases, the API formulas for collapse in tension may be needed as a supplement to the yield approach.

In Chapter 7, we developed a two-dimensional version of the von Mises yield surface. And, that is where the API based approach begins. The von Mises yield surface in two dimensions is given by the equation

$$\frac{\sigma_1 - \sigma_2}{Y} = \frac{\sigma_3 - \sigma_2}{2Y} \pm \sqrt{1 - \frac{3(\sigma_3 - \sigma_2)^2}{4Y^2}} \tag{8.22}$$

where $\sigma_1, \sigma_2, \sigma_3$ are the three principal stresses. In the absence of shear components, such as torsion, the three principal stresses are $\sigma_\theta, \sigma_r, \sigma_z$ and we may substitute them into equation (8.22) to get a convenient form for our use:

$$\frac{\sigma_\theta - \sigma_r}{Y} = \frac{\sigma_z - \sigma_r}{2Y} \pm \sqrt{1 - \frac{3(\sigma_z - \sigma_r)^2}{4Y^2}} \tag{8.23}$$

We purposely set the stress components in that order, so that the radial stress is the one we want to subtract from the other components. The radial stress is the negative value of the internal pressure (recall that yield due to internal or external pressure always occurs at the inner wall first). If it is zero, we can leave it out, but in any event, we should know its value, so for now we rewrite it with the internal pressure:

$$\frac{\sigma_\theta + p_i}{Y} = \frac{\sigma_z + p_i}{2Y} \pm \sqrt{1 - \frac{3(\sigma_z + p_i)^2}{4Y^2}} \tag{8.24}$$

Plotted it looks like Figure 8–2.

Now, here is how API uses this two-dimensional, or biaxial, formulation. API assumes that the tangential stress becomes an effective yield stress for collapse. So, we define an effective yield stress in collapse as

$$\tilde{Y} \equiv -\sigma_\theta \tag{8.25}$$

We may rewrite equation (8.24), adjusting the signs to account for the fact that the tangential stress is compressive (negative) in a collapse situation:

$$\tilde{Y} = Y\sqrt{1 - \frac{3}{4}\left(\frac{\sigma_z + p_i}{Y}\right)^2} + \frac{p_i - \sigma_z}{2} \tag{8.26}$$

This is essentially the API formula, except the API version assumes that the internal pressure is zero and, hence, uses the following formula:

$$\tilde{Y} = Y\sqrt{1 - \frac{3}{4}\left(\frac{\sigma_z}{Y}\right)^2} - \frac{\sigma_z}{2} \tag{8.27}$$

This formula is used to calculate a reduced yield value, $\tilde{Y}$. That reduced yield value then is used in the appropriate API collapse formula, from the earlier section on API collapse, to determine the reduced collapse value due to the tension. If there is internal pressure, then a correction factor is added to the reduced collapse pressure to account for the internal pressure:

$$p_{\text{clps}} = p_{\text{reduced}} + p_i\left(1 - \frac{2t}{d_o}\right) \tag{8.28}$$

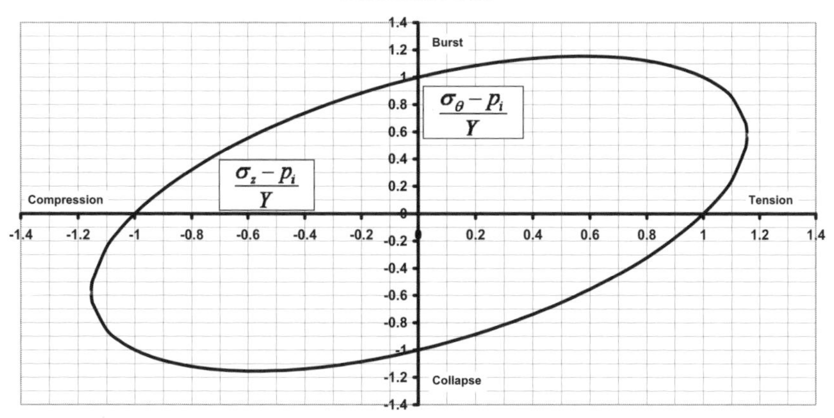

Figure 8–2 *Von Mises yield surface in two dimensions with principal stresses,* $\sigma_\theta, \sigma_r, \sigma_z$ *and* $\sigma_r = -p_i$.

Example 8–5

Using 7 in. 32 lb/ft N-80 casing from the basic casing design example and an axial design load at the bottom of that section of 42,000 lbf, we want to determine the reduced collapse strength of the casing. The published API collapse value with no tension is 8600 lbf/in.^{2}.

We calculate the reduced yield using equation (8.27):

$$\tilde{Y} = 80000\sqrt{1 - \frac{3}{4}\left(\frac{42000}{\frac{\pi}{4}\left(7.000^2 - 6.094^2\right)(80000)}\right)^2} - \frac{42000}{\frac{\pi}{4}\left(7.000^2 - 6.094^2\right)(2)}$$

$$\tilde{Y} = 77651 \text{ lbf/in}^2$$

We now have to determine which collapse formula to use. To do that, we need the value of d_o/t:

$$\frac{d_o}{t} = \frac{d_o}{0.5\left(d_o - d_i\right)} = \frac{7.000}{0.5(7.000 - 6.094)} = \frac{7.000}{0.453} = 15.453$$

We also need the API formula constants at the reduced yield strength. Without showing the calculations, we calculate them from the formulas shown in the section on API collapse formulas:

$$A = 3.062684$$

$$B = 0.065531$$

$$C = 1885.028$$

$$F = 1.993731$$

$$G = 0.042659$$

We start with the range limit for yield plastic collapse from equation (8.10) using those constants and $d_o/t = 15.453$:

$$\left(d_o/t\right) \le \frac{3.062684 - 2 + \sqrt{\left(3.062684 - 2\right)^2 + 8\left(0.065531 + 1885.028/77651\right)}}{2\left(0.065531 + 1885.028/77651\right)}$$

$$= 13.490$$

We see that our value of d_o/t is greater than that, so we must then check the formula for the upper range of the plastic collapse formula which is equation (8.12):

$$\left(d_o/t\right) \le \frac{77651\left(3.062684 - 1.993731\right)}{1885.028 + 77651\left(0.065531 - 0.042659\right)} = 22.672$$

Our d_o/t is within the range of the plastic collapse formula, so we use equation (8.11) to calculate the reduced collapse strength of the casing:

$$\tilde{p}_{\text{PC}} = 77651\left[\frac{3.062684}{15.453} - 0.065531\right] - 1885.028 \approx 8420 \text{ lbf/in}^2$$

This shows a reduction of 180 lbf/in.^{2} in the collapse resistance of the casing due to tension. It is not much in this case, but it might put our design below the minimum design factor, in which case, we might have to adjust the design to compensate. We did not add a pressure correction for internal pressure and use the actual external pressure as the collapse pressure, since in our example, the production casing was designed assuming no internal pressure.

This method is a bit tedious to do with manual calculations, but can be programmed to a spreadsheet calculation. And, there is also a table in API 5C2 (Table 4) that already calculated values of reduced collapse for certain values of tension.

8.5.3 Proposed API/ISO-Based Approach

The proposed new API/ISO formula for combined tension and compression begins something like the current API method as far as calculating a reduced collapse strength, in that it first uses an equation to calculate a reduced yield value. The equation for calculating the reduced yield is

$$\tilde{Y} = \frac{1}{2}\left(\sqrt{4Y^2 - 3(\sigma_z)^2} - \sigma_z\right) \tag{8.29}$$

It is the same as the equation we used to calculate the reduced yield strength in the current API method with no internal pressure, equation (8.26), although in a slightly different form. The reduced yield from this formula then is used in equation (8.19) to calculate a reduced yield collapse, which then is used in equation (8.17) to calculate a reduced collapse strength without internal pressure. If there is internal pressure, then a correction using equation (8.28) may be added. This internal pressure correction is called a simplified method in ISO 10400 and gives results to an accuracy of ±5% when $0 \le \sigma_z/Y \le 0.4$. Outside that range, one should refer to ISO 10400 for the more rigorous internal pressure correction method (currently listed in Appendix H of ISO/DIS 10400, 2004).

Example 8–6 Example of New Collapse Formulas

We apply these new formulas to determine the reduced collapse rating of the 7 in. casing in the previous example. In the section on proposed collapse formulas, we calculated the wall thickness (0.453 in.) and the collapse value without tension, which was 7650 lbf/in.^{2}, so we do not repeat those.

Now, we use an axial tension of 42,000 lbf to determine the reduced yield using equation (8.20):

$$\tilde{Y} = \frac{1}{2}\left(\sqrt{4Y^2 - 3\left(\frac{F_z}{A_t}\right)^2} - \frac{F_z}{A_t}\right)$$

$$= \frac{1}{2}\left(\sqrt{4(80000)^2 - 3\left(\frac{42000}{\frac{\pi}{4}(7^2 - 6.094^2)}\right)^2} - \frac{42000}{\frac{\pi}{4}(7^2 - 6.094^2)}\right)$$

$$= 77650 \text{ lbf/in}^2$$

Note again that this is exactly the reduced yield value we calculated using the historic API method (with round off). This method differs from the historic method in the manner in which the reduced yield is used once calculated. The reduced yield value is then used in equation (8.17). We may use the elastic collapse value of 16641 lbf/in.^{2} that we calculated before, but we must calculate a reduced yield collapse value using equation (8.19). We need not recalculate the elastic collapse value because it is independent of the yield strength:

$$\tilde{p}_{\text{yld}} = 2k_y Y\left(\frac{t}{d_o}\right)\left(1-\frac{t}{2d_o}\right)$$

$$= 2(0.870)(77650)\left(\frac{0.453}{7}\right)\left(1-\frac{0.453}{2(7)}\right)$$

$$= 8461 \text{ lbf/in}^2$$

Now, we plug this value and the elastic collapse value into equation (8.17):

$$p_{\text{clps}} = \frac{p_{\text{elas}} + p_{\text{yld}} - \left[\left(p_{\text{elas}} - p_{\text{yld}}\right)^2 + 4p_{\text{elas}}p_{\text{yld}}H_t\right]^{\frac{1}{2}}}{2(1-H_t)}$$

$$= \frac{16641 + 8460 - \left[(16641-8460)^2 + 4(16641)(8460)(0.164)\right]^{\frac{1}{2}}}{2(1-0.164)}$$

$$\approx 7460 \text{ lbf/in}^2$$

This formula gives a reduction in the collapse value of 190 lbf/in.^{2}, compared to a reduction of 180 lbf/in.^{2} using the current API formulas. But, the reduced collapse value with the new formula is 960 lbf/in.^{2} less than the reduced value with the API formula, because the new formulas lead to a lesser value of collapse in the absence of tension. All indications are that the newer formulas are better, but until (and unless) they are adopted, the old formulas remain the API standard.

8.6 Lateral Buckling

Much has been written over the years about lateral buckling of oil-field tubulars. Some of it has been good, some a bit misinformed, and some even has been ludicrous. Lateral buckling is called columnar buckling in most areas of structural engineering, but lateral buckling of oil-field tubulars differs from the common concepts of columnar buckling as it is understood by most structural engineers. In most structural applications where gravity is considered, the load on the column is at the top of the column as opposed to the bottom. In those cases, gravitational forces tend to contribute to the tendency of a column to buckle. In the case of oil-field tubulars, the loading is a bit different. Usually, the top of the column (casing or tubing) is fixed and the load is caused by a reactive force on the bottom. The reactive force may be due to the weight of some portion of the column resting on bottom, some pressure force on the bottom, or a combination of both. Many refer to lateral buckling of casing as simply buckling. However, collapse is a form of buckling called radial buckling. There is also axial buckling, in which the casing is crushed in an axial direction. And, we could include torsional buckling. This section is about lateral buckling. Lateral buckling occurs when the casing becomes unstable and displaces laterally disproportionately to the magnitude of a very small lateral force.

8.6.1 Stability

The best way to visualize the concept of stability is with a simple and commonly used illustration. In Figure 8–3, three balls are in equilibrium on three surfaces. In case (a), a ball rests at the low point on a concave surface. The ball is in static equilibrium; in other words, it will not move unless some force is applied to it. If we nudge the ball with some small force, it will move slightly then return to its original position as soon as the small force, called a perturbation, is removed. Also note that to move it further from its initial position requires an increasing force the farther it is moved. This ball is in a state of stable equilibrium. In case (b), the ball rests on a flat horizontal surface. It is also in a state of static equilibrium. If a small force is applied, the ball will move. It will continue to move with no requirement that the force be increased as in the first case. It will stop when the force is removed, or if its environment is frictionless, it will continue to move at constant velocity until another force is applied to stop it or it falls off the edge of the surface. It will not return to its original position, however. We call this case conditionally stable equilibrium or sometimes neutrally stable equilibrium. The third case, (c), also is in

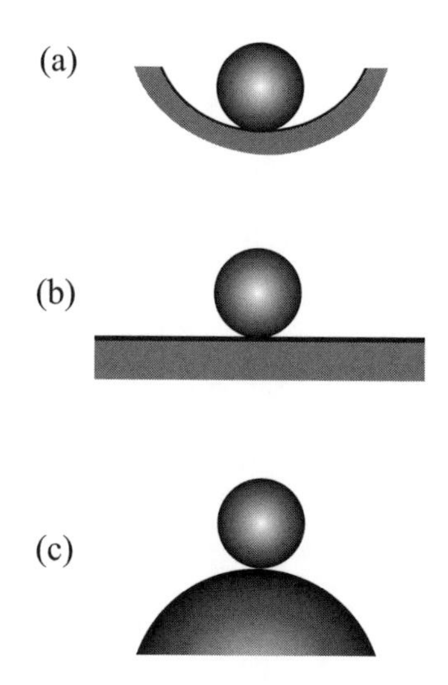

Figure 8–3 *Equilibrium states: (a) stable,(b) neutrally stable, (c) unstable.*

static equilibrium, though, from a practical standpoint, we might have a bit of trouble comprehending how someone could get a ball to balance on the high point of convex surface. Nevertheless, it is easy to understand that, if we apply even the smallest of perturbations to this ball, it will roll off the surface. Once it starts to move, no additional force is required to keep it rolling away from its static equilibrium point. We call this condition unstable equilibrium. This last condition is the type of instability that concerns us with buckling of casing.

How does lateral buckling occur? If you consider the type of lateral or columnar buckling shown in most engineering texts, you will see something like Figure 8–4(a). Typically, these are weightless columns with a vertical load applied at the top and the bottom either hinged or fixed. The initial buckling mode is in the form of a single curve, which is a portion of a sinusoidal curve. Other modes are possible (usually at higher loads), leading to sinusoidal-type configurations with increasing numbers of nodes. These additional nodes are mostly theoretical, because once the column buckles into a single curve, the other modes are not possible unless some constraints are applied. A perfect beam with a perfectly applied axial load (as is the case of our mathematical models) never buckles laterally unless some perturbation is applied. In other words, we could keep applying a load until the column yields in compression and deforms axially in a plastic regime, until it is just a lump of metal on the

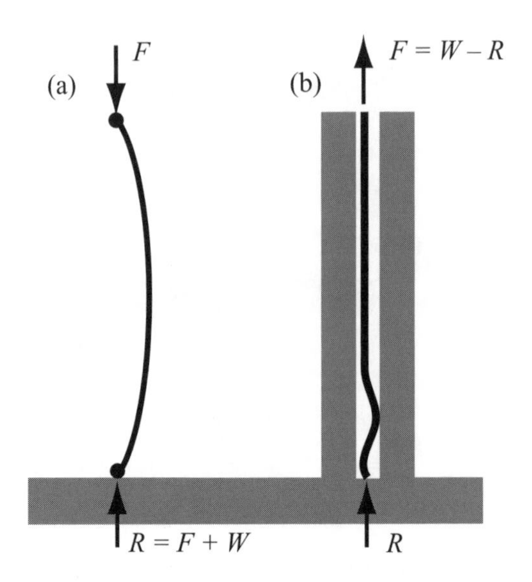

Figure 8–4 *Examples of column buckling: (a) structural columns, (b) casing.*

ground. For instance, the equation for the elongation or compression of an elastic tube is

$$\Delta L = \frac{2F + wL^2}{2\pi\left(r_o^2 - r_i^2\right)E}$$

Nowhere in that formula is any allowance for lateral buckling, because there is no inherent instability in the formula itself. Lateral buckling has to be determined in other ways. As to casing in a well bore, it usually is fixed on bottom with cement and has lateral constraint in the form of a bore-hole wall, Figure 8–4(b). Buckling in this case is affected by the weight of the casing, and we see that only the lower part of the casing is buckled, because the buckling is caused by axial compression at the bottom due to the weight of the casing. As the distance from the bottom increases, the axial compression decreases. When casing buckles in this manner, it initially may be in the shape of a sinusoidal curve, with decreasing frequency as the distance from the bottom increases, until a point is reached where there is no buckling. It may be even more extreme and form a helical shape in the well bore, with decreasing pitch as the distance from bottom increases.

It would seem intuitive, then, that for lateral buckling to occur in casing, it would have to be caused by a compressive load and some small perturbation. For many years, it was assumed that, if casing was hung in tension throughout the full string or at least the portion of the string above the top of the cement, then lateral buckling could not occur. That sounds intuitively simple, and that was the assumption up until papers by Lubinski (1951) and by Klinkenberg (1951) and several discussions of those papers.

The Woods Model

In 1951, in a discussion of a paper by Klinkenberg (1951), Henry Woods (1951) presented an example illustrating the neutral point in a simple way that became something of a classic. It essentially showed that, contrary to popular intuition, lateral buckling could occur in casing in tension (under certain pressure conditions). Much of what had been done previously was based on intuition. Woods, however, based his analysis on the theory of elastic stability, using a thought experiment as illustrated in Figure 8–5. A weightless tube is enclosed in a pressure device that has two chambers. The tube itself is fixed at the lower end and is free at the upper end but closed with a frictionless pressure seal in the top. There is also a frictionless pressure seal on the outside of the tube, between the upper chamber and lower chamber. The lower chamber, representing a well-bore annulus, has a pressure p_o. Inside the tube the pressure is p_i. The top chamber has a pressure of p_t. Woods said that the bottom chamber is large compared to the top chamber and the tube is long enough that its stiffness against bending is negligible. Also, the chambers are large enough that the pressures in each chamber remains constant. He then said, if one could apply a small lateral force to the tube in the large chamber such that the tube would deflect laterally by a small amount, then the top of the tube would slide downward some small amount δL. The following volume changes would occur:

$$\delta V_o = -\pi r_o^2 \, \delta L$$

$$\delta V_i = \pi r_i^2 \, \delta L$$

$$\delta V_t = \pi \left(r_o^2 - r_i^2 \right) \delta L$$

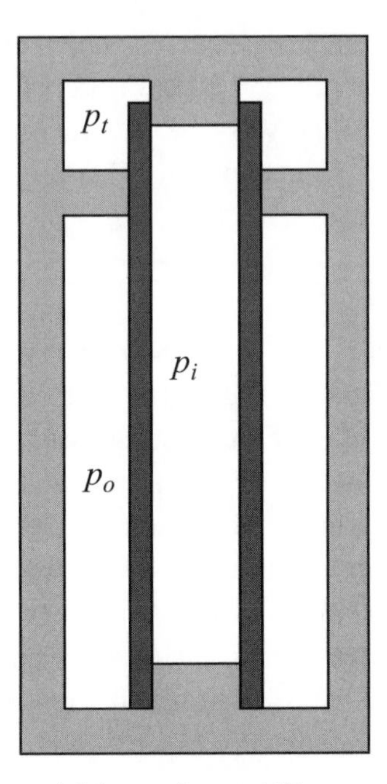

Figure 8–5 *The Woods model for casing stability.*

Since pressure is in each of the places where volume changes occur, a change in potential energy in each place is equal to the change in volume times the pressure, and the total change in potential energy, $\delta\Pi$ is given by

$$\delta\Pi = \delta V_o p_o + \delta V_i p_i + \delta V_t p_t$$

$$\delta\Pi = \pi r_o^2 \delta L p_o - \pi r_i^2 \delta L p_i - \pi\left(r_o^2 - r_i^2\right)\delta L p_t$$

The last term contains the pressure acting on that cross section. This is equivalent to the axial stress in the tube:

$$\sigma_z = -p_t$$

Note that the axial stress is a negative value, because the pressure is a compressive stress. So, we can substitute the axial stress into the equation:

$$\delta\Pi_p = \pi r_o^2 \delta L p_o - \pi r_i^2 \delta L p_i + \pi\left(r_o^2 - r_i^2\right)\delta L \sigma_z$$

For this system to remain stable, the change in potential energy must be zero or positive:

$$\pi r_o^2 \delta L p_o - \pi r_i^2 \delta L p_i + \pi\left(r_o^2 - r_i^2\right)\delta L \sigma_z \geq 0$$

$$\sigma_z \geq \frac{\pi r_i^2 p_i - \pi r_o^2 p_o}{\pi\left(r_o^2 - r_i^2\right)}$$

We could then write the stability condition in a number of ways, four of which follow:

$$\sigma_z \geq \frac{A_i p_i - A_o p_o}{A_o - A_i} \tag{8.30}$$

$$\sigma_z \geq \frac{r_i^2 p_i - r_o^2 p_o}{r_o^2 - r_i^2} \tag{8.31}$$

$$F_z \geq A_i p_i - A_o p_o \tag{8.32}$$

$$\sigma_z \geq \frac{1}{2}\left(\sigma_\theta + \sigma_r\right) \tag{8.33}$$

The last version was derived using the Lamé formulas. Recall that the sum of the tangential stress and the radial stress is constant through the wall of the pipe, so it makes no difference whether they are calculated at the inner or outer wall, as long as both are calculated at the same point.

What Woods's model illustrates is the difference between the change in potential energy by increasing the internal volume of the tube and decreasing the annular volume. If the defection results in an increase in potential energy, then the tube is stable, whereas if it results in a decrease in potential energy, then it is unstable. If there is no change in potential energy, then it is conditionally stable. Obviously, if the internal pressure is sufficiently greater than the external pressure, then the system is unstable, even if the axial stress is positive or in tension. Likewise with a sufficiently higher external pressure, the system is stable when the axial stress is compressive. This particular article by Woods turned around a lot of thinking about landing practices for casing. His results have been generally accepted.

General acceptance notwithstanding, there have been a number of observations about Woods' model as to whether it is applicable to real wells. One of the biggest points of contention is that the tube in Woods's model is in compression and can never be in tension in that model. This might, at first, appear intuitively obvious from a practical viewpoint, in that the pressure in the top chamber cannot cause the tube to be in tension. Remember though, that this is a thought experiment, and we can have a negative pressure even though that is difficult to visualize. Look at Figure 8–6, a schematic of a slightly different "test chamber." The results of this "version" of the Woods model are identical, showing the axial load in the tube can be either tensile or compressive. It is the change in potential energy that determines the stability.

One of the most important concepts to come from the Woods model is the concept of the neutral point as to lateral buckling. One interesting aspect of the equilibrium points derived from this model is in equation (8.32). That might look vaguely similar to something we saw in Chapter 2. If we rearrange it slightly, it shows that, at the neutral point,

$$F_z + \left(A_o p_o - A_i p_i\right) = 0 \tag{8.34}$$

Now, it should look familiar, because the left-hand side is exactly the effective load. So, at the neutral point for lateral buckling,

$$\hat{F}_z = F_z + \left(A_o p_o - A_i p_i\right) = 0 \tag{8.35}$$

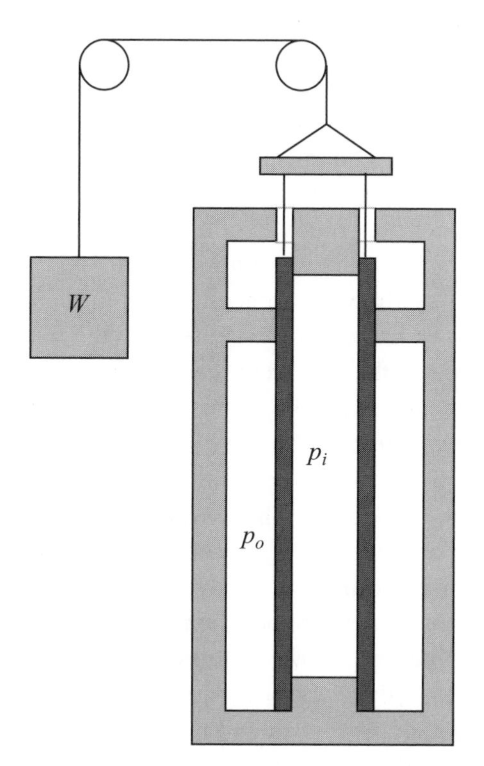

Figure 8–6 *An alternate version of the Woods (1951) model.*

The effective load is zero at the neutral point. It can be seen that the condition of stability is exactly the same as the neutral point of an effective load curve calculated using a buoyancy factor as opposed to a true axial load curve. The neutral point in buckling is the point at which the effective load curve goes from compression to tension; that is, the effective load is zero.

This worked fine for a long time in determining the number of drill collars needed to keep drill pipe in effective tension while drilling with a specified amount of weight on the bit. Then, in the 1960s, someone began looking at the true axial load as we calculated it in Chapter 2, using pressure and differences in cross-sectional area as opposed to weight in air multiplied by a buoyancy factor. From this, it was determined that the drill pipe was not all in tension but actually in compression above the drill collars. This actually was published in textbooks of the day, and many suddenly concluded that they had been "doing it wrong" all those years by using a buoyancy factor, so they used the true axial load and consequently added more drill collars. The rental tool companies thought this bit of contemporary engineering was a good thing. But some began to ask the question, if we had been doing it wrong all those years, why did it work? Then, some began to notice that, if you used the true axial load method, it showed that we needed drill collars to prevent buckling even when there was no weight on the bit. Did this mean we needed drill collars even when rotating off bottom? In terms of the true axial load, the neutral point for lateral buckling is not the point where the true axial load goes from tension to compression. All this was known since 1951 and possibly long before that, but along the way, someone got a bit confused. The fact is that, if a tube in a well has a density greater than the fluid in the well, then the hydrostatic force on the bottom of the tube and on cross-sectional area differences cannot cause the tube to buckle. If the pipe is closed on the end and its effective density, including the internal fluid, is greater than the fluid in the well bore, it still does not buckle due to the hydrostatic force on the bottom of the tube. The reason is that a hydrostatic force on the bottom can act only along the axis of the pipe, whereas our common examples of column buckling always have a load with a fixed direction. A hydrostatic end force cannot support a bending moment if it is always perpendicular to the pipe axis. Now, the reason for that tedious example of end moment due to hydrostatic pressure included in Chapter 2 was to show that, even though there is an end moment on a tube that is not vertical in a well bore, its effects essentially are negligible. Oddly, this "hydrostatic buckling" phenomenon is something that seems to reappear in the oil field from time to time. Many years ago, there were unbelievable discussions as to the maximum depth to which a wire line could be run in a well bore, because a vociferous few argued that, at some depth, the hydrostatic pressure on the cross section of the wire would cause it to buckle in the tubing. It was disproved then, but it resurfaced when floating drilling rigs were first built. Some argued there would be a depth limitation for free drill pipe because of the hydrostatic force on the bottom of the drill string as depths increased. It is hard to imagine it could ever come up again, but it is not the least uncommon, even today, to encounter someone in the oil field who is under the impression that hydrostatic pressure can cause lateral or helical buckling in a drill string or casing string. Certainly, in the case of a tubing string in a sliding seal assembly, it can happen, since the direction of the hydrostatic load is fixed, but not in the case of a free casing or drill string whose density is greater than the fluid in the well bore. So, all we need to determine whether or not a casing string will buckle is an effective load curve and a formula for the buckling load. And, the formula is where it starts to get complicated.

8.6.2 Lateral Buckling of Casing

Structural buckling has occupied the minds of engineers for centuries. Some engineers have and continue to base an entire career on the buckling of a single type of structural element. One would think by now all the problems would have been solved, yet buckling papers continue to be published on what seems like a frequency of several per week. It truly is a complicated topic, and there is always a different twist. Lateral buckling of tubulars in wells has been a considerable topic for many years also. It is a bit different from the structural engineer's typical columnar buckling, in that it is constrained by the walls of a bore hole. The first mode of lateral buckling is something like a single curve, which may become sinusoidal because of the bore-hole constraints and end conditions. In its most severe mode though, it becomes helical in shape. Most of us have seen a permanent helix in a recovered tubing string, work string, or tail pipe that was abused beyond the yield point.

For the most part, buckling of casing in well bores is not the problem it is with tubing or drill pipe. The primary reason for this is that the clearance between casing and the bore-hole wall is relatively small in most wells. So, even when it is buckled, in many cases, the degree of buckling is so small that there are few serious consequences. We are really serious about preventing drill pipe from buckling. Why? Because we experience drill pipe failure when the drill pipe buckles. Why does it fail? It fails because of the rotation while in a buckled configuration. Drill pipe seldom is harmed by buckling alone. It happens all the time in fishing operations, for instance. It happens in the slide drilling of some horizontal wells. But, unless the drill pipe is buckled to the point of yielding, it does not hurt it as long as it is not rotated. What about small drill collars? Drill collars do not buckle? Of course, they do. But when they do, they do not suffer the same frequency of failure as drill pipe, because the degree of buckling is limited by the annular clearance. (And, the connections usually are properly made up, which helps a lot.)

Casing is not drill pipe, and it is seldom rotated. So, what are the consequences of buckled casing? There are several. If the buckling is severe enough, the casing actually could yield or fail. This is rare and almost always a consequence of some geotectonic activity, like subsidence, fault movement, and so forth. And, it does happen in cases of high temperature fluctuations. Those who work in these types of environments are not likely to agree that damage is so rare. Less severe cases of buckling also can have serious consequences. One is the possibility of extensive casing wear in an intermediate string. Another is the difficulty of running and retrieving completion equipment in a buckled production string. So, for these reasons, we try to avoid it. If we are using a slip-type casing hanger, we have some control over the final axial load in the noncemented portion of the casing string. We may even be able to take into account possible thermal expansion and design our casing such that we can pull enough tension to avoid buckling. If we are using a mandrel-type hanger, there is not much we can do other than to try to support the casing with cement. When the top wiper plug is bumped at the end of the cement job, we have to live with whatever axial stress is in the noncemented portion. The final motion with a mandrel hanger always is downward. One of the most insidious forms of buckling can occur in a cemented section where there is an interval of bad cement and we have no means of controlling the axial load in the casing, no matter what type of hanger we have at the surface. Serious problems have arisen when casing in one of these sections of an intermediate string experiences increased temperatures from drilling and circulation in higher-temperature zones below the intermediate casing.

Buckling in a Vertical Well Bore

The published work on buckling of vertical structural columns could fill a small library. The written work on the buckling of tubulars in a vertical well bore could fill a shelf or two in that library. As interesting as all that may be to some, it is of little practical consequence. That is why this is the shortest section in this book. Real well bores seldom are vertical. If you do have a vertical well bore and the effective load is in compression, assume the casing will buckle; you do not need a formula.

Buckling in an Inclined Well Bore

It is a bit more difficult for casing to buckle in an inclined well bore, because gravity tends to hold it to the low side of the bore hole. It can start to move to away from the low side, but the farther it moves, the more it has to move up the side of the well bore. In one sense, it is like the ball resting in the low point of a concave surface, but that analogy is only for visualizing the gravitational effect; it does not tell us much about buckling. Here is a formula for buckling in a straight but inclined well bore (Dawson and Paslay, 1984):

$$F_{crit} = 2\sqrt{\frac{4EI\,\hat{w}\sin\alpha}{\Delta r}} \tag{8.36}$$

where

F_{crit} = critical axial buckling force

$\hat{w}$ = buoyed specific weight of casing

Δr = concentric radial clearance between pip and hole, $\left(r_{hole} - r_{pipe}\right)$

E = Young's elastic modulus

I = moment of inertial of tube cross section

α = borehole inclination angle

This formula has been used successfully for a number of years. Possibly the single biggest problem with it is that, almost everywhere the thing is published, the units are inconsistent. So, if you use oil-field units with the radii in inches, the buoyed specific weight of the pipe is in lbf/in. not lbf/ft as is usually stated. In SI units, all length measures are in meters. Note also that, as the inclination angle goes to zero (vertical well bore), the critical buckling load also goes to zero, implying that any compression in a vertical well will cause buckling.

This formula works pretty well for casing, but it does not take into account any well-bore curvature. Well-bore curvature was considered by He and Kyllingstad (1993) and then later by Mitchell (1999) in the course of resolving differing published formulations. This is Mitchell's solution:

$$F_{crit} = \frac{2EIR}{\Delta r}\left[1 + \sqrt{1 + \frac{\hat{w}\,\Delta r\,R^2 \sin\bar{\alpha}}{EI}}\right] \tag{8.37}$$

where $\bar{\alpha}$ is the average inclination angle over the short interval being considered. Later, Mitchell advanced his study to include the effects of couplings. In that paper (Mitchell, 2003), he recommended that the radial clearance be calculated with the coupling radius as opposed to the pipe body radius; that is, $\Delta r = r_{hole} - r_{cpl}$.

One should remember that all lateral buckling formulas are approximate, at the very best. Some have sweeping assumptions that may or may not be realistic. The two just discussed have been used extensively, and they are reasonable. They do not account for the effects of torsion, connections, and so forth. Mitchell continues to work in the area of tubular buckling in well bores and has taken some of those things into account. For any one interested in the subject, his papers on the subject are a good source.

8.7 Thermal Effects

So far we have not discussed temperature and how it affects casing design, except in brief. We now look at two aspects of temperature effects on casing, its magnitude and the change in magnitude.

8.7.1 Temperature and Material Properties

The engineering rule of thumb is that we do not consider temperature effects on most metals until the temperature exceeds 50% of the melting temperature of the metal; that is, $T \geq 0.5T_m$. That makes us pretty safe in terms of casing in all but the most extreme applications. API grades of steel are affected by temperature, but in most cases, they still retain at least 90% of their yield strength up to around 700° F (~370° C). Some charts regarding this are shown by Holliday (1969) to give you an idea. There is no universally accepted threshold temperature at which one should start down-rating the yield strength due to temperature, although some companies definitely have their own standards. I am not about to recommend a threshold temperature, but certainly at some point, one should definitely consider reducing the yield value of casing to around 90% or so. Geothermal and steam injection wells, for instance, not only have high temperatures but also large changes in temperature. Those types of wells are not considered here.

8.7.2 Temperature Changes

The thermal consideration in most wells is not the temperature per se but the change in temperature. Temperature change causes casing to expand or contract. When casing is run in the hole, the mud has been in a static condition for several hours before the casing reaches bottom. The temperature of the mud may or may not be close to an equilibrium state, depending on how long circulation has been static, but in most wells it is relatively close to static equilibrium. Once casing is on the bottom and circulation begins, it usually gets further away from the static thermal equilibrium state again. Much depends on the difference between the surface and down-hole temperatures. As circulation continues, the lower part of the hole normally is cooled below its static temperature and the upper part of the hole is warmer than its static temperature. Once the cement is in place and circulation ceases, temperatures begin to return to the static thermal equilibrium state. We normally assume that the cement sets at some time before normal static equilibrium is reached. There may be added axial compressive stress in the lower part of the casing as it warms, and there may be added tensile stress in the upper portion of the well as it cools. This amount of stress generally is ignored in most casing design, and in most cases, it likely is nowhere near any critical value. When we start to produce the well, though, the casing is exposed to a different thermal profile than it experienced before. Now, fluids from the formation travel up the hole and warm the upper part of the casing. No cooler fluids circulate downward from the surface to offset the warming. More of the casing in the upper part of the hole expands, and the axial stress change is toward compression. Whether it actually goes into compression or not depends on how much tension was in the pipe initially and how much the temperature increased.

We can show the effects of temperature change on uniaxial stress in casing with a one-dimensional version of Hooke's law:

$$\sigma = \sigma^o + E\varepsilon - E\alpha\Delta T \tag{8.38}$$

where

$\sigma =$ uniaxial stress

$\sigma^o =$ initial stress (before deformation or temperature change)

$E =$ Young's elastic modulus (material dependant)

$\varepsilon =$ uniaxial strain (change in length/original length)

$\alpha =$ coefficient of thermal expansion (material dependant)

$\Delta T = T_t - T_o =$ change in temperature from initial state to some time, t

In general, both the elastic modulus and the coefficient of thermal expansion are functions of temperature. Within a limited range, they are sufficiently constant that we assume them to be so here. With that assumption, this equation looks straightforward, and it is.

However, temperature effects are not always intuitive. For instance, we can say that a temperature change can cause a strain without a causing a stress, and it can also cause a stress without a strain. Look at Figure 8–7(a). A metal bar (or casing string) is suspended from the top end and free at the lower end. If we heat that bar by some amount, ΔT , then it is going to expand and get longer; that is, we induced a thermal strain in that bar. But, have we changed the stress? No, we have not. In this case, the measured uniaxial strain is equal to the thermal strain; that is, $\varepsilon = \alpha\Delta T$, so the stress

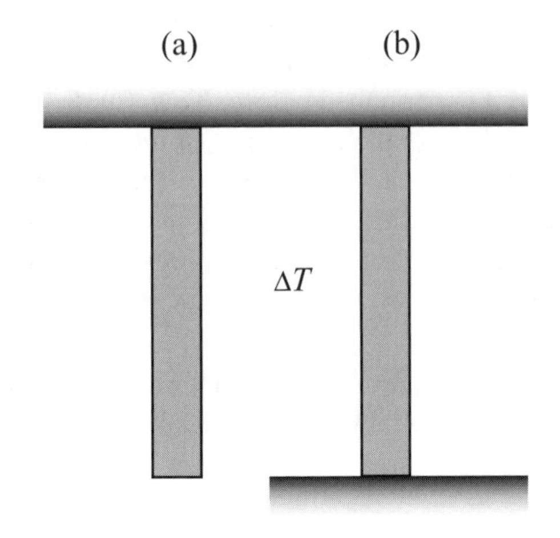

Figure 8–7 *Thermal effects: (a) suspended bar, (b) constrained bar.*

in the bar at any point has not changed, it still is equal to the initial stress, σ^o, which in this example is the body force due to gravity.

Now let us look at the bar in Figure 8–7(b). This bar is constrained at both ends. If we apply the same temperature change to this bar, it tries to expand, but it cannot (we will assume it does not buckle laterally). The bar has not gotten longer, so we have not caused any strain; that is, $\varepsilon = 0$. But, have we changed the stress in this bar? Absolutely. We changed it by the amount, $-E\alpha\Delta T$, a negative value, since the change in stress is compressive. What we see here is that the product of the coefficient of thermal expansion and the change in temperature, $\alpha\Delta T$, is something like an "effective strain." In the uniaxial case, it was relatively simple, and we can use this simple equation to calculate changes in axial stress in casing if we know the magnitude and the end conditions. It begins to get a bit more complicated in three dimensions:

$$\sigma_{ij} = \sigma_{ij}^o + C_{ijkl}\varepsilon_{kl} - \delta_{ij}E\alpha\Delta T \tag{8.39}$$

If you read the previous chapter, this should make sense to you. The temperature affects the stress only in the principal coordinate directions. For composite materials, this gets much more difficult, in that the thermal coefficient is not necessarily the same in all directions nor is the elastic modulus. One additional thing we might mention so that you do not go through life thinking that things are too simple. In the unconstrained bar, we said no stress is caused by heating the bar. That is true only if the bar is thin and heated slowly, so that the temperature is uniform throughout the bar. If it is thick and we heat it rapidly or locally, then we induce some amount of stress within the bar. And we actually can heat it locally so fast that the thermal stress cannot accelerate the mass of the rest of the bar fast enough to prevent the bar from yielding plastically before it can expand. This sort of thing can happen in hypersonic flight vehicles but, fortunately for us, not in casing.

In casing design, thermal effects usually lead to a situation of compression (note the negative sign in the Hooke's constitutive equation). That is something we are not accustomed to seeing in basic casing design, except in bending and bore-hole friction in inclined wells, which we cover in the next chapter. To determine the thermal effects in casing, we must know a number of things that we do not consider in most wells. The major thing we need to know, obviously, is the change in temperature. This can be measured in actual wells, but we also can use a heat transfer software model to estimate it. We must also know if the casing is free to move or not, and this we often do not know and cannot determine except at the wellhead and top of cement (where we assume it is not free to move). We already looked at the effect of changing pressure on axial stress when the pipe is constrained at the ends, and we might have to incorporate that into our thermal stress calculations, too. The best way to illustrate the thermal stress is with examples, where we can see the assumptions we must make along the way and how we might decide the question as to what additional data we require for a particular application. First, let us use the bar in Figure 8–7(b). Let us assume that, initially, the bar is at a constant temperature throughout its length. Let us also assume that it was hanging by its own weight before the lower end was constrained. Say, its cross-sectional area is 7.55 in.^{2}, its specific weight is 26 lbf/ft, and its total length is 10,000 ft (yes, it is 7 in. casing). The tensile force at the top is

$$F_o = w\ell = 26(10000) = 260000 \text{ lbf}$$

and, at the bottom, it is zero. If we set our z-coordinate axis at the top with a positive direction downward, then the axial stress at any point is

$$\sigma_z = \frac{F_o}{A_t} - \left(\frac{w}{A_t}\right)z$$

where A_t is the cross-sectional area of the bar (or tube). Now, if we apply a constant temperature change to this bar, we can calculate the axial stress at any point by

$$\sigma_z = \sigma_o + E\left(\varepsilon_z - \alpha\Delta T\right) = \frac{F_o}{A_t} - \left(\frac{w}{A_t}\right)z - E\alpha\Delta T$$

since $\varepsilon_z = 0$, because the bar is constrained. If we change to temperature in the bar by increasing it 100° F, the axial stress at the top is[7]

$$\sigma_z = \frac{F_o}{A_t} - \left(\frac{w}{A_t}\right)z - E\alpha\Delta T = \frac{260000}{7.55} - \frac{26}{7.55}(0) - 30\times10^6\left(0.69\times10^{-5}\right)100$$
$$\approx 13700 \text{ lbf/in}^2$$

At the bottom, the axial stress is

$$\sigma_z = \frac{F_o}{A_t} - \left(\frac{w}{A_t}\right)z - E\alpha\Delta T = \frac{260000}{7.55} - \frac{26}{7.55}(10000)$$
$$-30\times10^6\left(0.69\times10^{-5}\right)100$$
$$\approx -20700 \text{ lbf/in}^2$$

The bar is in compression at the bottom and in tension at the surface, although the tension at the top now is less than before the temperature change.

7. The coefficient of thermal expansion for casing is 0.69×10^{-5}/° F or 1.24×10^{-5}/° C.

Example 8–7

Suppose we have a well with a string of 7 in. 26 lb/ft L-80 casing in a vertical well. The top of the cement is at 10,000 ft, and the well is perforated in a zone at 14,000 ft. After the cement sets, the hook load is 275,000 lbf, and we calculate the true axial load at the top of the cement is 13,000 lbf. We pull an additional an additional 50,000 lbf on the casing above its hook weight and set it in a slip-type hanger. We run a shutin temperature survey in the well and find the temperature at the top of the cement is 220° F and its gradient is linear to a surface temperature of 70° F; that is, $T = 70 + 0.02h$, where h is the vertical depth. Below that point, we find that the formations are much hotter, and the temperature at 14,000 ft is 370° F. To keep this simple, let us say that our heat transfer model predicts that, with the anticipated production rate, the heat transfer will reach a near steady state with a temperature increase of 150° F uniformly along the casing string. We ignore any stresses caused by any temperature change between the time the cement set and the temperature survey was run and any effects from the possible expansion of fluids in the annulus outside the 7 in. casing, although expansion of trapped fluids is not something we can ignore in many cases (see Halal and Mitchell, 1993). We also assume that the pipe stays straight and does not buckle. The cross-sectional area of our casing is 7.55 in.^{2}.

We would like to determine the following.

1. The axial stress in the casing at the top of the cement.
2. The axial stress in the casing at the surface.
3. The amount of tension at the surface to avoid any compression in the casing due to the temperature increase during production.

Since the casing is constrained at the top and the bottom (wellhead and cement), there is no axial strain due to the change in temperature. The axial stress at the top is

$$\sigma_z = \sigma_o - E\alpha\Delta T = \frac{275000 + 50000}{7.55} - 30 \times 10^6 \left(0.69 \times 10^{-5}\right)150$$

$$\approx 12000 \text{ lbf/in}^2$$

and, at the top of the cement, it is

$$\sigma_z = \sigma_o - E\alpha\Delta T = \frac{13000 + 50000}{7.55} - 30\times10^6\left(0.69\times10^{-5}\right)150$$

$$\approx -23000 \text{ lbf/in}^2$$

The top of the casing is still in tension, but the bottom is in compression. What is the magnitude of the compressive force at the bottom?

$$F_z = \sigma_z A_t = -23000(7.55) \approx -173600 \text{ lbf}$$

Suppose we are concerned about buckling. What amount of tension should we pull so that the casing does not go into compression at the top of the cement with this amount of temperature change?

$$\frac{F_z + 13000}{7.55} - 30\times10^6\left(0.69\times10^{-5}\right)150 = 0$$

$$F_z \approx 221000 \text{ lbf (additional at the bottom)}$$

$$F_z = 275000 + 221000 = 496000 \text{ lbf (total hanging weight at surface)}$$

Assuming we use proprietary couplings with a higher tensile strength, the pipe body yield of this casing is only 604,000 lbf, leaving us with a tensile design factor of about 1.22, which is very low. If we down-rate the yield to 90%, as mentioned earlier, then we must look at other options. Additionally, we must consider whether our wellhead and conductor would support that amount of weight. We actually might have to live with some amount of buckling in the lower section of this casing string. In that case, we also have to determine the limit of our connections in compression near the bottom. There are no formulas for determining the compression strength of API connections. Some manufacturers of proprietary connections have compression strength data for their couplings, although that information usually is not published. See Jellison and Brock (2000) for a discussion of connections in compression. An alternative approach would be to bring the cement higher to a point where the casing tension is greater, but one also must consider that cemented casing is not necessarily immune to coupling failure in compression with large temperature increases.

The preceding is a simple calculation, and you will note, we made a lot of simplifying assumptions. We did not consider any inclination and friction in the bore hole that would resist pipe motion, and we assumed that the temperature change would be constant along the entire length of the string. However, we can use simple calculations like that to spot-check thermal stresses at various points to determine if we need a more in-depth investigation.

8.8 Closure

We covered a lot of ground in this chapter. Some topics we covered in more detail than others. This is not to slight any particular topic, but the material in this chapter easily could constitute a separate book. In the next chapter, we look at inclined and curved well bores. Much of what was covered in this chapter and the previous one will carry over into it and will be further explained as to how it applies in those circumstances.

8.9 References

API 5B1. (August 1999). "Threading, Gauging, and Thread Inspection of Casing, Tubing, and Line Pipe Threads." *API Recommended Practices 5B1.*Washington, DC: American Petroleum Institute.

API 5C2. (October 1999). *Bulletin on the Performance Properties of Casing, Tubing, and Drill Pipe.* API Bulletin 5C2. Washington, DC: American Petroleum Institute.

API 5C3. (October 1994). *Bulletin on Formulas and Calculations for Casing, Tubing, Drill Pipe, and Line Pipe Properties.* API Bulletin 5C3. Washington, DC: American Petroleum Institute.

Chater, E., and J. W. Hutchinson. (1984). "On the Propagation of Bulges and Buckles." *Journal of Applied Mechanics, Transactions of the ASME* 51: 269–277.

Dawson, R., and P. R. Paslay. (October 1984, Oct). "Drill Pipe Buckling in Inclined Holes." *Journal of Petroleum Technology*: 1734–1740.

Halal, A. S., and R. F. Mitchell. (1993). *Casing Design for Trapped Annulus Pressure Buildup.* SPE/IADC 25694. Richardson, TX: Society of Petroleum Engineers.

He, X., and A. Kyllingstad. (1993). *Helical Buckling and Lockup Conditions for Coiled Tubing in Curved Wells.* SPE 25370, Richardson, TX: Society of Petroleum Engineers.

Holliday, G. H. (1969). "Calculation of Allowable Maximum Casing Temperature to Prevent Tension Failures in Thermal Wells." ASME Petroleum Mechanical Engineering Conference, Tulsa, OK, September 21–22.

ISO/DIS 10400. (2004, draft). *Petroleum and Natural Gas Industries—Formulae and Calculations for Casing, Tubing, Drill Pipe, and Line Pipe Properties.* Geneva: International Organization for Standardization.

ISO 11960. (2001). *Petroleum and Natural Gas Industries—Steel Pipes for Use as Casing and Tubing for Wells.* Geneva: International Organization for Standardization.

Jellison, M. J., and J. N. Brock. (December 2000). "The Impact of Compression Forces on Casing-String Designs and Connectors." *SPE Drilling and Completion*: 241–248.

Klever, F. J., and T. Tamano. (2004). *A New OCTG Strength Equation for Collapse under Combined Loads.* SPE 90904. Richardson, TX: Society of Petroleum Engineers.

Klinkenberg, A. (1951). "The Neutral Zone in Drill Pipe and Casing and Their Significance in Relation to Buckling and Collapse." In *Drilling and Production Practice 1951.* Dallas [now Washington, DC]: American Petroleum Institute.

Kyriakides, S., C. D. Babcock, and D. Elyada. (1984). "Initiating of Propagating Buckles from Local Pipeline Damages." *Journal of Energy Resources Technology, Transactions of the ASME* 106: 79–87.

Lubinski, A. (1951). "Influence of Tension and Compression on Straightness and Buckling of Tubular Goods in Oil Wells." *Division of Production* 31, no. 4: 31–56.

Ludwick, P. (1909). *Elemente der Technologischen Mechanik.* Berlin: Springer Verlag.

Mitchell, R. F. (December 1999). "A Buckling Criterion for Constant-Curvature Wellbores." *SPE Journal* 4, no. 4: 349–352.

Mitchell, R. F. (2003). "Lateral Buckling of Pipe with Connectors in Curved Wellbores." *SPE Drilling and Completion*: 22–32.

Ramberg, W., and W. Osgood. (1943). *Description of Stress-Strain Curves by Three Parameters.* NACA Tech Note Number 902. Langley, VA: National Advisory Committee for Aeronautics [now NASA].

Tamano, T., T. Mimaki, and S. Yanagimoto. (1983). "A New Empirical Formula for Collapse Resistance of Commercial Casing." *Journal of Energy Resources Technology, Transactions of the ASME.*

Woods, H. B. (1951). "Discussion on Paper by Klinkenberg (1951)." In *Drilling and Production Practice 1951.* Dallas [now Washington, DC]: American Petroleum Institute.

Yeh, M. K., and S. Kyriakides. (1986). "Collapse of Inelastic Thick-Walled Tubes under External Pressure." *Journal of Energy Resources Technology, Transactions of the ASME* 108: 35–47.

CHAPTER 9

Casing in Directional and Horizontal Wells

9.1 Introduction

Many wells drilled each year are directional wells, and an increasing number of those are horizontal wells. In one sense, all wells are inclined to some degree, and the phenomena we are about to discuss affect them too, even though we usually ignore them when it comes to casing design for "vertical" wells. We are now ready to consider wells in which these phenomena cannot be ignored. What are they? There are principally two: bore-hole friction and bore-hole curvature.

Simply put, bore-hole friction affects the axial load in casing by resisting its motion. An upward motion of the casing increases the tension, and a downward motion decreases the tension. If the casing is rotated, this too adds a torsion load to the casing—something we never consider in basic casing design. Bore-hole curvature causes bending stresses in casing. This stress was ignored in most casing designs, that is, until horizontal wells began being drilled. Even then, it is surprising how many casing strings are still designed for horizontal wells that do not take bending into account, possibly because the bending stress magnitude is not understood to be significant. For example, a string of 7 in. K-55 casing run through the build section of a medium-radius well with a radius of curvature of 300 ft will have a bending stress amounting to over 50% of its yield strength, and that does not account for additional stress from friction, gravity, or pressure. Bending stress is not insignificant in horizontal wells.

In this chapter, we look at bore-hole friction and curvature and their effects on casing design. We also consider combined loads in directional wells with some examples of how we do the calculations.

9.2 Bore-Hole Friction

Friction is a resistance to motion between two bodies or media. We all studied it in basic physics or engineering courses and learned a so-called friction law for rigid bodies. It is not a law of physics at all, but you might not guess that from the way it is often presented in basic physics texts. Friction is quite complex by its very nature, and the simple friction relationship most of us learned does not hold up well in many real-life situations. However, it is a simple relationship, and it works well enough for numerous practical applications.

9.2.1 The Amonton-Coulomb Friction Relationship

The simple friction relationship is often referred to as the *Coulomb friction law* or a bit more accurately as the *Amonton-Coulomb friction law.* It was originally the outcome of two postulates by Amonton in 1699 and has been understood in its present form since about 1790, when Coulomb added a third postulate to it.

- Frictional force is proportional to the weight of the body being moved (Amonton, 1699).
- Frictional force is independent of the apparent contact area (Amonton, 1699).
- Frictional resistance is independent of the sliding velocity (Coulomb, 1790).

That is the simple friction relationship we all learned and is stated mathematically as

$$F \leq \mu N \tag{9.1}$$

It says that the frictional force is less than or equal to a friction coefficient multiplied by the normal contact force, *normal* meaning perpendicular to the contact surface. The relationship is necessarily an inequality, because the product of the friction coefficient and the normal contact force is equal to the frictional force only when the force opposite the friction force is equal to or greater than that product. In other words, if the force applied to generate motion of a body is less than μN, then the frictional force is equal to the applied force and not that product. Once the body is in motion, the friction force is equal to μN and independent of the applied force, as long as the motion is sustained. In 1699, Amonton was concerned with objects sliding on a level surface, so he used weight instead of contact force in his postulate.

What are the assumptions in that relationship? There are several, and they often are not mentioned in basic texts.

- The contact surfaces are smooth.
- The contact surfaces are dry (uh, oh!).
- The contact surfaces do not deform.
- The friction coefficient is a constant, that is, not affected by the heat generated.

We could add more, but that about covers the areas of our interest. We should discuss these limitations briefly.

When we require that the surfaces are smooth, we are talking about a matter of scale. On a microscopic scale, even smooth surfaces are not what we would consider smooth. There are numerous asperities that are elastically deformed, plastically deformed, fractured, melted, fused, and so forth, as two surfaces slide relative to one another. But these microscopic asperities are small compared to the entire surface area, and the distribution of the asperities on the smooth surfaces is relatively uniform. For example, if we could drag an object the size of an Egyptian pyramid through the Grand Canyon, we could not model the resisting force with Amonton-Coulomb friction because the "asperities" of the Grand Canyon are on a similar scale to the pyramid-sized object. However, if we had two geotectonic plates sliding over each other, we might model the friction with the Amonton-Coulomb relationship, because asperities the size of a large canyon or a pyramid are relatively small compared to the surface contact area and thickness of geotectonic plates. Casing sliding in a well bore is generally on a scale that allows the use of such a relationship.

What about deformation of the contact surfaces? Suppose we have a string of casing with LT&C couplings and a string of identical-weight casing with tapered integral connections. Which is going to slide in the hole (or out of the hole) easier? No question about it, the casing with the tapered connections will slide easier than the casing with the square shouldered couplings, even though both may have identical contact force. The friction relationship does not account for couplings gouging a borehole wall. But again, that is a matter of scale, since a long string of casing has many connections (asperities?) of uniform size. Their gross effect might be legitimately included in a friction coefficient. In other words, the coupling shape can be accounted for by the friction coefficient if we are considering numerous joints of casing.

What can we say about the effects of lubrication? Can the Amonton-Coulomb dry-friction relationship work for lubricated surfaces? In general, the answer is no, but again it depends on scale and the accuracy desired. If the lubrication is consistent, the dry-friction relationship can give reasonably practical results. By *consistent*, we mean that the friction coefficient does not vary significantly with contact force. The walls of a well bore usually are covered by a filter cake and the bore hole contains some type of liquid drilling fluid. This provides considerable lubrication as the casing slides along the well bore. If the casing couplings scrape the filter cake off portions of the wall as it slides, that changes the friction coefficient; in other words, the friction coefficient may increase in those areas. Another thing that may happen is that, with removal of some of the filter cake, the casing in contact with permeable formations may tend to be forced harder against the wall due to the difference in hydrostatic pressure in the well bore and formation. In this case, the contact force has increased. We may reasonably account for some of these things by lumping all we do not know into some average friction coefficient. The one problem we may encounter, though, is that the friction coefficient may vary with pipe diameter and weight.

The single question that most often arises about the Amonton-Coulomb relationship concerns the postulate that the frictional force does not depend on the apparent area of contact. If that is true, then why do racing cars have wide tires rather than narrow? The simple answer is that the postulate is true, but racing tires are much more complicated. Briefly, the crux of the matter is the last qualification mentioned, is the one regarding heat. For the most part, a narrow tire of the same composition gives the same frictional resistance on a given race car as a wide one—at low speeds. At high speeds, several things come into play, and heat is a major one. Wide tires distribute the heat over a larger area, and the heat is dissipated to the atmosphere more quickly. There is also the matter of wear or ablation of the tire. A narrow tire with a much smaller area of compound in contact with the track wears in its radial dimension much faster than a wider tire. In fact, a narrow tire might make only a few laps before it would have to be replaced. On a wet track, racers have to switch to grooved tires, which actually reduce the contact area. This is a matter of hydrodynamics, where the narrower contact areas tend to cut through the lubricating layer of water, whereas the slick tires tend to "float" on top of it. A volume could be written on racing tire performance. That is totally off the subject, but it comes up in almost every discussion of the Amonton-Coulomb friction relationship, so that is why we address it. Still on the issue of heat, a good example of a breakdown of the simple friction relationship is automobile brakes. This is a practical example of heat affecting the friction coefficient significantly. Brakes are said to "fade" when they get hot, in other words the friction coefficient is reduced with heat. That is true in most cases of friction. Fortunately, we generally need not be concerned about the heat of friction affecting the motion of casing in a bore hole. It can affect casing wear, however.

Most of the preceding discussion on the limitations of the Amonton-Coulomb friction relationship has to do with linearity. In other words, the assumption of the relationship is that the friction coefficient is a constant. As long as the friction coefficient is not a function of the contact force, equation (9.1) is simple and easy to use. Also, we easily can see that, if we change our mud properties to reduce the friction coefficient by 25%, for instance, then we also reduce the frictional force by 25%. Likewise, if we reduce the contact force by some amount, we also reduce the friction force by the same amount. Or, if we change our casing design by increasing the wall thickness of the casing in a horizontal section of a well, we increase the friction force in that part of the hole proportionally.

Before we leave this friction relationship, there is one other important point to cover. When most of us learned the simple version of friction, we were taught that two friction coefficients apply to a particular problem, a static friction coefficient, μ_s , being the larger, and a kinetic friction coefficient, μ_k , being the smaller. That means the force to initiate motion is greater than the force required to sustain the motion once initiated. Which do we use for casing design? It would seem obvious that a static friction coefficient is more realistic, since motion of the pipe ceases and initiates for each connection in the casing string. This might not seem worth the concern, if we are considering the running process, but if we are going to reciprocate the casing while cementing or if we encounter an obstruction in the bore hole before reaching bottom and have to pull the casing out of the hole, it is a serious concern. The caveat about static friction coefficients is that they problematic except for rigid bodies. We see an example of that on the rig weight indicator all the time. When pipe is picked up off bottom we see the weight indicator increase gradually until some maximum point is reached, then it drops back to a slightly lower value. That pretty much seems to confirm what we have been taught, but reconsider what we really observe. There is more to this than what we might at first think. When the driller starts to pick up the pipe, we see it moving through the rotary as the weight indicator reading increases. Some of the pipe already is moving before we reach the peak load. And, before the final peak is reached, a lot of the pipe is moving. The pipe is not a rigid body like the simple objects we encountered in basic friction applications. Once the entire string of pipe is in motion, the situation is fairly simple, since it is moving more or less as a rigid body and an approximate kinetic friction coefficient pretty much predicts the resistance to motion. But, what about the initiation stage? What is going on there? Obviously, if some of the pipe is in motion, we cannot assume a single static friction coefficient and apply it to the entire string. This brings up a basic flaw in the Amonton-Coulomb friction relationship. Suppose we have a rigid body with a total weight, W, resting on a flat surface, as in Figure 9–1. We apply a gradually increasing force, P, to the body, and the friction force, F, resists motion.

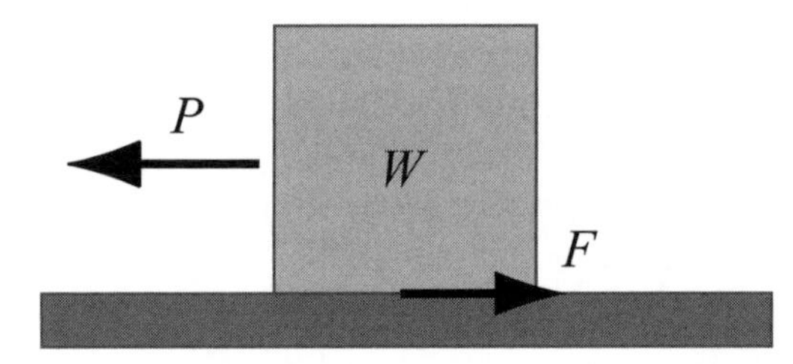

Figure 9–1 *A rigid body of weight, W, at rest on a flat surface.*

As P increases so does the value of F, which is equal to P, until it reaches its maximum value, $F = \mu_s W$. As P continues to increase, the value of the friction force decreases suddenly to its maximum dynamic value of $F = \mu_k W$. It remains at that value as long as $P \geq F = \mu_k W$. This is shown in Figure 9–2.

An extensible body does not behave like that. Suppose now, we model the same body with four pieces instead of one and connect them with weightless springs to simulate extensibility. The total weight is the same, but each segment now weighs one fourth the total. This is illustrated in Figure 9–3.

We stipulate that initially the segments are static, and there is no load in the connecting springs. In this case, as we apply a force, P, we see that initially the only body acted on until motion is initiated is the one on the left. No force is transmitted to the next segment until the first one moves. As the static friction in the first segment is overcome, the friction force drops back to the kinetic value. As we continue to increase the

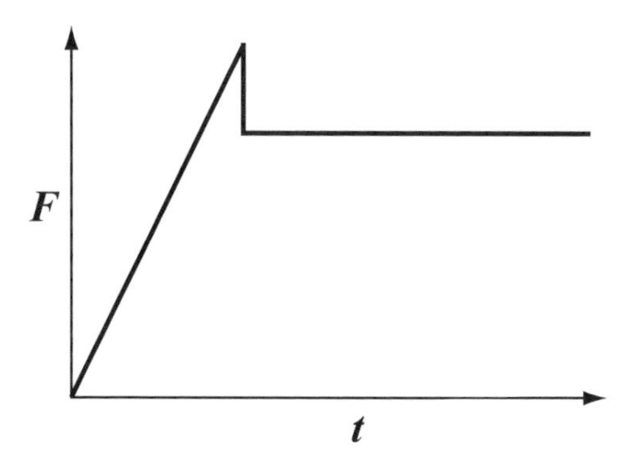

Figure 9–2 *Static friction force increases until motion is initiated, then changes to a kinetic friction force.*

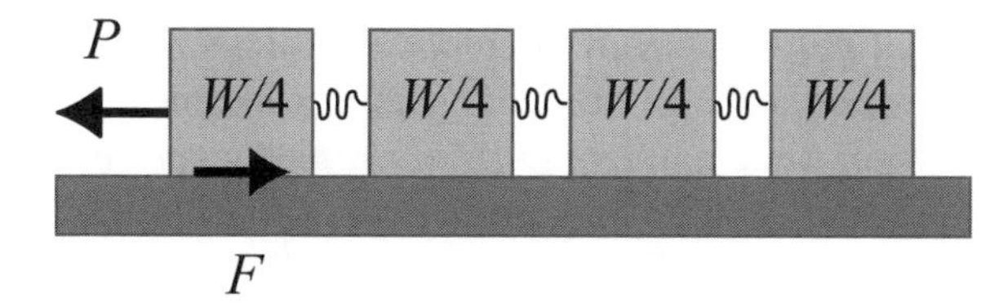

Figure 9–3 *Extensible body model.*

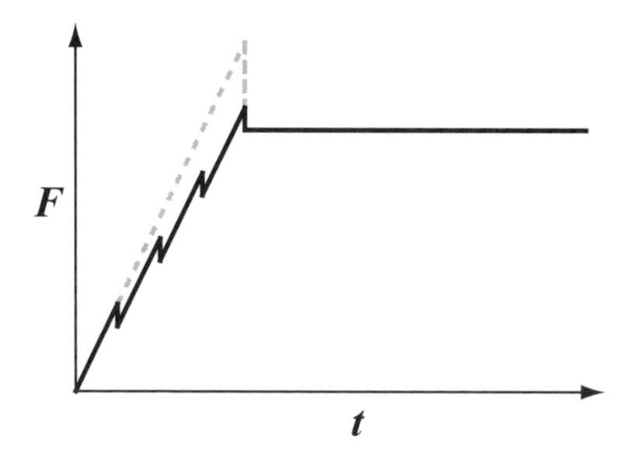

Figure 9–4 *Friction force in a segmented body.*

force, P, the same thing takes place successively in each of the other segments, but when the final segment is set in motion the total difference between the maximum static friction force and the maximum dynamic friction force is only one fourth that of the previous rigid body example. The friction force for the segmented example is shown in Figure 9–4.

Now, suppose we divide the body into an infinite number of segments attached by weightless springs. We could say that the difference between the static friction and dynamic friction disappears altogether, and it does, at least in this model. However, some would argue that the weight of each segment goes to zero; and while that argument may at first seem valid, it does not affect the limit. Since the static friction coefficient is greater than the kinetic friction coefficient, we can show the friction force at any time as the sum of the segments in motion times the kinetic friction coefficient plus the weight of one static segment times the static friction coefficient.

$$\mu_k < \mu_s \rightarrow F = \mu_k \frac{W}{n}(n-i) + \mu_s \frac{W}{n} \tag{9.2}$$

where n is the total number of segments, $n-i$ is the number of segments in motion, and i is the number of static segments. We can take the limit of the friction force as the number of segments goes to infinity:

$$\lim_{n\to\infty} F = \mu_k W \lim_{n\to\infty}\left(\frac{n-i}{n}\right) + \mu_s \lim_{n\to\infty}\left(\frac{W}{n}\right)$$

$$\lim_{n\to\infty} F = \mu_k W(1) + \mu_s(0)$$

$$\lim_{n\to\infty} F = \mu_k W \tag{9.3}$$

That may be a bit over simplified, but what we have shown is that, for an extensible body, a static friction coefficient does not exist in the context of the linear Amonton-Coulomb friction relationship. This is not to say that a static friction coefficient does not exist even though all bodies are extensible to some degree, but that there are serious deficiencies in the linear Amonton-Coulomb friction relationship in regard to extensible bodies. It also points out that, for casing in a well bore, we cannot assume some value for a static friction coefficient, as with a rigid body, and use it with any degree of predictable accuracy.

So, as to calculating casing loads with friction, we are pretty comfortable with a kinetic friction coefficient. We know that it will take more force to initiate motion, but to determine a static friction coefficient for a particular well generally is not addressed because it is not possible. All we can truthfully say is that we require some amount of force greater than the kinetic friction force to initiate motion in a casing string that has come to rest. Another thing that we may notice in many well bores is that the initiating force often increases with time. That usually is a sign of differential sticking, which in terms of friction translates into increased contact force rather than a different friction coefficient. In many such cases, the friction coefficient becomes a catchall for all those things we cannot quantify otherwise.

There is another important concept to understand when dealing with directional wells: There is some critical inclination angle at which a body is at static equilibrium with movement impending. In engineering terms, this is known as the *angle of repose*, except in that case the angle is measured from horizontal. Since we measure inclination from the vertical, we simply call it a *critical inclination angle* (see Figure 9–5), meaning that any casing in the well where the inclination is greater than this value will not slide under its own weight; that is, it has to be pushed into the hole.

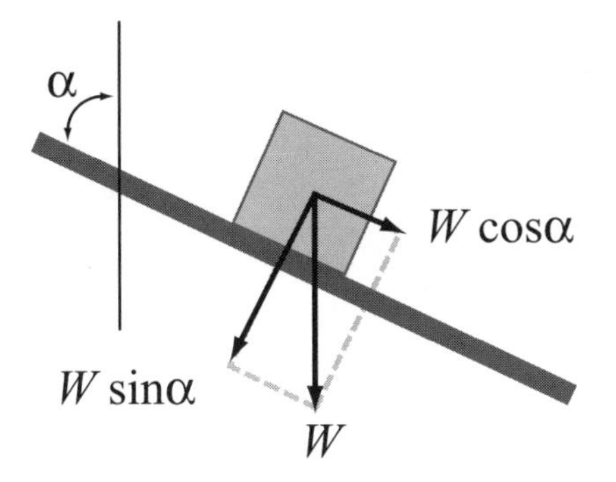

Figure 9–5 *Critical inclination angle.*

The critical inclination angle depends on the friction coefficient, and it can be derived as follows:

$$W\cos\alpha_c - \mu W\sin\alpha_c = 0$$

$$\tan\alpha_c = \frac{1}{\mu}$$

$$\alpha_c = \tan^{-1}\frac{1}{\mu} \tag{9.4}$$

Since we know the friction coefficient only approximately, this formula at best gives us an idea as to where in our well bore the casing will cease to slide due to its own weight. For example, if the friction factor is 0.4, then the critical inclination angle, $\alpha_c = 68^{\circ}$. We know, then, that we cannot expect the casing to slide due to its on weight and must be pushed anywhere the inclination exceeds $68°$. All of us have seen similar limiting values in relation to wire line logging tools.

9.2.2 Calculating Bore-Hole Friction

A number of commercial friction software models on the market calculate bore-hole friction, probably to the point that hardly anyone cares how they work as long as they give reliable results. One does not really have to have software, since the calculations can be done manually, although it is a tedious process. In addition to the tedium, doing bore-hole friction calculations manually is error prone, in that if an error is made, it carries through to all the subsequent calculations. One can fairly easily program one of the models into a spreadsheet and get results as accurate as any commercial software, although the commercial software has numerous options to make life much easier.

Some assumptions are common to all the current models for bore-hole friction:

- The Amonton-Coulomb friction relationship is valid.
- The tubular string is a rigid body in translation.
- The tubular string is a rigid body in rotation.
- The tube has no bending stiffness.
- The tube is in contact with the bore hole everywhere.

The first assumption already has been discussed adequately. The second, third, and fourth assumptions are referred to collectively as the *soft string* assumptions. None of them is true, but that needs explanation. If the entire string is in translation or rotation, it does not matter if it is extensible or not because we are primarily interested in the value of friction when the entire string is in translation or rotation and not some intermediate values when only a portion of the string is in motion. The fourth assumption is that bending the pipe around a curve does not add to the contact force. In other words, the contact force required to bend the tube is ignored. This is reasonable for drill pipe in most cases, and even for casing in most situations, so we can generally accept this assumption. Some of the commercial software now includes contact force due to bending stiffness. The fifth assumption, regarding discontinuous contact, might be questioned, however. Without better data than we currently get regarding the bore-hole path and shape, it is not possible to determine where the pipe is in contact with the bore-hole wall and where it is not. This is not a serious limitation, though, because of the first assumption in the list. Since the Amonton-Coulomb friction relationship is independent of the area of contact, the portion of the pipe that is not in contact is accounted for in the contact force where the pipe is in contact. From a practical standpoint, the current torque-and-drag software works quite well.

Most of the commercial software is based on the model of Johancsik, Frieson, and Dawson (1984), which is a difference equation that uses the buoyed specific weight of the pipe in its calculations. The result is an effective axial load. Another model was formulated by Sheppard, Wick, and Burgess (1987). It is in the form of a differential equation and it ultimately produces the true axial load, with the effective axial load as an intermediate step. One might question why the commercial models use the former instead of the latter. It was published first, it is easy to program, and not many people use the true axial load for casing design in directional wells, because the unfortunate truth is that many do not know the difference. When programmed, the second model actually does the same thing in terms of buoyed specific weight and results in an effective axial load, but it also shows how to take that result one step further to calculate the true axial load. The issue is inconsequential from a practical standpoint, in that both models produce the effective axial load and one goes a step further to produce the true axial load. The true axial load formulas of Sheppard et al. may also be used with the Johancsik et al. model results to determine the true axial load. The differential equation of Sheppard et al. is not solvable in closed form, except in the case of a single plane, that is, constant azimuth, and the assumption that the bore-hole curvature is that of a segment of a circle between survey points. (A circle-segment well path between surveys is the basic assumption of the minimum curvature method currently used in directional drilling.) However, the single-plane, or 2-D, closed-form solution is of little practical use, because it can be used only for idealized single-plane well paths and it results in two closed-form equations. This is because, at points where the contact is on the high side of the hole, the gravitational force subtracts from the contact force, and when the contact is on the low side of the hole, the gravitational force adds to the contact force. An incremental calculation method must be used to determine when to switch from one solution to the other. A number of tests were done using the closed-form solution to test the accuracy of numerical techniques for solving Johancsik et al.'s difference equation and Sheppard et al.'s differential equation. A simple incremental method was used in Johancsik et al.'s equation. An Euler method, a second-order Taylor series, and a fourth-order Runge-Kutta method were used for solving Sheppard et al.'s differential equation. As it turns out, the simple incremental method gave almost identical results with the difference equation as the more sophisticated fourth-order Runge-Kutta method for the differential equation, when the same number of increments were used. The Euler method never approached the order of accuracy of the other methods, even with twice the number of increments. The net result is that a simple incremental solution to either equation gives acceptable results, as long as one uses a sufficient number of subintervals between survey points. The following is the differential equation of Sheppard et al.:

$$\frac{\mathrm{d}\hat{F}}{\mathrm{d}s} = \hat{w}\cos\alpha \pm \mu\left[\left(\hat{F}\frac{\mathrm{d}\alpha}{\mathrm{d}s} + \hat{w}\sin\alpha\right)^2 + \left(\hat{w}\frac{\mathrm{d}\beta}{\mathrm{d}s}\sin\alpha\right)^2\right]^{\frac{1}{2}} \tag{9.5}$$

where

$\hat{F}$ = effective axial load

$\hat{w}$ = buoyed specific weight of casing (effective specific weight)

s = coordinate along the casing central axis

μ = friction coefficient

α = inclination angle

β = azimuth angle

The first term on the right is the gravitational contribution to the axial load, and the second term is the frictional contribution, that is, the friction coefficient multiplied by the contact force. The ± sign is determined by whether the pipe motion is into the well (negative) or out of the well (positive). Two things to note about this equation are that the axial load appears on both sides of the equation and the contact force always is positive. In a straight section of bore hole, the axial load is dependent on the contact force, but the contact force is not dependent on the axial load. In that case, the axial load disappears from the contact force term. In a curved bore hole, the axial force is dependent on the contact force as in a straight section , but also the contact force is dependent on the axial load. The differential equation may be solved numerically as an initial-value problem using a second-order Taylor method or a fourth-order Runge-Kutta method. As previously mentioned, an Euler method does not give very good results, even with a significantly greater number of increments. An incremental formulation of the Sheppard et al. equation gives equally good results as the more sophisticated Taylor and Runge-Kutta techniques with sufficient number of increments. The initial condition at the bottom of the hole is $\hat{F}(0)=\hat{F}_o$ which accounts for any weight set on bottom. The incremental form is

$$\hat{F}_n=\hat{F}_o+\sum_{i=1}^{n}\left[\left(s_i-s_{i-1}\right)\hat{w}_i\cos\left(\frac{\alpha_{i-1}+\alpha_i}{2}\right)\pm\mu_i N_i\right] \tag{9.6}$$

where the contact force is given by

$$N_i=\left(s_{i-1}-s_i\right)\left\{\begin{array}{l}\left[\hat{w}_i\sin\left(\frac{\alpha_i+\alpha_{i-1}}{2}\right)+\hat{F}_{i-1}\left(\frac{\alpha_i-\alpha_{i-1}}{s_{i-1}-s_i}\right)\right]^2\\+\left[\hat{F}_{i-1}\left(\frac{\beta_i-\beta_{i-1}}{s_{i-1}-s_i}\right)\sin\left(\frac{\alpha_i+\alpha_{i-1}}{2}\right)\right]^2\end{array}\right\} \tag{9.7}$$

In this last equation, the angle measurements must be in radians. (When angles appear in formulas outside of trigonometric functions, they are almost always in radians.) The numbering of the nodes starts at the bottom of the hole with node 0 and proceeds to the top as seen in Figure 9–6.

Using these formulas one may write a simple spreadsheet program to do the calculations of bore-hole friction for casing design.

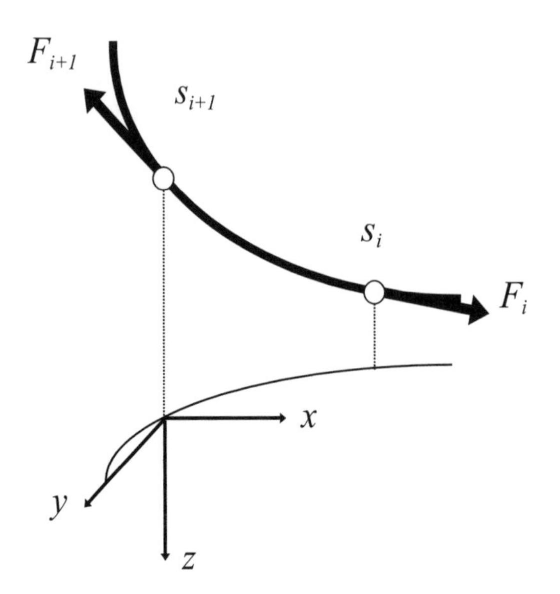

Figure 9–6 *Node numbering system for bore-hole friction calculations.*

To convert the values of effective load at any node to true load, one may use the formula from Chapter 2:

$$F_i = \hat{F}_i - (p_o A_o - p_i A_i) \tag{9.8}$$

There are times when we might consider rotating the casing while cementing. This is not often done in highly inclined or horizontal wells because of the significant amount of friction. However, if it can be done at a torque less than the maximum recommended makeup torque of the connections, then there is a possibility of doing it. After all, getting a good cement job in a horizontal well is difficult enough as is, and everything we could do to displace the mud is worthwhile. The differential form of the torsion equation is

$$\frac{\mathrm{d}\tau}{\mathrm{d}s} = r\mu \left[\left(w \sin\alpha + F \frac{\mathrm{d}\alpha}{\mathrm{d}s} \right)^2 + \left(F \frac{\mathrm{d}\beta}{\mathrm{d}s} \sin\alpha \right)^2 \right]^{\frac{1}{2}} \tag{9.9}$$

where r is the radius of the casing. This is also a one-dimensional differential equation with the boundary condition, $\tau(0)=\tau_o$ at the bottom of the tube, although that would be zero for most casing strings. We could solve this initial-value problem using a second-order Taylor method or a fourth-order Runge-Kutta as previously mentioned. We could just as easily cast this equation in an incremental form, where it becomes

$$\tau_n = \tau_o + \sum_{i=1}^{n} r_i \mu_i N_i \tag{9.10}$$

and the normal contact force is calculated by

$$N_i = \left(s_{i-1} - s_i\right)\left\{ \begin{aligned} &\left[\hat{w}_i \sin\left(\frac{\alpha_i + \alpha_{i-1}}{2}\right) + \hat{F}_{i-1}\left(\frac{\alpha_i - \alpha_{i-1}}{s_{i-1} - s_i}\right)\right]^2 \\ &+ \left[\hat{F}_{i-1}\left(\frac{\beta_i - \beta_{i-1}}{s_{i-1} - s_i}\right)\sin\left(\frac{\alpha_i + \alpha_{i-1}}{2}\right)\right]^2 \end{aligned} \right\} \tag{9.11}$$

In this case, the axial load used to calculate the normal contact force is not itself a function of the contact force, so it may be calculated separately for use in equation (9.11) from

$$\hat{F}_i = \hat{F}_o + \sum_{k=1}^{i} \left(s_{k-1} - s_k\right) \hat{w}_k \cos\left(\frac{\alpha_{k-1} + \alpha_k}{2}\right) \tag{9.12}$$

An incremental approach such as this and the one for sliding friction are used in most commercial software. And, as previously mentioned, numerical techniques do not give better solutions when applied directly to the differential form as long as a sufficient number of increments are used.

The drawbacks to using these models for casing design are the lack of actual friction coefficients and the idealized well plan as opposed to the actual hole as drilled. There are no tables of values in which to look up friction coefficients for bore holes. There are some average values for water-based drilling fluids:

- 3.0–4.0 for an open hole.
- 2.0–2.5 for a cased hole.

If one were using an oil-based mud, those values might be reduced by 30–50%. In practice, measurements of the actual hook load are made in the field, and the values are plugged into a commercial torque-and-drag model, which iteratively finds a friction coefficient that gives results matching the field measurements. This is of great benefit during drilling operations but of little use if the casing is being designed before the well is drilled, unless one has data from previously drilled wells. Even if we have the correct friction coefficient, the next problem facing the casing designer is the well path.

In a conventional L-shaped horizontal well, for instance, the vertical portion of the well plan is exactly that—vertical! There is no friction in that portion of the hole according to the models. And the rest of the hole also is totally smooth with no wobble. The actual hole is quite different, there is friction in the "vertical" section, and none of the rest of the well is quite as smooth as the planned well path. How do we deal with that? One possibility is to use the data from a similar well with some possible adjustments. Some commercial software have a way to impose some tortuosity on the planned well path in the form of a sinusoidal curve some type of random "noise" curve. Both are good, but they require some experience to know how much tortuosity to use. One other method is to add an inclination of a few degrees to the "vertical" section of the plan and use a "high" friction coefficient for casing design. Figure 9–7 shows the calculated true axial load for upward motion of 5½ in. casing in a well plan and the true axial load in this well as it was actually drilled.

For most wells, it is not critical what method is used, as long as it is recognized that one cannot use a perfect well path plan to design a casing string for a real well. Long-reach and high-pressure wells may require a lot more planning and even considerable effort to drill the well as nearly smooth and close to the planned path as possible.

The result of the friction calculations just discussed is a load curve similar to Figure 9–7 that we would employ for designing our casing string for tensile loads. In that particular curve, the friction is calculated for an upward motion of the casing string once on bottom, similar to what we would expect if reciprocating the casing while cementing. We can plot

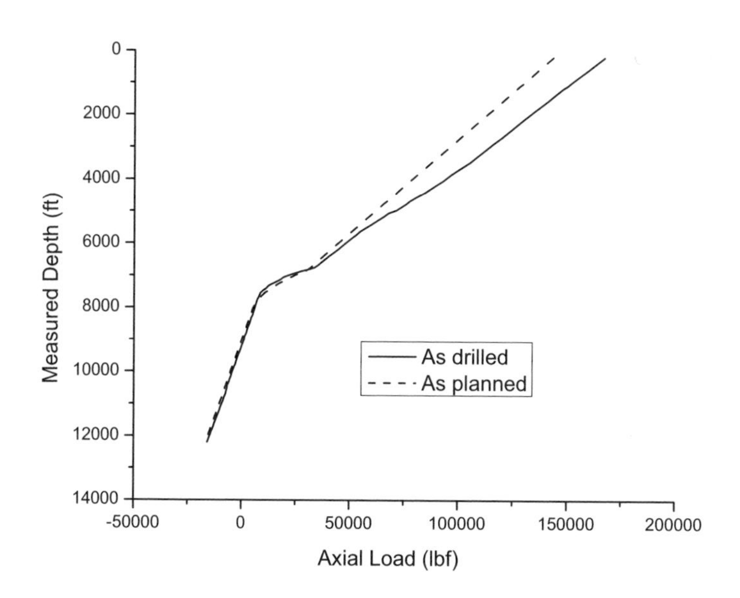

Figure 9–7 *Drag friction distribution for 5½ in. casing pulled off bottom in a horizontal well, as planned and as drilled.*

these curves either in terms of effective axial load or true axial load, depending on how we intend to design our casing string. If we expect to rotate the casing while cementing, we would also want a plot of the rotating torque from friction, so that we might verify whether or not the connections of our string will permit rotation. Maximum recommended makeup torque generally is the limiting factor in rotating most casing strings. The main point here is that, in highly deviated wells, we do not use the simple vertical well assumptions we used in Chapter 5. That having been said, there is one other important point: The friction curves do not account for bending stresses in the pipe due to bore-hole curvature. That is the next topic.

9.3 Curvature and Bending

Curved well bores add stress to casing, and often that added stress is quite significant. A general lack of understanding of this has led to casing failures in the build sections of a number of horizontal wells. We now examine the effects of bore-hole curvature and the resulting bending stresses in casing.

But, before we get into particulars, we need to see a couple of formulas, because we encounter them time and again in working with curved well bores. Curvature is the change in angle with respect to the distance along the path, or mathematically,

$$\kappa \equiv \frac{d\theta}{ds} \tag{9.13}$$

where

κ = curvature
θ = reference angle (inclination angle in 2-D well bore)
s = distance along cuved path (well bore length measurement)

The curvature may be expressed in two forms: as curvature, meaning a change in angle as just shown, or a radius of curvature, R:

$$R = \frac{1}{|\kappa|} \tag{9.14}$$

There is a mathematical quirk here, in that the curvature can be either positive or negative but the radius of curvature always is positive, since a negative radius is meaningless.

Curvature is measured in units of reciprocal length, that is, L^{-1}. In oil-field parlance, curvature is called by a colorful term, *dog-leg severity*, in reference to a crooked hole. It usually is measured in degrees per 100 ft or degrees per 10 m. To convert from curvature as used in oil-field terminology to radius of curvature and vice versa, we need a formula. In oil-field units this is

$$\kappa = \frac{18000}{\pi R}$$

$$R = \frac{18000}{\pi \kappa} \tag{9.15}$$

where

κ = curvature (dogleg severity), degrees/100 ft
R = radius of curvature, ft

In SI units, the conversion formula is

$$\kappa = \frac{1800}{\pi R}$$

$$R = \frac{1800}{\pi \kappa} \tag{9.16}$$

where

κ = curvature (dogleg severity), degrees/10 m
R = radius of curvature, m

There are two other versions of curvature in metric units, degrees per 30 m and degrees per 100 m. The former is numerically approximately the same as degrees per 100 ft and was used for a number of years but is fading from popularity. When using degrees per 30 m, the numerator is 5400. The latter measure, degrees per 100 m, was common about 20 years ago but does not see much use today.

We employ two conventions in referring to curvature here:

- Although we often use radius, R, to quantify curvature, the descriptions *small* or *large values of curvature* refer to the values of κ. Hence, a large radius refers to a small curvature and vice versa.
- Measure of curvature in well bores always is assumed to be taken at the central axis of the bore hole.

9.3.1 Simple Bending

Calculating bending stresses can be a formidable undertaking in general. Even the planar bending problem is a two-dimensional elasticity boundaryvalue problem, and several assumptions usually are adopted so that a simple solution may be obtained. These ad-hoc assumptions are known variously as *Euler-Bernoulli beam theory, planar beam theory*, or simply just *beam-bending theory*. No theory really is involved, but a merely set of a-priori assumptions about the way a beam deforms in bending that allows for an analytic solution to the more complicated boundary-value problem. For the case of tubes (see Figure 9–8), such as casing, these are the typical assumptions:

- The tube initially is straight.
- The tube cross section is symmetric about the central longitudinal axis.
- All cross sections normal to the longitudinal axis before bending remain normal to the axis after bending.
- The central longitudinal axis (neutral axis) experiences no axial strain
- The tube radius is small compared to the length.
- The bending deflections are small in comparison to the length, so that the radius of the tube remains constant in all directions.

The result of these assumptions is an equation for the axial strain:

$$\varepsilon_s = -y\frac{\mathrm{d}\theta}{\mathrm{d}s} \tag{9.17}$$

where y is a coordinate in the bending plane with origin at the neutral axis (center), θ the angle in the plane of curvature, and s an axial coordinate along the neutral axis of the tube. Substituting this into a one-dimensional constitutive equation (Hooke's law) gives us the axial bending stress:

$$\sigma_\mathrm{b} = \sigma_s = E\,\varepsilon_s = -E\,y\frac{\mathrm{d}\theta}{\mathrm{d}s} \tag{9.18}$$

It is obvious that the maximum stress occurs at the point where y is equal to the outside radius of the pipe, r_o. But we might want to determine the stress at the inner wall also in cases of internal pressure, so we will just leave off the subscript with the understanding that $r_i \le r \le r_o$. The term

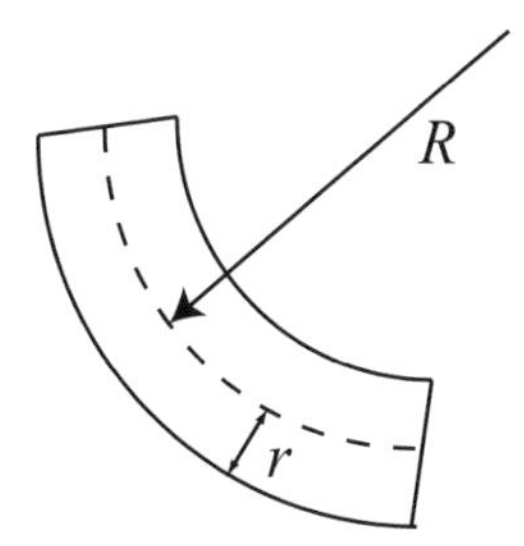

Figure 9–8 *Simple bending of a tube in a single plane.*

$d\theta/ds$ is the curvature of the bent tube, which is the reciprocal of the radius of curvature, R . So, in practical form, the equation becomes

$$\sigma_b = \pm E \frac{r}{R} \tag{9.19}$$

where

σ_b = bending stress, (+) for tension, (-) for compression
E = Young's elastic modulus
r = radius of pipe where stress is determined (i.e. inside or outside)
R = radius of curvature of borehole path

It is important that the units used are consistent. In oil-field units, the radius of the pipe usually is in inches and the radius of curvature of the bore-hole path usually is in feet, so they must be converted to the same unit (it does not matter which). In SI units, both measures should be in meters. Young's modulus usually is in units of lbf/in.^{2}, kPa, or MPa, and the bending stress is in the same units.

It is necessary to remember the assumptions of this formula before using it, especially with tubes. As a tube bends, its cross section tends to ovalize rather than remain circular. The pipe radius in the bending plane is reduced as the cross section becomes ovalized, and the formula no longer is valid. Since there is no easy way to determine the point at which the shape is too ovalized to use of the formula, the tendency is to ignore it, since it will overpredict the maximum bending stress when the pipe is slightly ovalized. That makes the formula possibly a bit conservative in casing design. For long and medium radius of curvature wells, it seems to work well for all but larger-diameter, thin-wall pipe. For short-radius wells, it should be used with caution, and again, it would depend on the pipe diameter and wall thickness. That may seem to avoid the specific, but for certain, we can say that it becomes meaningless if the yield point is exceeded.

9.3.2 Effect of Couplings on Bending Stress

One limitation of the simple bending formula, as we typically apply it, is that it assumes the casing is in contact with the bore-hole wall along its entire length and its curvature is the same as that of the bore hole. This does not account for an amount of standoff due to the couplings. The coupling standoff allows for local bending with a smaller radius of curvature than that of the bore hole, therefore possibly a higher bending stress due to axial tension or compression loading in the pipe. In the tensile case, the couplings usually are tangent to the bore-hole wall, so that if the pipe between couplings is not in contact with the bore-hole wall, then the tension tends to straighten the joint between couplings. The result is that the greatest bending stress in the joint is in the pipe body near the couplings, until the pipe makes contact with the bore-hole wall, then the maximum bending stress in the pipe may be at some other point. In the compression case, the casing between couplings is forced toward the bore-hole wall in a compressive mode and the bending stress is higher near the couplings. Once pipe body contact with the bore-hole wall is made in compression, the point of maximum curvature and bending stress may be at some other point in the tube. This is illustrated in Figure 9–9.

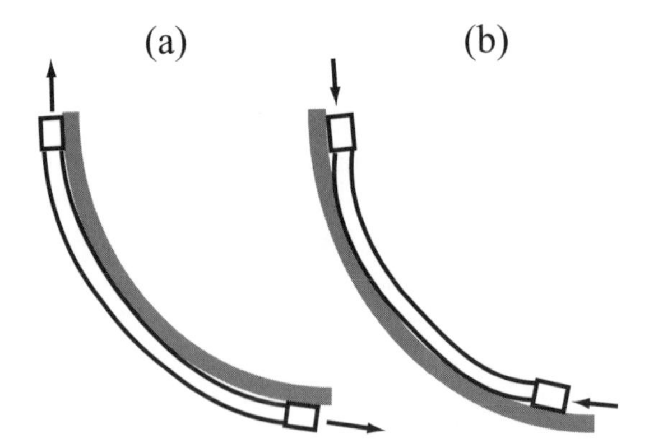

Figure 9–9 *Effects of couplings in bending: (a) tension, (b) compression.*

An equation for determining the maximum bending stress with connections was derived for the tension case (Lubinski, 1961, 1977), and later derivations were done for both tension and compression (Paslay and Cernocky, 1991). Lubinski's equation initially was developed to account for the standoff of drill-pipe tool joints and the effects this had on the fatigue of drill pipe rotating in tension in curved well bores. His equation later began to appear in conjunction with casing design in curved wellbores. His equation (and all that follow in this discussion) assumes that the bore-hole curvature is constant and in a plane between casing couplings, the coupling length is small compared to the length of the joint, the couplings are in contact with the well-bore wall, and the couplings are tangent to the well-bore curvature at the point of contact. Also assumed is that, since the coupling is relatively small in length, its entire length is in contact with the wall and it does not bend. These are reasonable assumptions. Although not in Lubinski's original form, his equation from 1961 can be written as follows:

$$\sigma_{\mathrm{b}} = \lambda E \frac{r_o}{R} \tag{9.20}$$

which is essentially the same as our previous bending equation, equation (9.19), except for the factor, λ, which has been called a *bending-stress magnification factor.* Note that there is no $\pm$ in this equation, since Lubinski's equation is valid only in tension. Lubinski's bending stress magnification factor is

$$\lambda = \frac{\varphi}{\tanh(\varphi)} \tag{9.21}$$

where

$$\varphi \equiv \frac{\ell}{2}\eta \tag{9.22}$$

and

$$\eta \equiv \sqrt{\frac{|F_s|}{EI}} \tag{9.23}$$

where

ℓ = joint length between couplings
F_s = axial load (tension in this case)
E = Young's modulus of elasticity
I = moment of inertia (second area moment of tube cross section)

The second area moment of the tube cross section about an axis passing through the center of the tube perpendicular to its longitudinal axis is

$$I = \frac{\pi}{4}\left(r_o^4 - r_i^4\right) = \frac{\pi}{64}\left(d_o^4 - d_i^4\right)$$

The units of length should be in inches for oil-field units and in meters for SI units. Note also that we used the absolute value of the axial load in this formula. In this particular equation, we are talking about tension, but later we use the same quantities for equations in which the axial load is in compression, a negative value, and the square root would be a complex number. We choose this slight modification so that we may use the same nomenclature in both the tension and compression states. And, in that context, we could resume use of the $\pm$ sign in equation (9.20), once the equations for λ are given for the compression case.

The limitation of this equation of Lubinski's is that the casing does not contact the bore-hole wall between the couplings. Simply stated, if there is contact, the equation no longer is valid. That notwithstanding, his equation appeared in various places to show that coupling standoff is important in casing design, but without mention of the contact limitation. One cannot just plug numbers into Lubinski's equation without understanding its limitations. One always must determine the valid range for a specific application.

The maximum displacement from a straight configuration to contact by the midpoint of the casing between couplings is

$$y_{max} \approx \frac{\ell^2}{8R} + \omega \qquad (9.24)$$

This gives the approximate maximum displacement of the midpoint of an initially straight pipe deflected to the point of contact with the bore-hole wall. The second term, ω , is the standoff due to the coupling, defined as

$$\omega \equiv r_{cpl} - r_{csg} = \frac{d_{cpl} - d_{csg}}{2} \tag{9.25}$$

where

$\omega =$ standoff

$r_{csg} =$ outside radius of casing

$r_{cpl} =$ outside radius of coupling

$d_{csg} =$ outside diameter of casing

$d_{cpl} =$ outside diameter of coupling

As previously mentioned, Paslay and Cernocky (1991) did additional work in this area. They solved the problem in both tension and compression and cast it in a slightly different format, which lends itself to computer programming. The tension case results in a formula for each of the three possibilities: no contact, point contact, and wrap contact. The first two are self-explanatory, and the wrap contact is reached when the curvature of the pipe in contact with the bore hole begins to follow the curvature of the bore hole. Since there are both tension and compressive axial-loading possibilities, the pipe can take six modes of deformation:

Mode 1. Tension, no contact.

Mode 2. Tension, point contact.

Mode 3. Tension, wrap contact.

Mode 4. Compression, no contact.

Mode 5. Compression, point contact.

Mode 6. Compression, wrap contact.

Paslay and Cernocky (1991) derived equations for all six modes, plus four equations necessary to define the transition between modes (two for tension and two for compression). We present their results with very little explanation, and one should read their paper for a full understanding of the derivations.

Mode 1. Tension, No Contact

The bending-stress magnification factor for this mode is

$$\lambda = \varphi \frac{\cosh \varphi}{\sinh \varphi} \tag{9.26}$$

This equation is equivalent to Lubinski's equation, equation (9.21). The nomenclature for this equation and the following remain the same as previously defined.

Mode 2. Tension, Point Contact

Point contact begins when the tension is such that

$$1 - \cosh \varphi + \left(\frac{\varphi}{2} - \frac{2\eta \omega R}{\ell} \right) \sinh \varphi = 0 \tag{9.27}$$

This nonlinear equation in φ must be solved numerically for values of tension, F_s, contained in η and φ to establish the value of tension at which contact is established. The equation for the bending-stress magnification factor for point contact is

$$\lambda = \frac{\varphi \left[(\sinh \varphi - \varphi) - \left(\frac{\varphi}{2} + \frac{2\eta \omega R}{\ell} \right) (\cosh \varphi - 1) \right]}{2(\cosh \varphi - 1) - \varphi \sinh \varphi} \tag{9.28}$$

Mode 3. Tension, Wrap Contact

Wrap contact begins when the curvature in the casing first begins to equal that of the bore hole where the two are in contact. Wrap contact begins when the magnitude of the tension is such that

$$\frac{\varphi}{2}(\cosh \varphi + 1) - 2\sinh \varphi + \left(\frac{2}{\varphi} - \frac{2\eta \omega R}{\ell} \right) (\cosh \varphi - 1) = 0 \tag{9.29}$$

This nonlinear equation must be solved numerically to determine the value of the axial tension at which the wrap contact begins. The equation for the bending-stress magnification factor in wrap contact is

$$\lambda = \frac{\frac{\xi^2}{2} + \eta^2 R\omega(\cosh\xi - 1) - \xi(\sinh\xi - \xi)}{\xi\sinh\xi - 2(\cosh\xi - 1)} \tag{9.30}$$

where we must first solve the following nonlinear equation numerically for ξ:

$$\frac{\xi^2}{2}(\cosh\xi + 1) - 2\xi\sinh\xi + (2 - \eta^2 R\omega)(\cosh\xi - 1) = 0 \tag{9.31}$$

Paslay and Cernocky state that the solution of interest is in the range

$$\varphi \le \xi \le \eta R \cos^{-1}\left[\frac{R}{R+\omega}\right] \tag{9.32}$$

Mode 4. Compression, No Contact

The equation for no contact in compression is.

$$\lambda = \frac{\varphi}{\sin\varphi} \tag{9.33}$$

Mode 5. Compression, Point Contact

When the compression load in the casing reaches a magnitude at which point contact is made, the following condition is satisfied:

$$1 - \cos\varphi - \left(\frac{\varphi}{2} + \frac{2\eta\omega R}{\ell}\right)\sin\varphi = 0 \tag{9.34}$$

It must be solved numerically to obtain the axial load at which point contact occurs.

The Paslay and Cernocky bending magnification factor for point contact in compression is a bit more complicated, in that there are three possibilities as to where the maximum bending-stress occurs. The maximum could occur at the coupling, the midpoint of the joint, or under some circumstances, at another location in the pipe. One must determine the first two, then determine if the third possibility exists and, if so, its magnitude. Once those are calculated, the maximum of the three is the bending-stress magnification factor.

At the coupling,

$$\Lambda_{cpl} = \frac{\varphi\left[\left(\frac{2\eta\omega R}{\ell} - \frac{\varphi}{2}\right)(1-\cos\varphi) + \varphi - \sin\varphi\right]}{2(1-\cos\varphi) - \varphi\sin\varphi} \tag{9.35}$$

At the midpoint,

$$\Lambda_{mid} = \frac{\varphi\left[-\left(\frac{2\eta\omega R}{\ell} - \frac{\varphi}{2}\right)(1-\cos\varphi) + \varphi\cos\varphi - \sin\varphi\right]}{2(1-\cos\varphi) - \varphi\sin\varphi} \tag{9.36}$$

Two additional quantities are needed to determine the possible third point:

$$\Gamma = \frac{\varphi\left[\left(\frac{2\eta\omega R}{\ell} - \frac{\varphi}{2}\right)\sin\varphi + 1 - \cos\varphi\right]}{2(1-\cos\varphi) - \varphi\sin\varphi} \tag{9.37}$$

$$\Omega = \tan^{-1}\left(\frac{\Gamma}{\Lambda_{cpl}}\right) \tag{9.38}$$

Then, the third possible bending-stress moment is

$$\Lambda_s = \Lambda_{\text{cpl}} \cos\Omega - \Gamma \sin\Omega \qquad \Leftrightarrow \qquad 0 < \Omega < \varphi \tag{9.39}$$

If Ω is outside the valid range, then Λ_s does not exist and the maximum bending-stress moment will be either at the midpoint or the coupling. The maximum bending-stress factor is the maximum absolute value of the three possibilities:

$$\lambda = \max\left(\left|\Lambda_{\text{cpl}}\right|, \left|\Lambda_{\text{mid}}\right|, \left|\Lambda_s\right|\right) \tag{9.40}$$

Mode 6. Compression, Wrap Contact

When the compression load in the casing reaches a magnitude such that wrap contact begins, the following condition is satisfied:

$$\frac{\varphi}{2}(1+\cos\varphi) - 2\sin\varphi + \left(\frac{2}{\varphi} - \frac{2\eta\omega R}{\ell}\right)(1-\cos\varphi) = 0 \tag{9.41}$$

As before, this must be solved numerically to determine the axial compressive load at which wrap contact begins.

The bending-stress magnification factor for wrap contact in compression has two possibilities: The maximum is at the coupling or some other point between the coupling and the midpoint. Since the casing curvature at the midpoint is the same as the bore-hole curvature, the bending magnification factor there is unity.

At the coupling,

$$\Lambda_{\text{cpl}} = \frac{\frac{\xi}{2} - \eta^2 \omega R(1-\cos\xi) + \xi(\xi - \sin\xi)}{\xi\sin\xi - 2(1-\cos\xi)} \tag{9.42}$$

where we must first solve the following nonlinear equation numerically for ξ:

$$\frac{\xi^2}{2}(1+\cos\xi) - 2\xi\sin\xi + \left(2 - \eta^2\,\omega R\right)(1-\cos\xi) = 0 \tag{9.43}$$

Paslay and Cernocky do not give a range for the solution, but it appears the range given in equation (9.32) might be at least a starting point.

We then calculate two more quantities:

$$\Gamma = \frac{\left(\frac{\xi}{2} - \eta^2 \omega R\right)\sin\xi + \xi\left(1-\cos\xi\right)}{\xi\sin\xi - 2\left(1-\cos\xi\right)} \tag{9.44}$$

and

$$\Omega = \tan^{-1}\left(\frac{-\Gamma}{\Lambda_{\text{cpl}}}\right) \tag{9.45}$$

Then, we calculate a stationary value of the bending-stress factor at some location in the pipe, if it exists:

$$\Lambda_s = \Lambda_{\text{cpl}}\cos\Omega - \Gamma\sin\Omega \quad \Leftrightarrow \quad 0 < \Omega < \xi \tag{9.46}$$

If Ω is outside the specified range, then Λ_s does not exist. Paslay and Cernocky recommend assigning a value of unity in that case, if it is used in a computer program. A value of unity is equivalent to saying there is no bending-stress magnification at that point. Then, the maximum bending-stress magnification factor for this joint of pipe is the maximum of the absolute values of Λ_{cpl} and Λ_s :

$$\lambda = \max\left(\left|\Lambda_{\text{cpl}}\right|, \left|\Lambda_s\right|\right) \tag{9.47}$$

Comments on Bending-Stress Magnification

As can be seen, calculation of the bending stress magnification factors of Paslay and Cernocky (1991) is not exactly something that can be done manually. These formulas may be programmed for computer implementation, but the programming is far from trivial. Paslay and Cernocky were interested primarily in drill-pipe fatigue, and they mention solving the transition equations for the standoff quantity that we labeled ω. In that context, the equations are linear and solved easily. That may be of some use in selecting a drill-pipe string with various options as to tool joint dimensions, but in most casing design, the standoff, ω, is a fixed quantity, and our primary interest in the transition equations is in the value of the *axial load* that determines the transition point from one mode of contact to another, so that we might apply the appropriate equation for the bending-stress magnification factor, λ. In terms of the axial load, which is contained in the variables, η and φ, these equations are nonlinear and must be solved numerically. All the nonlinear equations that must be solved numerically have been recast here to avoid singularities in the numerical solutions, so they may not exactly resemble those of Paslay and Cernocky. However, some of these equations have local minima and maxima and multiple roots, so they are best solved with a bracketing technique, such as the bisection method. It might be of some help to see a plot of some of the nonlinear equations that determine the contact mode transition points, and an example is shown in Figure 9–10. This figure is for a specific casing size and bore-hole curvature so it will vary for different casing sizes and borehole curvature.

The two equations that define the transition between tensile modes are relatively easy to solve with a simple bracketing method, such as a bisection method. However, for small values of curvature, κ, (or large radius, R), there is a range for which contact is physically impossible and will result in an infinite root. That condition can be stated as

$$\omega > \sqrt{R^2 + \left(\frac{\ell}{2}\right)^2} - R \tag{9.48}$$

Another characteristic of the tensile modes is that, when the curvature is small, the value of the tensile load at which point contact occurs is far greater than the joint strength of the casing, so there is no point in searching for a root if it lies beyond the joint strength.

The two transition equations for compression exhibit especially bad behavior, as equations with trigonometric functions often do over a wide range of values. As previously mentioned, the equations of Paslay and Cernocky have been recast here to avoid numerical singularities, but they still produce multiple roots. The first roots for these two equations tend to lie close to the origin for large curvature and further away for smaller

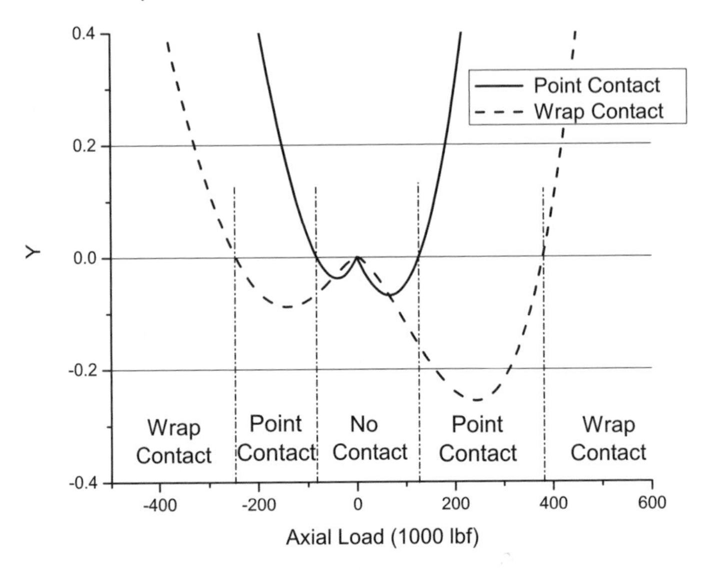

Figure 9–10 *Typical behavior of transition equations for compression and tension. The compression equations usually have additional roots farther to the left.*

values of curvature. From a computational standpoint, this means that a combination procedure that first brackets the root starting very near the origin, then proceeds to locate the root within the bracket, is a good approach. Perhaps, someone with an inclination toward such things might even define a valid range based on the mechanics of the problem and save us the effort. The physical meaning of the additional roots to these equations does not appear to have been explored, but Mitchell (2003) has done work in lateral buckling of drill pipe in curved well bores and shows that the Paslay and Cernocky equations underpredict the bending-stress magnification in cases where compressive loading causes lateral buckling. The reason for the higher bending-stress magnification in those cases is that the lateral buckling is out of the plane of the well-bore curvature to which the Paslay and Cernocky equations are confined.

It is does not seem to have been mentioned in any of the discussions on the bending-stress magnification factor as to the nature of the axial load used for actual calculations. Simple bending, as described by equation (9.19), is independent of axial loading, but the bending-stress magnification factor, λ, is not. That raises the question as to whether we use the true axial load or the effective axial load to calculate λ. We use the effective axial load. (And if you do not believe that, then it falls on you to prove otherwise.)

If one actually calculates the bending-stress magnification factors for a particular casing design, one may be alarmed at their magnitudes, which easily may range between 1 and 4, yet for some reason, this process rarely is considered in actual casing design. One excuse might be that it is not something that can be calculated easily, but surely if casing failures actually occur because of bending-stress magnification, then everyone would have a computer program to calculate it. The most probable reason that we do not recognize problems caused by bending-stress magnification is likely because, in most casing strings, the highest value of tension occurs near the surface, where the curvature often is relatively small. It is not unusual to see bending-stress magnification factors of over 5 in such instances, but the bending stress due to bore-hole curvature is so small that, when multiplied by a large magnification factor, it still is only a small percentage of the yield strength of the casing, especially when we are inclined to use relatively large design factors in tension loading. This is not to say that it can be ignored, but that it does not seem to cause problems in most wells (or at least that we recognize as such). Bending-stress magnification should certainly be considered for casing design in deeper wells and any well where the combined loading may be close to the yield of the casing.

We definitely do not attempt an example calculation here, but Figure 9–11 shows an example of the bending stress magnification factors for an actual horizontal well. And, as would be expected from our discussion, they are highest near the surface.

These factors were calculated to check a previously designed casing string for the specific well bore in which it would be run. While the results might tend to be cause for apprehension due to the large magnitudes, the overall effect in this well is negligible, as will be seen in a following section of this chapter devoted to combined loading in directional wells. The combined loading for this well is shown with and without bending-stress magnification in that section in Figure 9–12. It can be seen there that the only significance of the bending-stress magnification is in the build section of the well, as one would intuitively expect.

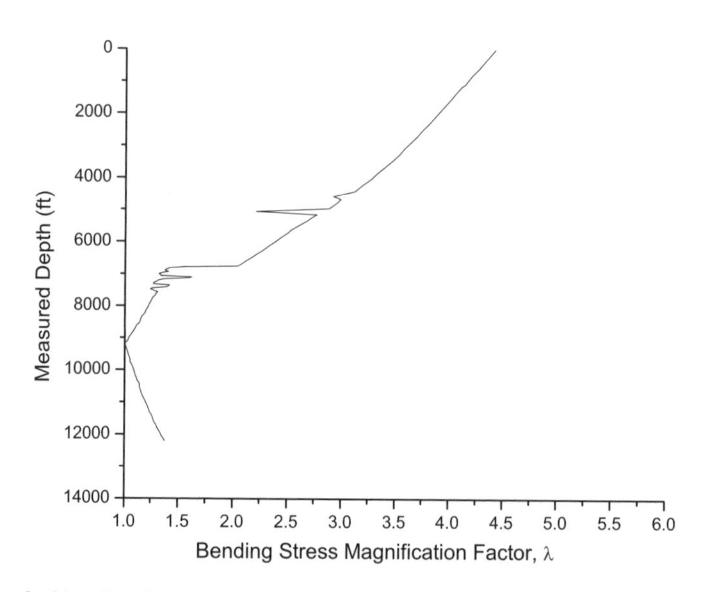

Figure 9–11 *Bending stress magnification factors for a horizontal well.*

9.3.3 Effects of Bending on Coupling Performance

There are no standards on coupling performance in bending other than what some companies and manufacturers established for themselves. The exception to that is the API formula for bending strength of 8-rd thread couplings, which appears in API Bulletin 5C3 and ISO 10400 (2004 draft). This formula predicts the maximum tensile load that can be placed on a coupling in a curved well bore. Actually, there are two formulas and one has to be tried first to verify if it is the correct one; if not, then the other is used. The two formulas with their validation criteria are

$$F_{max} = 0.95 A_{min} \left[U - \left(\frac{140.5 d_o \kappa}{U - Y} \right)^5 \right] \quad \Leftrightarrow \quad \frac{F_{max}}{A_{min}} \geq Y \tag{9.49}$$

$$F_{max} = 0.95 A_{min} \left[\frac{U - Y}{0.644} + Y - 218.15 d_o \kappa \right] \quad \Leftrightarrow \quad \frac{F_{max}}{A_{min}} < Y \tag{9.50}$$

where

F_{max} = maximum axila load in bending, lbf

$A_{min} = \frac{\pi}{4}\left[\left(d_o - 0.1425\right)^2 - d_i^2\right]$ = cross sectional area under last full thread, in^2

κ = well bore curvature, °/100 ft

d_o = outside diameter of pipe, in.

d_i = inside diameter of pipe, in.

Y = minimum yield strength of pipe, lbf/in^2

U = minimum ultimate strength of pipe, lbf/in^2

These formulas are full of conversion factors that are not worth sorting out. Even the new ISO 10400 (2004 draft) does not bother to present them in SI units, if that gives you pause for thought. There is some disagreement as to the worth of these formulas, and for that reason, no SI version is included here either. In a bending situation, the threads on the convex side of the pipe are subjected to the axial load plus the maximum tensile bending stress; and on the concave side, the threads are subjected to the axial load less the maximum compressive bending load. The threads perpendicular to the maximum and minimum are subjected to only the axial load in the pipe due to tension or compression (neglecting pressure). One way that combined bending and axial load was handled historically amounted to multiplying the maximum bending stress times the cross-sectional area of the tube, adding it to the axial load, then comparing that sum to the joint strength of the casing. The API formulas probably are better than that but not by much. The best recommendation that could be made here is no recommendation at all, as neither method inspires much confidence. One thing that should be understood about the 8-rd thread in a bending situation, though, is that it is considered a poor choice by most operators. V-shaped threads have a tendency to "jump out" or override one another because the shape is conducive to this, and possibly more so in the presence of a thread lubricant. Most operators of horizontal wells elect to use a thread with a more squared profile as opposed to a V-shaped profile, because it lessens the possibility of jump-out. A buttress thread has a square contact in tension but is somewhat tapered in compression.

A buttress-type thread has performed successfully in medium-radius wells for many operators. Most proprietary threads are a better choice, and in critical wells, the proprietary threads that tend to interlock are a much better, although more expensive, solution. Some proprietary thread manufacturers publish bending performance data for their connections, and these can be quite useful. Until some meaningful standards are published, the API formulas for 8-rd threads are a starting point.

9.4 Combined Loading in Curved Well Bores

We examined combined loading in the previous chapter. Perhaps, the most common occurrence of failure due to combined loading is in horizontal wells or highly deviated wells. Primarily, this is caused by the addition of the bending stress, which many do not realize is quite significant in magnitude. Our approach to casing design for these wells is to use conventional load curves for burst and collapse. Then, the tension design is based on a load curve that includes the effects of both gravity and friction, so it will be quite different from casing hanging in a vertical well. A typical curve like this is seen in Figure 9–7. John Greenip (1989) illustrated a simple method for designing casing strings in highly deviated wells, using torque-and-drag curves like we discussed in the previous section on bore-hole friction. He went a step further to include bending stress, which he converted to an equivalent axial load by multiplying the bending stress by the cross-sectional area of the tube so that it might be considered part of the tensile load. That equivalent axial load is added to the axial load from the bore-hole friction curves in the appropriate places to constitute a tensile load curve for casing selection. The procedure produces an adequate design for most directional wells and has been used successfully by numerous operators. Here, we assume that we designed a casing string by that method or something similar, and we now are ready to check it for the effects of combined loading, especially in the build section.

Although none of the single loads, such as tension, bending, or burst, may exceed the yield strength of the pipe individually, it is quite possible that the combination of the loads may exceed the yield strength. And there are no handy charts to adequately show the effects of the combined loads often encountered in highly deviated wells. It is easy to check the combined loads at critical points manually to determine whether we need to make adjustments to a design or not. We also could do this with a spreadsheet and check the entire string. Next is an example of a horizontal well in which there was a casing problem. The operator had drilled the vertical part of the hole and a build section to an inclination of 90°. The company was running 7 in. casing, and about halfway through the build section, the casing hit an obstruction, which is not unusual in a horizontal well, if the drilled cuttings are not sufficiently cleaned out of the bore hole.[8] The operator took immediate action. He put a circulating head on the casing, then tried to wash through the obstruction. At first unsuccessful, the driller slacked off the brake even further. As hard as it may be to understand, he set the entire string weight on the obstruction. The shoe plugged and the internal pressure at the surface rose to 3000 lbf/in.^{2} before he was able to shut off the pumps and pick up the string off bottom to relieve the pressure. The operator pulled the casing out of the well and found one of the joints in the build section had buckled and crushed. Here are the data and calculations for this example.

Example 9–1

Data at 6000 ft (TVD):

Casing OD = 7.0 in.

Casing ID = 6.366 in.

Casing grade = J-55 (Y = 55,000 lbf/in.^{2}).

Young's modulus = 30×10^6 lbf/in.^{2}

Axial compression = –122,000 lbf (from torque and drag estimate).

Radius of curvature = 300 ft.

External pressure = 3000 lbf/in.^{2}.

Internal pressure = 6000 lbf/in.^{2}.

Determine if combined loads will yield pipe.

1. Calculate the axial stress:

$$\sigma_s = \frac{-122,000}{\frac{\pi}{4}(7.0^2 - 6.366^2)} = -18331 \text{ lbf/in}^2$$

8. The build section of horizontal wells is notoriously difficult to clean because cuttings tend to migrate down the low side of the hole at inclination angles between 45° and 60°, even at relatively high circulating rates.

2. Calculate the bending stress:

$$\sigma_b = -\frac{30\times10^6(7.0/2)}{12(300)} = -29167 \text{ psi}$$

Note that we had to get the radius of curvature into the same units as the pipe radius.

3. Calculate the pressure effect using at the outside wall using the Lamé equations:

$$\sigma_r = -3000 \text{ lbf/in}^2$$

$$\sigma_\theta = \frac{-3000(3.5^2 + 3.183^2) + 2(6000)(3.183)^2}{3.5^2 - 3.183^2} = 25694 \text{ lbf/in}^2$$

4. The maximum axial stress at the external wall is the preceding axial stress plus the bending stress:

$$\hat{\sigma}_s = -18331 - 29167 = -47498 \text{ lbf/in}^2$$

5. Since there is no torque, these are principal stress components and can be plugged directly into the von Mises criterion:

$$\Psi = \sqrt{\frac{1}{2}\left[(-3000-25694)^2 + (25694+47498)^2 + (-47498+3000)^2\right]}$$

$$\Psi \approx 63900 \text{ lbf/in}^2$$

$$63900 > 55000 \rightarrow \text{yield}$$

Clearly, this combined load value is well above the yield strength of the casing. This casing string, in fact, did fail. The operator was not aware that the combined loading could be that significant. In this case, the operator was lucky to get the casing out of the hole. Others have encountered similar circumstances and found the lower part of the casing string missing when they pulled the string out of the hole. In this case, we calculated the combined load at the outer wall of the casing, assuming that, since the maximum bending stress occurs at the outer wall, that would be the critical location. But, the maximum stress due to pressure occurs at the inner wall, and if we calculate it at the inner wall, we find that the combined load there would be slightly less at 63,700 $lbf/in.^2$. In this case, it was not obvious as to whether the maximum combined load would occur at the inner wall or outer wall, but it did appear obvious that, since the pipe was in compression, it would occur on the concave side of the curve. One should be careful about such assumptions, however. While the maximum combined load typically occurs on the concave side when the pipe is in compression and on the convex side when it is in tension, that is not true in general because of the influence of the pressure.

It is relatively easy to program the Lamé equations and the von Mises yield criterion in a spreadsheet to calculate the combined loading for a well for use in casing design. From a directional survey, one needs the measured depths, true vertical depths, and radius of curvature between survey points. From a torque-and-drag model, one needs the true axial loads for motion in both directions, so that one may check for the worst-case scenario. Additionally, one needs the pipe dimensions and specific weights, the fluid densities, applied pressures, and so forth. One does not actually need both directional and friction software to get this data, as it all can be programmed into a single spreadsheet. The addition of a bending-stress magnification factor would be a bit cumbersome in a spreadsheet though, but it can be done. One convenient way to look at combined loading in casing design is in a plot of the combined load as a percentage of pipe yield stress. Figure 9–12 shows the combined load in a simple L-shaped horizontal well as a percent of the yield of the casing. This is a rather simple case, where the entire string is a single weight and grade of casing, 5½ in., 17 lb/ft, N-80. The operator plans to do multiple hydraulic fracture treatments in this well at high rates and pressures. The combined load is calculated for a burst scenario, where the frac treatment might screen out with a full column of frac fluid and proppant such that the pressure equalizes at maximum surface treatment pressure.

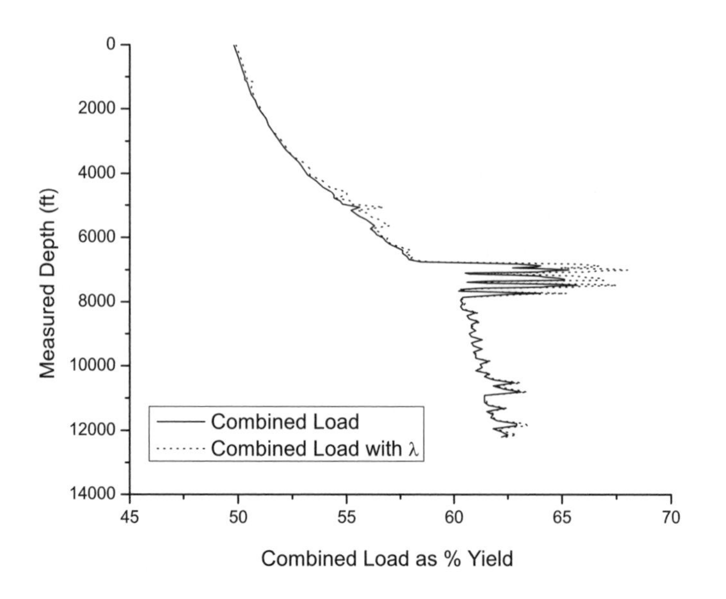

Figure 9–12 *Combined loading in a horizontal casing string.*

A plot like this figure provides an easy way to visualize the effects of combined loading. If necessary, a casing string could be modified to reduce the combined loading effects. One last precaution though, combined loading using a von Mises yield criterion does not account for connection strength nor collapse prior to yield, as mentioned in previous chapters.

9.5 Closure

Directional wells and horizontal wells place additional loads on casing that are not present in vertical wells. We looked at the effects of friction and curvature in this chapter and saw the extent of these types of loadings. We also examined the combined effects of tension or compression, pressure, and bending in these types of wells to get a feel for the relative significance. What one should get from this chapter is an understanding of the loads in these types of wells and a certain amount of comfort in being able to check conventional designs for these types of wells.

9.6 References

API 5C3. (October 1994). *Bulletin on Formulas and Calculations for Casing, Tubing, Drill Pipe, and Line Pipe Properties.* Washington, DC: American Petroleum Institute.

ISO/DIS 10400. (2004, draft). *Petroleum and Natural Gas Industries—Formulae and Calculations for Casing, Tubing, Drill Pipe, and Line Pipe Properties.* Geneva: International Organization for Standardization.

Greenip, J. F., Jr. (December 1989). "How to Design Casing Strings for Horizontal Wells." *Petroleum Engineer International.*

Johancsik, C. A., D. B. Friesen, and R. Dawson. (June 1984). "Torque and Drag in Directional Wells—Prediction and Measurement." *Journal of Petroleum Technology*: 987–992.

Lubinski, A. (February 1961). "Maximum Permissible Dog-Legs in Rotary Boreholes." *Journal of Petroleum Technology*: 175–194.

Lubinski, A. (March–April 1977). "Fatigue of Range 3 Drill Pipe." *Revue de l'Institut Français du Pétrole* 32.

Mitchell, R. F. (2003). "Lateral Buckling of Pipe with Connectors in Curved Wellbores." *SPE Drilling and Completion*: 22–32.

Paslay, P. R., and E. P. Cernocky. (1991). *Bending Stress Magnification in Constant Curvature Doglegs with Impact on Drillstring and Casing.* SPE 22547. Richardson, TX: Society of Petroleum Engineers.

Sheppard, M. C., C. Wick, and T. Burgess. (December 1987). "Designing Well Paths to Reduce Drag and Torque." *SPE Drilling Engineering.*

CHAPTER 10

Special Topics

10.1 Introduction

This last chapter is a place to discuss a few topics that are not exactly related to casing design but nevertheless are related to the use of casing. The first topic is casing wear, which is significant when drilling takes place below the casing string. A method is shown as to how to determine where the wear will be the most severe. Given that knowledge, one can know where it is necessary to use pipe protectors and possibly even modify a casing design to include thicker wall pipe.

The second topic is expandable casing, which is a relatively new product. It has seen some good success and increasing use. We discuss briefly the applications and limitations as they currently are understood.

Finally, the topic of drilling with casing and liners is discussed. Whether this topic belongs in a book on casing, a book on drill strings, or a book on drilling procedures is an open question. What we attempt here is a brief overview of the topic, primarily as it relates to casing rather than the specific rig requirements and drilling techniques.

10.2 Casing Wear

Casing wear is a serious problem in many intermediate strings and some surface strings. It often is the reason that an intermediate string cannot be used as a production string in combination with a production liner. Reduced wall thickness or a hole in the pipe can be disastrous. There is no good way to repair badly worn pipe so that it will contain higher internal pressures other than to run a new string inside it with the accompanying reduction of internal diameter. Hence, it is quite important to prevent or minimize casing wear.

The primary mechanism for casing wear is the rotation of drill pipe, although the tripping of the drill pipe also contributes to the wear but to a lesser degree. Two things are necessary for the wear to occur, and these are fairly obvious: contact force and movement of the drill pipe (rotation and sliding). The rate of wear depends on a number of things, such as

- Magnitude of contact force.
- Rotation speed.
- Lubricity of the drilling fluid.
- Relative hardness of the drill pipe tool joints and casing.
- Presence of abrasives.

Of course, the total amount of wear depends on all these plus the time duration during which wear occurs. Typically, we measure the amount of wear as a percentage of reduction in the wall thickness, with 100% meaning that the wall thickness is completely worn through. Reduction of wall thickness is a linear measure and therefore somewhat misleading. The amount of metal removed under a specific set of conditions generally is a linear function of time, but the reduction of wall thickness is not. Figure 10–1 illustrates why it is not.

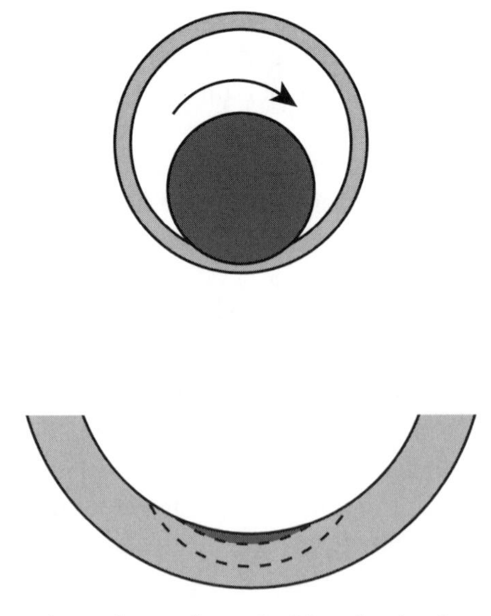

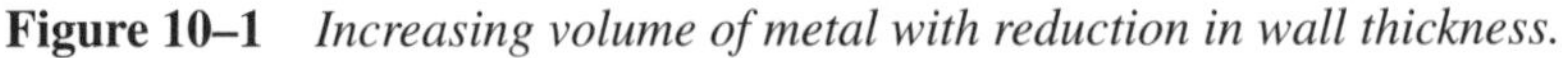

Figure 10–1 *Increasing volume of metal with reduction in wall thickness.*

It is easy to see that, as the tool joint wears into the wall of the casing, more volume of metal must be removed in relation to the amount of penetration. So, while the rate of metal removed may be linear with time or cumulative revolutions, the reduction in actual wall thickness is not. We can see that, initially, the wall thickness reduction is quite fast, but as it progresses, it becomes much slower because of the increasing volume of metal that must be removed for a corresponding reduction in wall thickness.

Prevention of casing wear is of utmost interest in most wells. Historically, most of what was known about preventing wear came from common sense and experience. We have long known that rough hard-banded tool joints can wear a hole in casing as quickly as a mill. Even heat galling can take place when the lubricity of the mud is low and the contact force is high. And, no matter what precautions are taken, if there is sand in the mud, all wear mechanisms are accelerated. Even with rubber pipe protectors, the presence of sand causes wear, since the sand grains can become imbedded in the rubber itself. So, assuming we know to keep abrasives to a minimum and hard-banded tool joints out of the casing while rotating, where do we install the pipe protectors to reduce wear? It was once thought that we could make a plot of dog-leg severity (hole curvature) to determine where the critical wear areas were. Historically, this proved unreliable. In general, casing wear is not a function of the magnitude of the dog-leg severity. The worst wear in casing typically occurs nearer the surface rather than deeper and often where the magnitude of the dog-leg severity is typically less than 1° or 2° per 100 ft, as opposed to deeper in the well where the dog-leg severity might exceed 4°/100 ft, for example. Another approach that proved more useful is a plot of the difference between successive dog-leg severity measurements. While that is a much better indication of the areas of most severe wear, it too can be grossly misleading in some parts of the well. Until wear began to be studied more seriously, that remained the only tool readily available to most operators for determining the best location for pipe protectors. Most operators just ran them on every joint or so in the upper half of the casing as a precaution.

A lot of work has been done to try to quantify wear in casing, and software is available to predict the amount of wear. The results of such predictions have been mixed at best and, in many cases, have been totally unreliable. The difficulty in quantifying wear is in quantifying all the variables that affect the process. In other words, one has to know pretty accurately the time spent rotating, the penetration rate, the lubricating properties of the mud, the rotation speed, the type and concentration of solids and abrasives in the drilling fluid, and so forth. However, this is not a dismissal of such software by any means. While it has proven relatively poor at quantifying actual casing wear, it is extremely good at predicting where the critical wear areas in a casing string are located. For any given mud system and amount of rotating time, the areas of most severe wear are those areas that suffer the greatest amount of contact force between the tool joints and casing. That contact force is quite easy to quantify, at least to the accuracy needed.

An investigation was done several years ago with this type of software run post priori on several wells that had experienced holes worn in the casing. Good drilling data were available, as well as caliper logs that had been used to locate the holes, and of course, directional surveys, which are essential for use of the software. The results were a bit disappointing. In none of these particular wells did the software predict a hole in the casing, even though each actually had a hole worn through the casing. In fact, the worst wall thickness loss predicted in any of the wells by the software was slightly more than 50%. But, the important point again is that where the software predicted the worst wear to occur was exactly where the holes were (see Figure 10–2). In addition to the wear curves, the contact force curves were plotted for comparison (Figure 10–3), and in fact, the wear curves and contact force curves were almost identical, except of course for the scale. The conclusion of that particular study was that, while the software was not very good at quantifying the amount of wear, it was excellent for determining the critical wear areas. It also was found that a contact force curve by itself was adequate for predicting where pipe protectors were needed while drilling below those strings of casing. And that, ultimately, is what we want to know. because we cannot know with certainty the exact properties of the mud system, abrasives content, and rotating time prior to drilling. However, we do know the shape of the hole and the planned well path below the casing well enough to predict the amount of contact force on the casing.

Differential dog-leg severity essentially is the difference between the dog-leg severity at one point and that at the previous point. While it often gives a similar plot to the two above, it gives misleading results near the bottom of the casing string because it cannot account for the reduction in contact force.

Contact force can be calculated from the bore-hole friction formulas in this text (Chapter 9), and they easily can be programmed into a spreadsheet. Most commercial torque-and-drag software also generates a contact force curve. But, to use the contact force for determining the need for pipe protectors, one must have directional survey data, and in the case of

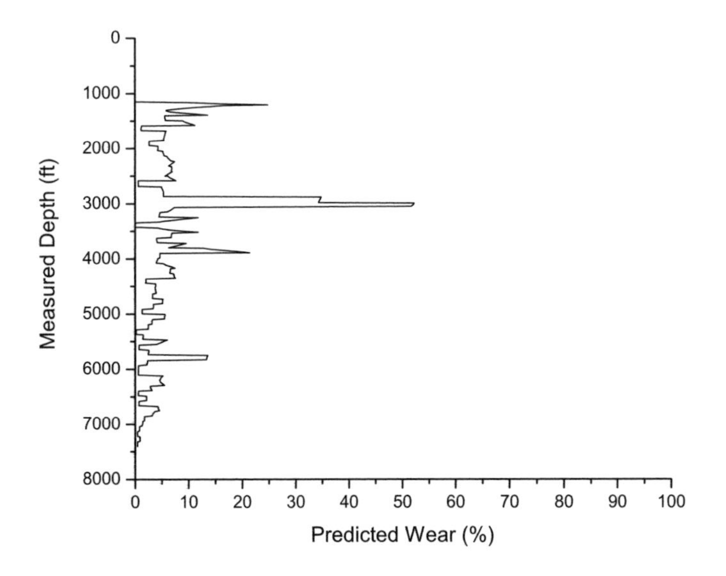

Figure 10–2 *Results from casing wear software, showing the predicted amount of wear in a particular well. This casing string had a hole in it at about 3000 ft.*

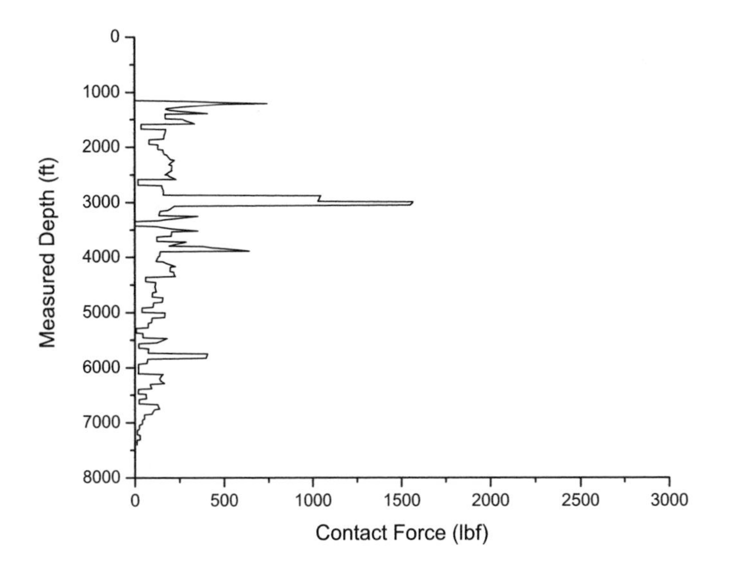

Figure 10–3 *Contact force curve for the same well.*

vertical wells, this may not be available. Many companies feel that. if a well requires an intermediate string for over-pressured reservoirs below, then it should also have a gyro survey run in the intermediate casing in the event it becomes necessary to drill a relief well to kill a blowout. In those cases, the gyro survey can serve both purposes. There really is no reason for not being able to determine where the casing wear will be most severe and where pipe protectors should be placed in a drill string. Given that knowledge and common sense as regard to wear mechanisms, casing wear should not be a severe problem.

10.3 Expandable Casing

Two problems with casing sizes sometimes arise in the drilling of wells, both of which can increase the well costs and possibly prevent a well from reaching its objective:

- Unanticipated conditions that require an additional casing string of casing after the well has been started.
- Known conditions that require multiple casing strings for a well before it has been started.

In the first case, the sizes and depths already are selected, and one or more strings may be set before the need for an additional string arises. Unexpected bore-hole stability or pressure problems may require an additional string that was not originally planned for. Another problem of similar nature is the possibility that a planned casing string may stick before reaching its planned depth and thereby necessitate an additional sting. In these cases, an additional casing string or liner must be set and the final casing string at total depth will be smaller than desired, unless some contingency was included in the original plan to allow for such an event. The second case is becoming more common in some areas, especially where depleted zones may be present. Typically, we think in terms of a surface string, an intermediate string, and a production string, possibly with a liner somewhere in that mix, but basically three or four strings. Occasionally, we may even find it necessary to run five or six strings, counting liners. In recent times, however, we are seeing wells that require 7 or even as many as 10 strings of casing to reach an objective. A conventional approach to this problem requires some very large well bores and casing to reach the total depth with a final casing string size that allows for adequate production. In each of these cases, size and clearance become serious problems.

One answer to these problems is expandable casing. This is a type of casing with connections that can be run through conventional casing (or other expandable casing) then expanded to a larger diameter than a conventional string run through that same size pipe. While this is a relatively new technology, it has seen some good success in numerous applications. However, it is not necessarily a panacea, as there are drawbacks, too.

10.3.1 Expandable Pipe

Expandable casing is not your typical casing product. First of all, it must be ductile enough that it can be expanded without rupturing and still have sufficient strength to function properly. We discussed plastic material behavior in Chapter 7, so we do not rehash that here, but this is exactly the type of behavior that takes place with this type of material. Consequently, it does not come in standard API grades, weights, and so forth. Likewise, there are no published standards of performance properties but rather those are set by the manufacturer. Most expandable pipe is not seamless pipe, since the wall thickness has to be much more uniform than most seamless pipe. It is manufactured from flat plate steel that has been precisely rolled to within close tolerances. Now, seamless pipe can be used and is being used in some expandable applications, but the wall thickness must be very carefully determined to within close tolerances. You can imagine the results of the expansion process if the wall thickness is not uniform before the expansion; that is, most of the expansion takes place in the portions where the wall thickness already is the thinnest. Additionally, the connections must be expanded, since it must be run as individual joints. When you consider the amount of expansion of the pipe body and the threaded connections, you come to appreciate the technology of the process, in that it is not nearly as simple as it might first appear. Obviously, for the performance properties of the expanded pipe to be reasonable, the expansion process must be uniform.

10.3.2 Expansion Process

Two basic processes are used for expanding pipe, and they essentially are the same two processes that have been around for more than 40 years, since the first internal casing patches were introduced. Of course, they have seen considerable improvement since that introduction. One process involves a swaging operation in that an internal swage mandrel is run with the expandable casing, and it expands the pipe from the bottom up as it is pushed or pulled through the tube. This typically is a hydraulic process. The other process employs a roller-type device that expands the casing from the top down, using a tapered device with rollers that expand the casing as the device is rotated with a work string. One thing that must be kept in mind, though, is the elastic unloading discussed in Chapter 7 about elastic-plastic behavior. If we expand a tube plastically, it always exhibits some amount of elastic shrinking from its plastic state. This means that it has to be expanded to a slightly larger diameter than its final diameter to account for the elastic unloading once the expansion tool is removed.

The swaging process uses a mandrel with a circular cross section that is in the expandable casing as it is run. The mandrel can be either a solid piece or of a type that allows for retraction and retrieval through a smaller diameter. In its expanded position, it is pumped or pulled through the casing from the bottom upward and expands the casing as it moves upward. This may be accomplished hydraulically or with a mechanical pulling force, as long as the casing is not allowed to move as it is being expanded. It is a positive type of expansion, in that, for the mandrel to pass through, the casing must expand to the diameter of the mandrel. Advantages of this process are that it imparts a true hoop stress to the casing being expanded. If the wall thickness of the casing is uniform and the material is isotropic, then the hoop stress is uniform. The expanded tube should be round. The disadvantages are that the swaging process induces an axial stress in the pipe as it is being pumped or drawn through and may require a special coating internally to reduce the friction. If the mandrel is a one-piece device and for some reason it cannot be pulled through the entire expandable casing, then it cannot be removed unless it can be milled up. This may not be a problem with retractable-type mandrels, though they lack the simplicity of a single piece mandrel.

The roller-type process was in use long before the expandable casing patch was introduced, over 40 years ago. It historically was used to try to restore partially collapsed casing. The roller process is simple, in that the process starts from the top and expands the casing as it is rotated downward into the expandable or collapsed pipe. It has the advantage that it can be removed at any time, replaced, and resume the operation where it stopped. The historic problem with rollers is that they do not work very well, at least in the fixed version. A roller device does not induce true uniform hoop stress in a tube, because it contacts the casing at only a finite number of points, usually three or four. The old roller-type casing patches typically failed because they never were round in cross section once expanded, because the rollers had only three contact points. Expanders with four rollers were introduced and had better success than three rollers but still never were as successful as the swage-type process. The use of casing rollers to restore partially collapsed casing historically enjoyed limited success primarily because of the point contact with the casing wall and the elastic unloading between contact points. Figure 10–4 shows a typical swaging expansion process.

10.3.3 Well Applications

At the beginning, we mentioned the unanticipated well problem as a possible application for expandable casing. For someone who has been involved in drilling operations for a period of time, this usually is the first application that comes to mind. Expandable casing could be used in such an application as a temporary means of getting past some troublesome zone. Originally, the availability and lead time required for the expandable pipe to be a readily available solution for this type of problem was limited. The expandable casing had to be ordered and available as a backup for a particular well before the actual need arose. This changed in time and is seldom a limitation now. Another drawback to expandable casing as an unplanned contingency string is the cementing issue. If the expandable string is to be reliably cemented, then the hole in which it is to be placed must be either underreamed to a larger diameter than the bit that will pass through the casing above it or it must be drilled initially to a larger diameter as with a bicenter bit. These are not necessarily amenable to unanticipated situations that may arise and require an additional casing string. As it currently stands, expandable casing is a planned part of the casing program and for that it has proven quite successful.

The cementing process in regard to expandable casing is a bit different from conventional casing cementing. The usual procedure is to displace the cement prior to expanding the casing. This requires that the casing expansion be completed before the cement begins to harden. The expandable casing can be reciprocated and even rotated during the displacement process, so in that respect, it is no less effective than a conventional liner cementing job. The biggest differences may be the cement near the top of the liner and whether or not one wants cement in the annulus above the liner before the expansion process begins. As the casing is expanded, the mud and cement in the annulus must be displaced somewhere, and it goes into the annulus between the running string and the previously set casing. If cement actually is displaced into this space above the expandable liner, then there is a considerable discomfort factor until the expansion is complete and this cement can be circulated out of the well bore. For the most part though, since the expandable casing is used for a temporary drilling liner, it is not critical to have cement all the

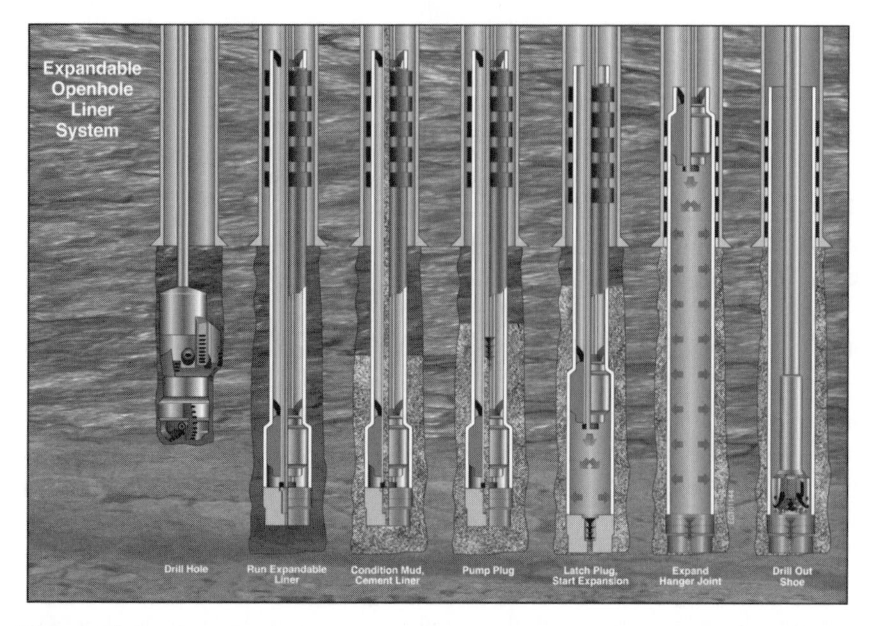

Figure 10–4 *Running, cementing, and expanding process (courtesy of Enventure Global Technology).*

way to the top of the liner. The process has been successful in numerous applications, but it is a cause for concern. Most expandable casing is run as a liner and the final part of the expansion process is the expansion of the overlap in which some type of elastomer seals on the outside of the expandable casing seal against the casing through which it has been run. Once that seal has been established there is no way to displace cement into the annulus short of perforating and squeezing.

10.3.4 Collapse Considerations

The collapse rating of expandable casing is usually less than what one is accustomed to in similar sizes of conventional casing. This is mostly due to the thinner wall of the expanded tubes as compared to API tubes. The thinner wall is the tradeoff we accept for the larger internal diameter which in turn is the primary reason we choose the expandable tube. In the discussion on plasticity, we mentioned that a material that strain hardens in plastic tension may gain yield strength in that direction, but in the process, it loses yield strength in compression. Also, if the casing wall does not expand uniformly, then the collapse strength is less than if it had expanded uniformly. With conventional casing, we can inspect its wall thickness and eccentricity before it is run in the hole. With expandable casing, there is no way to know with certainty the final wall thickness and eccentricity, if any, until after it is in the hole and expanded. It can be seen in some of the commercial videos that show the expansion process on the surface that the final expanded tube has some amount of curvature in it. The causes of this curvature could be attributed to variations in wall thickness, residual stress, or anisotropic hardening. This is not said to denigrate the expandable casing but to point out that one must understand that expandable casing is not the same thing as conventional casing. It has different properties, one of which is a reduced collapse strength, and that must be considered in any particular application and casing design.

Consequently, the most obvious practical application for expandable casing is as an intermediate string or liner to be utilized during the drilling of a well that eventually will be cased with conventional casing. In those applications, it can be invaluable. For instance, Figure 10–5 shows a conventional casing program for a particular application.

By utilizing expandable casing in the same well the program can be modified as seen in Figure 10–6. The advantage is readily apparent, in that the total depth now can be reached with the same size conventional casing and smaller casing at shallow depths, if expandable casing strings are used in the well plan. There are several possible variations. While this may not be applicable for the most common wells drilled in the world, it represents a considerable advantage in those costly wells that do fall outside the common category.

10.4 Drilling with Casing and Liners

Drilling with casing is not new. It was routine practice in some shallow fields between 50 and 60 years ago to attach a bit to a small-diameter production string, drill a few hundred feet, and leave it on bottom once total depth was reached. Outside that limited sphere, it generally was recognized that, if casing could be used as a drill string without making bit trips, there would be some obvious advantages over conventional drilling. The earliest general thinking was that the ability to drill with casing would

- Reduce or eliminate trip time.
- Eliminate the need for a separate drill string.

The big drawback to the whole idea was that bits did not last very long, and unless there was either some dramatic improvement in bit life or some way to retrieve and replace the bit without tripping the casing, there was no reason to consider the process except in the limited context of very

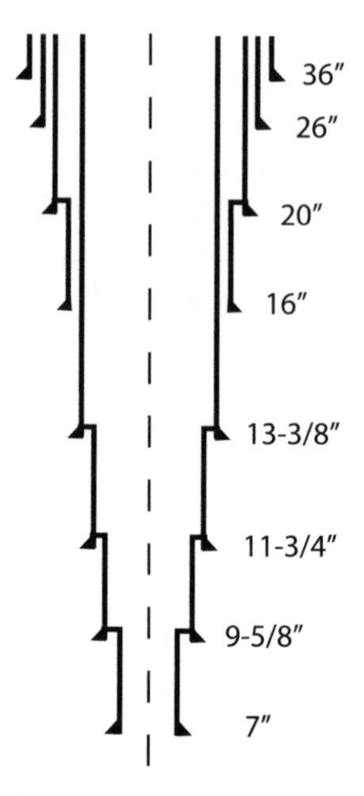

Figure 10–5 *Conventional casing program.*

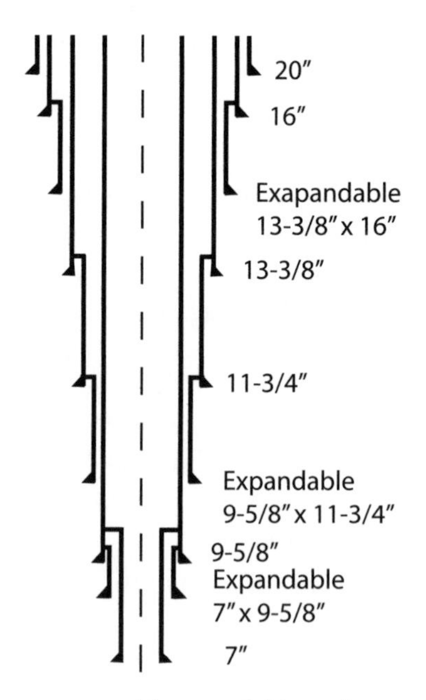

Figure 10–6 *Casing program with expendable casing.*

shallow wells. In recent years, improvements in technology along those lines and the development of rigs specifically built for drilling with casing brought about a considerable change. As a result, the two benefits have begun to be realized, but so too have many other benefits that are in the details of the process. Some are obvious, some are not. Before compiling a list, it might be best to describe the process, so that the advantages become more apparent.

10.4.1 Drilling Methods with Casing

Some of the shallow wells drilled with casing in the 1940s and early 1950s in northern Louisiana were drilled "on the cheap" so to speak. Typically, a reconditioned bit was welded to the pin end of a joint of 4½ in. casing, and the well was drilled to a depth of a few hundred feet. The drill string/casing string consisted of about 15–20 joints of pipe and was cemented as soon as total depth was reached. The bit was left on the bottom of the string. It was a crude process, based almost entirely on its economic advantage.

The current processes of drilling with casing are considerably more advanced. There essentially are four approaches:

1. Rotation drilling with conventional bits.
2. Rotation drilling with expendable drill-through bits.
3. Rotation drilling with wire line retrievable bits.
4. Slide drilling (with or without rotation) using retrievable downhole motors and bits.

The first case essentially is the same technique as the early methods, except the technology is several orders of magnitude more advanced. However, the general concept is the same, in that the bit is a permanent part of the casing string, and if it fails before the total depth is reached, then the casing must be tripped to replace it. This particular method typically is used for a final production string of casing or a final production liner. Subsequent deepening of a well drilled with this procedure requires a window or section milling operation and a sidetrack, although this method is unlikely to be used if deepening is a consideration.

The second method is slightly different from the first, in that the bit does not form a permanent barrier on the bottom of the casing string. The bit is designed so that it can be drilled through with another bit that will pass through the casing. These expendable bits are drag-type bits with

PDC cutters set in a drillable matrix material. This method also has a particular disadvantage in common with the first method, in that if the bit fails before the required depth is reached, then the casing must be tripped to replace it. Typically, this method is used for surface casing strings or relatively short sections, where bit life is not a significant factor.

The third method is the heart of the new technology and the true significant step that makes the drilling with casing process truly viable as an alternative to conventional drilling. Current tools allow the bit to be retrieved and replaced by wire line, so that dull bits may be replaced with new or different types of bits. During the retrieval and replacement process, it is also possible to keep the casing in motion to prevent sticking, if that is a potential problem.

The fourth method of using retrievable down- hole motors has been used primarily in drilling with liners but also is the method by which directional wells may be drilled with casing. With this method, both the motor and the bit (as well as directional tools) may be retrieved and replaced using wire line.

The methods just described have a number of variations, but that is the essence. These methods are necessarily accomplished with a top-drive rig, although the drilling with a liner could be done with a conventional rotary rig. The best results are with rigs specifically designed and built for drilling with casing. A number of these rigs already exist and more are likely to be built as the process gains wider acceptance. In addition to conventional and special bits for these applications, special underreamers have been developed for drilling below a cemented casing string to provide more clearance while drilling and for cementing.

10.4.2 Casing as a Drill String

The casing used in drilling obviously undergoes stresses not normally considered in conventional casing design. Primarily, the difference is in the addition of torsion-induced shear stress. While it is common practice to rotate liners while cementing, it is not that common to rotate full casing strings, because of the high torque required to do so. The torque required to rotate most full casing strings exceeds the maximum recommended makeup torque of the connections, so torsional stress in the casing body itself seldom is even considered.

Connections

When casing is rotated for drilling, the first consideration is the connections. Casing connections usually are thought of in terms of one-time makeup and are never intended to be tripped in and out of a well bore. Anyone who ever had to pull a full casing string out of a well can attest to this. Standard API 8-rd threads are especially bad, in that a good percentage of them gall in such a process. This is true of many interference types of threads in general, but any type of thread that is not properly lubricated with approved thread lubricant can experience galling problems. The basic premise for drilling with casing is that the casing will not be tripped, but in reality, a trip always remains a possibility, even though remote. The main concern with many types of casing connections, then, is the possibility of overtorqueing them to the extent of causing damage. This is especially significant in connections that do not have a shoulder. So the main concern with torque for most drilling with casing applications is to not damage the connections. One solution is to use only shouldered connections. This can be a rather expensive solution. Another possibility, and one that has proven effective, is the use of stop rings, in other words, inserts in the couplings that allow the pins to make up into the coupling to a specified depth and no farther. This represents a sound, inexpensive alternative to shouldered connections. While such rings have been used in ST&C and LT&C 8-rd connections for some casing rotation applications, these connections are not considered good choices for drilling with the casing. For drilling applications, buttress threads with stop rings have proven a much better choice.

Body Strength

We discussed the effects of combined loading in previous chapters. The only thing to add as far as drilling is concerned is torsion. So, by knowing the internal and external pressures and axial load, we can calculate the radial, tangential, and axial stress components, just as we did in Chapter 7. We also considered axial stresses due to bending in curved well bores and gave the formula for calculating the additional shear stress component due to torsion. All those stress component values are substituted into the von Mises yield formula to determine the combined load, which we compare to the yield strength of the casing. In most cases, a string adequate for the normal application as casing also should be adequate for drilling as far as the casing body goes.

However, one additional aspect of casing design that is unique to drilling with casing is the matter of casing collapse. One normally would consider that collapse occurs only near the bottom of a casing string, but that may not always be the case when the string also is used for drilling. Suppose that, while drilling a well with casing, we experience a gas kick.

How do we kill the kick? We do it the same way we do it with drill pipe: We mix a kill-weight mud and circulate the gas bubble out, holding enough back pressure on the choke to keep the annular bottom-hole pressure slightly above the formation pressure. No problem there, because that is fairly routine. But, what do we normally do when we get the kill-weight mud to the bottom of the drill string? We normally stop the pump to be sure that the surface drill string pressure goes to zero. This is how we check to see if the "kill-weight" mud is actually "kill weight"; it is our first and best opportunity to check this. Now consider that our drill string is casing instead of drill pipe, so the internal-volume to annular-volume ratio is quite different than in a conventional scenario. In this case, we likely already have gas to the surface by the time kill-weight mud is at the bottom of the casing string. Think about the surface pressures. The pressure on the inside of the drill string/casing string is effectively zero, and the annular pressure is the confined gas pressure at the surface. The casing at the surface also has the maximum tension load of the entire string at that time. Is the combined tension and collapse strength of the casing string sufficient in this case? This is something that should be considered in our casing design. We are not accustomed to thinking in terms of casing collapse at the surface in a situation such as this, but this is something we must consider now, because such a casing collapse could be disastrous. We likely have a "wet buckle," as described in Chapter 8, and in that event, we no longer can circulate the well from the bottom.

In one of our examples in Chapter 4, we considered that a gas kick while drilling below the 9⅝ in. intermediate casing could give us a maximum gas pressure of 7190 lbf/in.^{2} at the surface (Figure 4–6). Later, in Chapter 5, we calculated the buoyed axial load at the top of the 7 in. production casing run through that intermediate string at 341,000 lbf. The collapse pressure of that 7 in. 29 lb/ft P-110 production casing is 8530 lbf/in.^{2}. If we include the effects of the axial tension, the reduced collapse value of the casing is 7260 lbf/in.^{2}. If we used that string of casing for drilling and experienced a well control event like we just described, that would put us within 70 lbf/in.^{2} of the minimum collapse strength of the casing. If we were to pull an additional 15,000 lbf tension in the casing while trying to keep it from sticking and there is a good chance the production casing would collapse at the surface. This is a rather extreme example of a gas kick that fills the entire annulus, which we would like to think would not happen. Whether or not we actually could have such an event is debatable, but it does represent a worst-case scenario. This is something that at least should be considered in our casing design, if the casing is also going to serve as a drill string because we are not accustomed to considering collapse in casing design anywhere except near the bottom.

10.4.3 Casing Wear and Fatigue

Earlier in this chapter, we discussed casing wear, but we were considering wear from the inside as opposed to the outside. The rotation of a casing string while drilling obviously is going to cause some amount of wear on the outside of the string. We could use techniques like calculating the contact force to determine where the wear will be the most significant, and that should prove reliable, just as before, but the difference is that, in this case, the casing string is not static. Such a method would tell us at what depths in the bore hole the wear will be greatest, but it will not tell us where in the casing string the wear will be the greatest, because the casing string is not static in relation to the bore hole. We probably could develop some rotation time and contact force correlations, but that would require data we may or may not have. Certainly, this can be developed in the future, but for now, the assumption is that any casing in the open hole is subject to wear and must be protected. Currently, this is accomplished with wear rings that are crimped onto the casing with hydraulic tools specifically designed for that purpose. Practical experience so far shows that this is effective.

Fatigue of casing in drilling operations has certainly not received the amount of study that the drill pipe has. The principles essentially are the same, except for the different types of connections. Casing fatigue as related to drill string harmonics has been observed in practice. Well-bore curvature in directional applications might cause some fatigue problems, too, although that has not been seen in practice so far. Increasing applications of drilling with casing will lead to further developments in this area.

10.4.4 Cementing

Cementing a casing string that has been used to drill the bore hole in that hole should be relatively easy. Special float equipment has been developed to use while drilling or to be inserted into the string once drilling has been completed. The casing string can be both reciprocated and rotated during cementing to aid in displacement of the mud, which is something generally not possible in conventional cementing operations. Crimped-on centralizers have been developed and are run with the casing initially. These are rigid with curved blades to aid in mud displacement. It appears the only thing that cannot be run are sidewall scrapers and wipers. With good-quality mud, centralizers, and pipe movement by reciprocation and rotation, adequate mud displacement should not be a problem in cementing. Liners used as drilling strings may be cemented in a similar manner. Liner hangers have been developed so that the liner may be hung and cemented immediately after drilling.

10.4.5 Advantages and Questions

Having discussed the casing and liner drilling operations, it seems that a list of advantages is now in order and most of these should already be obvious:

- Trip related
  - Time savings.
  - No drill collar handling.
  - Less wear on equipment.
  - Less chance of swabbing.
  - Less chance of surge fracturing formations.
- Crew related
  - Less crew exposure to pickup and lay-down operations.
  - No derrick man for numerous trips.
  - No casing crews.
  - No casing elevator or stabbing board with which to contend.
- Drilling related
  - Lower circulating rates and pressures.
  - Better cuttings transport (reduced annular volume).
  - Better control of equivalent circulating density.
  - Overall better well control while drilling.
  - Less chance of differential sticking in most cases due to pipe movement.

One other interesting phenomenon observed in some actual drilling cases is a reduced tendency for lost circulation where it has been a known problem in the same area with conventional drilling. This has been attributed to the formation of filter cake or compaction of mud filter cake into formation pore spaces at the bore-hole wall. A filter cake or plugged pore spaces cannot make a formation stronger per se, but as discussed briefly in Chapter 3, the fracture pressure of a formation is lower if the fracturing fluid can enter the pore spaces prior to fracture. It appears that what has been observed in these cases is that the drilling fluid is unable to enter the pore spaces prior to fracture. While it is possible that the rotating casing more thoroughly "packs" the pores on the formation face with mud filter cake, it also opens the question as to how much damage is done to the filter cake in tripping the drill pipe. It is also possible that effect of the trip surges on the formation are not always apparent during the trip itself. This remains an interesting phenomenon for further investigation.

Despite the advantages of drilling with casing and liners, obviously questions remain unanswered or only partially answered. They are related mostly to equipment development, reliability, availability, and so forth; and there are questions regarding specific issues such as

- Fatigue failures of casing.
- Actual casing wear.
- Cementing reliability.
- Differential sticking tendencies of larger-diameter pipe.
- Consequences of trips with casing if the necessity should arise.

These are questions that will be answered in time and as the technology develops.

It should be apparent from this discussion that we have not examined the economic benefits of drilling with casing or liners, and that is not a subject for this particular book. A fair portion of what has been done so far has been experimental and developmental in nature. However, definite economic benefits are being demonstrated in certain locales, and there is no reason to think that they are limited to any particular area. Those benefits must be evaluated on an individual basis.

10.5 Closure

This brings us to the end of this chapter and the end of this book. I resist the temptation to say more.

Index

D

P

Q

R

S